SITE CARPENTRY LEVEL 2

DIPLOMA & NVQ

Stephen Jones

SERIES EDITOR: MARTIN BURDFIELD

HODDER EDUCATION
PART OF HACHETTE UK

Orders: please contact Bookpoint Ltd, 130 Milton Park, Abingdon, Oxon OX14 4SB. Telephone: +44 (0)1235 827720. Fax: +44 (0)1235 400454. Lines are open from 9.00am to 5.00pm, Monday to Saturday, with a 24-hour message-answering service. You can also order through our website www.hoddereducation.co.uk

If you have any comments to make about this, or any of our other titles, please send them to educationenquiries@hodder.co.uk

British Library Cataloguing in Publication Data
A catalogue record for this title is available from the British Library

ISBN: 978 0 340 98176 4

First Edition Published 2009
Impression number 10 9 8 7 6 5 4 3 2 1
Year 2012, 2011, 2010, 2009

Illustrations by Oxford Designers & Illustrators

Hachette UK's policy is to use papers that are natural, renewable and recyclable products and made from wood grown in sustainable forests. The logging and manufacturing processes are expected to conform to the environmental regulations of the country of origin.

Cover photo © Arthur Tilley / Photodisc
Typeset by Pantek Arts Ltd
Printed in Italy for Hodder Education, an Hachette UK Company, 338 Euston Road, London NW1 3BH

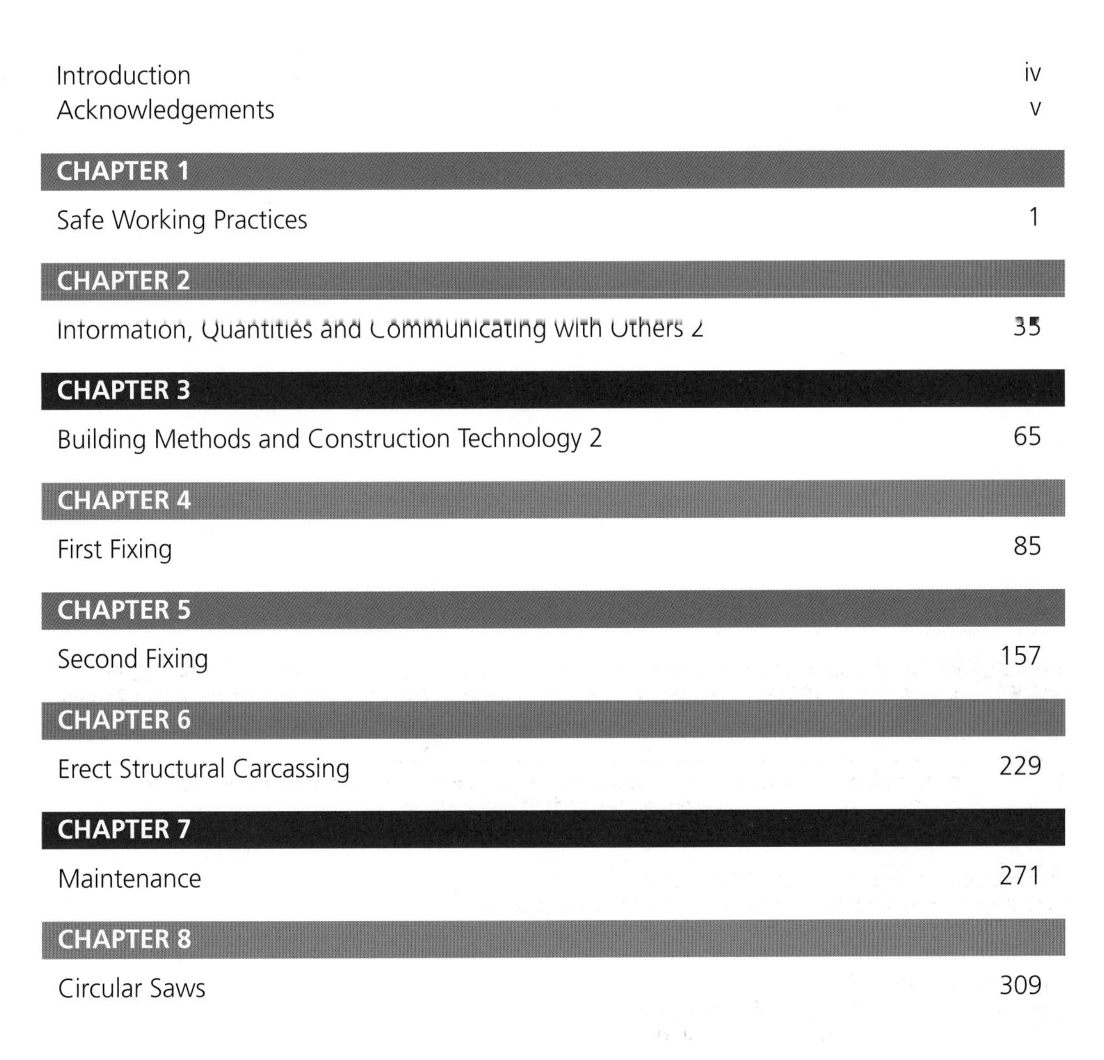

CONTENTS

INTRODUCTION

This book and its supporting resources have been produced to build on prior knowledge from Level 1 and to guide apprentices and students through their academic studies, whilst working towards completion of the Level 2 Diploma in Site Carpentry.

Each of the eight chapters is specifically designed to focus entirely on the units contained within the Occupational Standards set by the Awarding Body (City and Guilds), but is not exclusive to people studying in this area. Hundreds of fully coloured illustrations, current regulations and good working practices are demonstrated throughout this book, with editorial input from a senior examiner to ensure factual accuracy and theoretical rigour.

Frequently asked questions occur in appropriate places throughout each chapter to clarify terminology and expand on specific detail in direct and clear language that readers can understand. At the end of each topic, activities are used to re-enforce students' understanding and also as a source of revision before completion of the unit. Multiple choice questions at the end of each chapter will also enhance students' prior understanding in preparation for their final assessment of the core and occupational units. Trade secrets are introduced to this book to develop students' knowledge from the perspective of experienced craftsmen working in the industry. Information contained within these segments will also help to bridge the transition between the knowledge and simulated assessments completed at their training organisation, through to employment within the construction industry.

A supporting website for this book – www.hodderplus.co.uk/carpentry – contains sample video footage of technical skills and electronic versions of multiple-choice questions.

Website username: carpentry2
Website password: construction

ACKNOWLEDGEMENTS

I would like to thank Stephen Halder and the team at Hodder Education for the opportunity to write this book, and for their continued support. Thank you to Martin Burdfield for his expertise and guidance whilst devoting his time to work with me on this project.

Most of all, special thanks and love to my wife Rebecca for believing in me, and my beautiful family: Daniel, Rachel and Jessica for sparing me the time to complete this book.

The authors and publishers would like to thank the following for use of photographs in this volume:

Figure 1.1 Reproduced under the terms of the Click-Use Licence; Figure 1.3 ©RichVintage/istockphoto.com; Figure 1.11 CSCS Ltd; Figure 1.12 © Paul Gibbings - Fotolia.com; Figure 1.19 © Stas Perov/istockphoto.com; Figure 1.27 © Dan Wilton/istockphoto.com; Figure 1.31 Andrew Howe/Getty Images; Figure 1.32 Image Source/Construction Photography; Figure 1.34 © Rob Fox/istockphoto.com; Figure 2.7 -Vladimir-/istockphoto.com; Figure 2.8 © Branko Miokovic/istockphoto.com; Figure 2.20 © George Peters/istockphoto.com; Figure 2.21 Rex Features; Figure 2.22 © Roger Milley/istockphoto.com; Figure 2.30 Ryan McVay/Gettt Images; Figure 4.11 © Steven Miric/istockphoto.com; Figure 4.21 thumb/istockphoto.com; Figure 4.28 © Construction Photography/Corbis; Figure 4.29 Rex Features; Figure 4.36 Neil McAllister/Alamy, Rex Features; Figure 4.43 Jon Woodfine; Figure 4.48 Ryan McVay/Gettt Images; Figure 4.58 © marc fischer/istockphoto.com; Figure 4.60 ©digitalskillet/istockphoto.com; Figure 4.61 © Jim Jurica/istockphoto.com; Figure 4.115 © Kristian Septimius /istockphoto.com; Figure 5.102 © Steven Miric/istockphoto.com; Figure 5.131 © M. Eric Honeycutt/istockphoto.com; Figure 6.16 © Bill Noll/istockphoto.com; Figure 6.17 © Emrah Turudu/istockphoto.com; Figure 6.23 © jounruh/istockphoto.com; Figure 6.53 © Christopher Dodge - Fotolia.com; Figure 7.1 © Mandy Hartfree-bright/istockphoto.com; Figure 7.9 Coppa Gutta Ltd/www.coppagutta.co.uk; Figure 7.12 reproduced with kind permission by BSI; Figure 7.13 Jon Woodfine; Figure 7.29 Either Jet Tools & Machinery Ltd; Figure 7.36 Benson Ind Ltd bensontoolssales@aol.com; Figure 7.37 Benson Ind Ltd bensontoolssales@aol.com; Figure 7.54 Tarmac Ltd/www.tarmac.co.uk

All other photographs taken by the author.

SAFE WORKING PRACTICES

CHAPTER 1

LEARNING OUTCOMES

By the end of this chapter you should have developed a knowledge and understanding of:

- health and safety regulations;
- accident, first aid, emergency procedures and reporting;
- identifying hazards on construction sites;
- health and hygiene;
- safe handling of materials and equipment;
- basic working platforms;
- working with electricity;
- using appropriate personal protective equipment (PPE);
- fire and emergency procedures.

INTRODUCTION

The aim of this chapter is for learners to be able to recognise situations that may put themselves and others at risk through work activities. It also highlights the duty holders' responsibilities to conform to current health, safety and welfare law in the United Kingdom. In addition, this chapter explains the relevant health and safety legislation and good working codes of practice recommended by the Health and Safety Executive (HSE).

HEALTH AND SAFETY LEGISLATION

Pre-1974, the construction industry suffered an exceptionally high number of accidents and deaths occurring in the workplace. During that period there were various legislations loosely controlling activities in places of work. Following the Robens Report the government passed a primary piece of law known as the Health and Safety at Work Act (HASAWA) 1974. The Act is an umbrella piece of legislation that facilitates a number of other laws and regulations underneath it. Its aim is to promote and encourage high standards of health and safety in all places of work. When the HASAWA was introduced it superimposed many older laws and legislations with others being phased out or replaced with new regulations and supporting 'codes of practice'.

The text below gives an overview of the Health and Safety at Work Act and some of the regulations under its enabling umbrella.

HEALTH AND SAFETY AT WORK ACT 1974 (HASAWA)

GENERAL DUTIES OF EMPLOYERS TO THEIR EMPLOYEES:

1. to ensure the health, safety and welfare at work of all his/her employees;
2. to provide and maintain equipment and systems of work that are safe and without risks to health;
3. to make arrangements for ensuring safety and absence of risks to health in connection with the use, handling, storage and transport of articles and substances;
4. to provide information, instruction, training and supervision as is necessary to ensure the health and safety at work of their employees;
5. to provide safe access and exit for employees to their place of work;
6. to provide a safe working environment with adequate welfare facilities;
7. to prepare and revise a written statement of his/her health and safety policy.

GENERAL DUTIES OF EMPLOYEES AT WORK

It shall be the duty of every employee while at work:

1. to take reasonable care for the health and safety of themselves and of other persons who may be affected by his/her acts at work;
2. to comply with their employer or any other person under any of the relevant statutory legislations, to cooperate with them so far as is necessary to enable that duty or requirement to be performed or complied with;
3. not to interfere with or misuse things provided for health, safety or welfare.

GENERAL DUTIES OF MANUFACTURERS, DESIGNERS AND SUPPLIERS:

1. to ensure that the article/substance provided will be safe without risks to health and safety at all times when it is being used, handled, processed, stored and transported;
2. to provide information about the use for which the article/substance is designed, tested, dismantled or disposed of, to ensure health and safety;
3. to make sure that the article/substance is designed and constructed so that it will be safe without risks to health and safety.

HEALTH AND SAFETY EXECUTIVE (HSE)

The Health and Safety Executive was appointed as part of the Health and Safety at Work Act 1974. Its role is to control the risks and exposure to hazards in the workplace by providing health and safety legislation, information and enforcement of the law.

Health and safety inspectors may visit workplaces to investigate reported occurrences, either from the local authority or directly for the HSE. Health and safety inspectors have many powers, including the following:

- they may enter any premises at any reasonable time;
- if necessary, use the police to prevent an obstruction of their duties;
- examine and investigate;
- take photographs, measurements, samples and recordings;
- destroy or dismantle dangerous items of equipment or machinery;
- seize equipment and render it harmless if it poses an imminent danger to health and safety;
- retain documentation to carry out their investigations;
- take possession of materials;
- take written statements and declarations from employers and employees;
- issue improvement or prohibition notices.

FREQUENTLY ASKED QUESTIONS

What are prohibition and improvement notices?

A 'prohibition notice' is a ban imposed by the health and safety inspector. A prohibition notice may be issued on a particular item of equipment if it is considered to pose a risk of serious personal injury. The equipment will not be permitted to be used until the fault has been rectified.

An 'improvement notice' requires the duty holder to put right an item of equipment or method of work within a specified period of time.

Failure to comply with the Health and Safety Executive and the law may initiate the prosecution of the duty holder. In a court of law, those found to be guilty of compromising the health and safety of others may have heavy fines imposed upon them, a jail sentence or even both.

REPORTING INJURIES, DISEASES AND DANGEROUS OCCURRENCE REGULATIONS 1995 (RIDDOR)

It is a legal requirement to report injuries, diseases and dangerous occurrences in the workplace to the Health and Safety Executive. This allows them to investigate the report, advise the employer and reduce the likelihood of it happening again. The HSE will need to be informed immediately of the following events, known as 'reportable incidents':

- dangerous occurrences or near misses;
- injuries resulting in employees being absent from work for three days or more;
- injuries to members of the general public;
- major injuries;
- work-related deaths;
- work-related diseases.

FREQUENTLY ASKED QUESTIONS

▶ How can you inform the Health and Safety Executive of an incident at work?

There are several methods that can be used to contact the HSE. The quickest methods are as follows.

1. Complete and submit an 'F2508 incident report form' online. A copy of the report will be sent from the HSE for your records, to comply with RIDDOR requirements.
2. Telephone the incident contact centre (ICC). The operator will ask you questions in order to complete the report form before sending a copy for your records.

Alternatively, the HSE can be informed via email or post. In the event of a fatality or serious injury out of working hours, the duty officer will need to be informed.

The Reporting Injuries, Diseases and Dangerous Occurrence Regulations apply to all work activities, but not all incidents are reportable. Employers and people in control of premises have a duty under the regulations to report incidents to the HSE as soon as possible after the event (certain incidents up to ten days). Employees also have a responsibility to inform their employers if they have witnessed a dangerous occurrence, had an accident at work or been certified by a doctor as having a reportable work-related disease. This will then allow the employer to complete a report and pass the information provided to the Health and Safety Executive. Duty holders have a responsibility to keep records of any reportable incidents at work for a minimum of three years after the event.

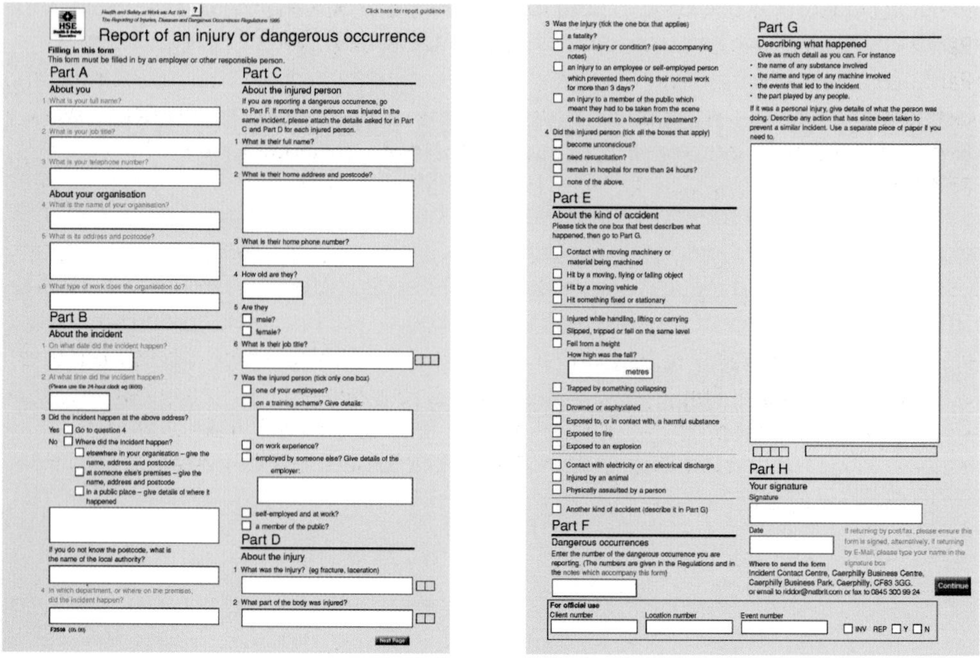

HSE
Health and Safety at Work etc Act 1974
The Reporting of Injuries, Diseases and Dangerous Occurrences Regulations 1995
Click here for report guidance

Report of an injury or dangerous occurrence

Filling in this form
This form must be filled in by an employer or other responsible person.

Part A

About you
1 What is your full name?
2 What is your job title?
3 What is your telephone number?

About your organisation
4 What is the name of your organisation?
5 What is its address and postcode?
6 What type of work does the organisation do?

Part B

About the incident
1 On what date did the incident happen?
2 At what time did the incident happen? (Please use the 24-hour clock eg 0600)
3 Did the incident happen at the above address?
Yes ☐ Go to question 4
No ☐ Where did the incident happen?
☐ elsewhere in your organisation – give the name, address and postcode
☐ at someone else's premises – give the name, address and postcode
☐ in a public place – give details of where it happened
If you do not know the postcode, what is the name of the local authority?
4 In which department, or where on the premises, did the incident happen?

F2508 (05 00)

Part C

About the injured person
If you are reporting a dangerous occurrence, go to Part F. If more than one person was injured in the same incident, please attach the details asked for in Part C and Part D for each injured person.
1 What is their full name?
2 What is their home address and postcode?
3 What is their home phone number?
4 How old are they?
5 Are they
☐ male?
☐ female?
6 What is their job title?
7 Was the injured person (tick only one box)
☐ one of your employees?
☐ on a training scheme? Give details:
☐ on work experience?
☐ employed by someone else? Give details of the employer:
☐ self-employed and at work?
☐ a member of the public?

Part D

About the injury
1 What was the injury? (eg fracture, laceration)
2 What part of the body was injured?

Next Page

3 Was the injury (tick the one box that applies)
☐ a fatality?
☐ a major injury or condition? (see accompanying notes)
☐ an injury to an employee or self-employed person which prevented them doing their normal work for more than 3 days?
☐ an injury to a member of the public which meant they had to be taken from the scene of the accident to a hospital for treatment?
4 Did the injured person (tick all the boxes that apply)
☐ become unconscious?
☐ need resuscitation?
☐ remain in hospital for more than 24 hours?
☐ none of the above.

Part E

About the kind of accident
Please tick the one box that best describes what happened, then go to Part G.
☐ Contact with moving machinery or material being machined
☐ Hit by a moving, flying or falling object
☐ Hit by a moving vehicle
☐ Hit something fixed or stationary
☐ Injured while handling, lifting or carrying
☐ Slipped, tripped or fell on the same level
☐ Fell from a height
How high was the fall? metres
☐ Trapped by something collapsing
☐ Drowned or asphyxiated
☐ Exposed to, or in contact with, a harmful substance
☐ Exposed to fire
☐ Exposed to an explosion
☐ Contact with electricity or an electrical discharge
☐ Injured by an animal
☐ Physically assaulted by a person
☐ Another kind of accident (describe it in Part G)

Part F

Dangerous occurrences
Enter the number of the dangerous occurrence you are reporting. (The numbers are given in the Regulations and in the notes which accompany this form)

Part G

Describing what happened
Give as much detail as you can. For instance
- the name of any substance involved
- the name and type of any machine involved
- the events that led to the incident
- the part played by any people.

If it was a personal injury, give details of what the person was doing. Describe any action that has since been taken to prevent a similar incident. Use a separate piece of paper if you need to.

Part H

Your signature
Signature
Date
If returning by post/fax, please ensure this form is signed, alternatively, if returning by E-Mail, please type your name in the signature box

Where to send the form
Incident Contact Centre, Caerphilly Business Centre, Caerphilly Business Park, Caerphilly, CF83 3GG.
or email to riddor@natbrit.com or fax to 0845 300 99 24

Continue

For official use
Client number | Location number | Event number | ☐ INV REP ☐ Y ☐ N

FIGURE 1.1 HSE Report of an injury or dangerous occurence form

CONSTRUCTION (DESIGN AND MANAGEMENT) REGULATIONS 2007 (CDM)

The new CDM Regulations have been revised in recent years and therefore supersede the CDM Regulations 1994 and the Construction (Health, Safety and Welfare) Regulations 1996, to combine the two laws into one single regulation. The aim of the CDM Regulations is to:

- reduce accidents and raise awareness of health and safety in the industry;
- coordinate the correct people at the correct time to manage health and safety on site;
- centre on effective planning and managing health and safety risks in the industry.

The CDM Regulations place legal responsibilities on almost everybody involved in the construction process, including:

- CDM coordinators – the client is responsible for appointing a competent person to manage the process of health and safety on sites that last for more than 500 person days or lasting more than 30 days; this person is known as the 'Construction (Design and Management) Coordinator';
- clients – the source of funding for a concept and project;
- subcontractors – are usually employed by the principal contractor to complete portions of the contract;
- designers – this is a broad term used for architects, engineers and quantity surveyors, or anybody else involved in the preparation of drawings, schedules and specifications, etc.;
- principal contractors – the main contractor appointed by the client to complete a project;
- workers – this term is used for anyone that is involved in the maintenance, alteration, construction or demolition of a building or structure.

The CDM Regulations 2007 are divided into five parts:

- Part 1 deals with the application of the Regulations and definitions;
- Part 2 covers general duties that apply to *all* construction projects;
- Part 3 contains additional duties that apply *only* to notifiable construction projects, i.e. those lasting more than 30 days or involving more than 500 person days of construction work;
- Part 4 contains practical requirements that apply to *all* construction sites;
- Part 5 contains the transitional arrangements and revocations.

PROVISION AND USE OF WORK EQUIPMENT REGULATIONS 1998 (PUWER)

In general, these regulations require any equipment and machinery provided at work to be safe for its intended use. Employers also have duties to ensure that adequate training, instruction, and supervision in the use and maintenance of the equipment is given to employees. The Provision and Use of Work Equipment Regulations is covered in further detail in Chapter 8, Circular Saws.

ACTIVITIES

Activity 1 – Health and safety regulations

Read through the following questions and answer them as fully as you can to help you develop your underpinning knowledge of this subject area.

1 What do the initials HASAWA stand for?
2 What organisation enforces health and safety legislation in the construction industry?
3 List three powers of the Health and Safety Executive.
4 List three reportable incidents under the RIDDOR.
5 List the key responsibilities of employers under the Construction (Design and Management) Regulations (CDM).

MANUAL HANDLING OPERATIONS REGULATIONS 2002 (MHO)

Manual handling injuries occurring in the workplace account for an alarming proportion of reportable occurrences. The Health and Safety Executive (HSE) has reported that, between 2001 and 2002, over a third of reported injuries were caused by handling, moving or lifting objects. In most cases employees were absent from work for an average of 20 days a year.

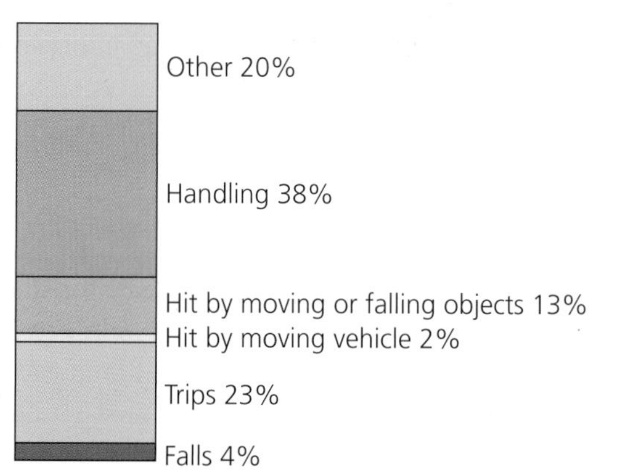

FIGURE 1.2 Injuries reported to the HSE during 2001/2 resulting in a minimum of three days' absence from work

The majority of handling accidents and injuries are caused as a result of the following:

- people using awkward postures to move or lift items;
- heavy manual labour;
- manually handling materials.

The Manual Handling Operations Regulations 2002 place the following duties on employers and employees.

FIGURE 1.3

EMPLOYERS' DUTIES:

1. if possible avoid their employees having to undertake any manual handling operations which pose a potential risk to their health and safety;
2. carry out assessments to record significant risks;
3. as much as reasonably possible, take steps to remove or reduce the risk of injury;

4. provide employees with adequate information to carry out manual handling operations. Employees are normally informed of the risks posed through manually moving or lifting objects in a 'risk assessment'.

The assessment should be reviewed if there is reason to suspect that it is no longer valid or there have been significant changes in the MHO.

FREQUENTLY ASKED QUESTIONS

What is a 'method statement'?

A 'method statement' is a written document, normally completed by a competent person appointed by an employer. They are used to detail safe methods of work to reduce or eliminate the risks highlighted in risk assessments.

EMPLOYEES' DUTIES:

1. employees are responsible for complying with the safe systems of work enforced by their employer;
2. use the equipment provided by the employer;
3. report hazards encountered;
4. make sure that their activities do not endanger others.

RISK ASSESSMENTS

SAFETY METHOD STATEMENT	
Description of work	Re-manufacture of MDF panel products
Risks	• Irritation to the skin, eyes, nose ot throat • Possible explosion
Company training and safety method statements to be observed	• Housekeeping • Manual handling • Using power tools • Using wood-working machines
Requirements to be observed	• Always wear gloves when handling MDF • Eye protection and dust masks should be worn • Wash and brush down at the end of your shift and before eating drinking or going to the toilet, to ensure all dust has been removed • Use barrier cream • Do not smoke
First aid procedures	• In the event of skin irritation or discomfort, please seek traetment from your first aider or consult a doctor

FIGURE 1.4 Sample method statement

Manual handling of items should always be a last resort, after every other possible method has been discounted. While carrying out a manual handling 'risk assessment' you should consider the following questions.

1. Will the operation involve twisting repetitively?
2. Can you avoid having to move the object?
3. Could a 'lifting aid' be used?
4. Will the lifting operation require more than one person?
5. Is the load harmful, bulky, heavy, stable, etc.?
6. Can the load be divided into smaller parts to reduce the weight?
7. Are there risks imposed by the environment, e.g. weather conditions, lighting, floor surface (uneven, slippery, etc.)?
8. What personal protective equipment (PPE) will be required to undertake the task safely?
9. How far will employees be expected to carry the load?
10. How long will employees be expected to repeat the manual handling operation?

RDJ contractors

RISK ASSESSMENT

Activity: manually handling plasterboard

Participants: site operatives

Does the activity involve any potential risk?

Yes / ~~No~~ (delete as appropriate)

Please indicate the level of risk?

~~Low~~ / Medium / ~~High~~ (delete as appropriate)

Please provide details of action taken to protect against potential risks:

1 Wear safety boots

2 Wear gloves

3 Use panel carrier (lifting aid)

4 No more than two sheets to be carried at one time – between two operatives

5 Ensure correct lifting techniques are used

Signature of risk assessor: D Smith

Date of assessment: 8 January 2009

Action to be taken by: 21 March 2009

Date of reassessment: 3 April 2009

FIGURE 1.5 Sample risk assessment

MANUAL HANDLING OPERATIONS

The Health and Safety Executive (HSE) has recommended maximum lifting weights for manual handling stable objects with both hands.

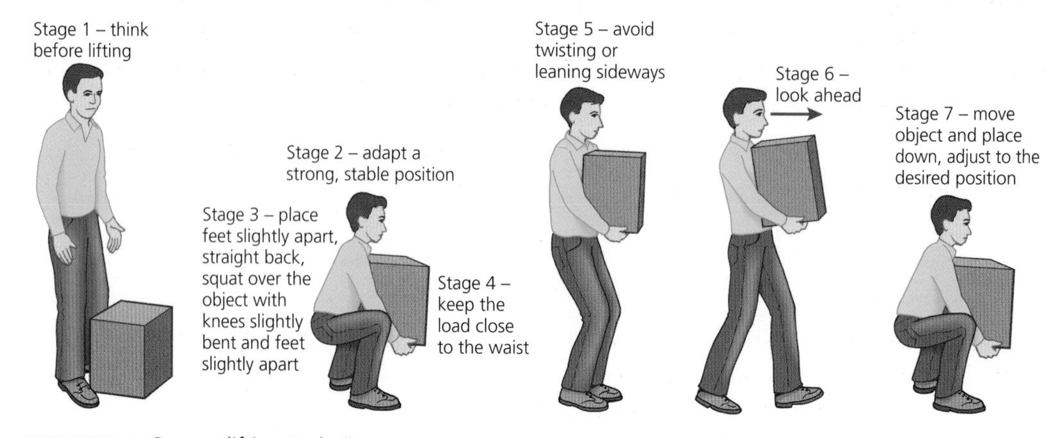

FIGURE 1.6 Correct lifting technique

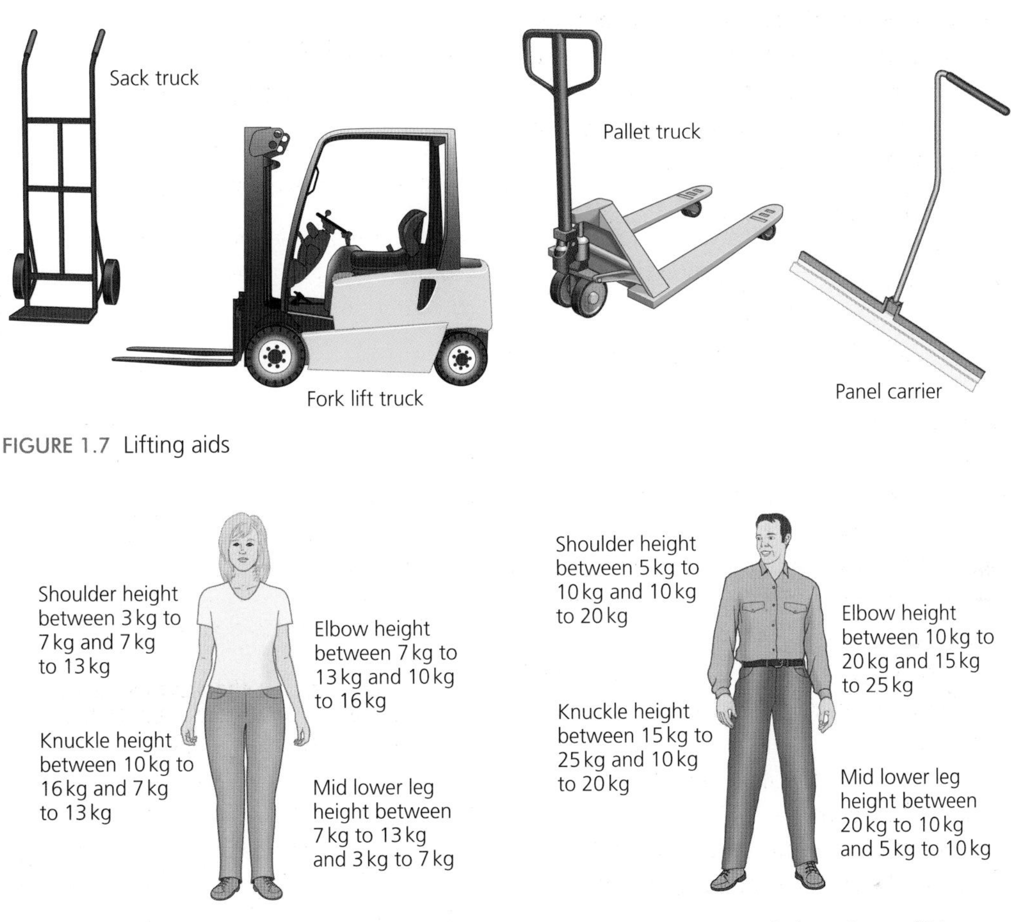

FIGURE 1.7 Lifting aids

FIGURE 1.8 Recommended maximum lifting weights for a female

FIGURE 1.9 Recommended maximum lifting weights for a male

A SUMMARY OF EMPLOYERS' AND EMPLOYEES' DUTIES

It would be impractical to expect employees to be able to remember all the current health and safety workplace law in the UK, but they should have at least a good general understanding of their responsibilities. Employers also have legal obligations to protect every employee's health, safety and welfare while they are at work. In general employers *must* inform and protect employees at work, while employees have a legal duty to report issues of health and safety to their employer or safety representative.

The following lists give a brief overview of the current responsibilities employers and employees have to comply with workplace legislation.

EMPLOYERS' RESPONSIBILITIES TO INFORM EMPLOYEES OR SAFETY REPRESENTATIVES OF THE FOLLOWING:

1. arrangements to appoint competent people to satisfy health and safety law;
2. changes to health and safety at work;
3. provide information about health and safety planning;
4. provide information on the risks and dangers at work, and the measures in place to reduce or eliminate them.

EMPLOYERS' DUTIES/OBLIGATIONS:

1. appoint a suitably qualified person to deal with matters of health and safety;
2. carry out risk assessments, record significant findings, and arrange health and safety measures;
3. draw up a 'health and safety policy' to bring issues of health and safety to the attention of employees for companies with five or more personnel;
4. eliminate or control the use of substances hazardous to health;
5. ensure machinery, plant and equipment is safe to use;
6. ensure materials and substances are transported, stored and used properly;
7. ensure safe methods/systems of work;
8. ensure suitable safety measures are in place to protect against electrical equipment, flammable and explosive substances, noise and radiation;
9. ensure that the health, safety and welfare arrangements are satisfied in the workplace;
10. inform the Health and Safety Executive of all reportable accidents, diseases and dangerous occurrences;
11. make sure that the equipment provided for work activities is suitable for its intended use, correctly used, and regularly serviced and maintained;
12. provide adequate emergency procedures;
13. provide adequate information, training, instruction and suitable supervision in matters of health and safety;
14. provide adequate lifting aids to avoid unsafe manual handling;
15. provide and maintain adequate safety signs and notices;
16. provide first aid facilities;
17. provide personal protective equipment and clothing free of charge to employees;
18. provide welfare facilities for employees, e.g. toilets, drying rooms, washing facilities;
19. provide a safe working environment without risks.

EMPLOYEES' DUTIES/OBLIGATIONS:

1. act responsibly and take care for their own health and safety in the workplace;
2. cooperate with supervisors, managers and employers in matters of health and safety;
3. not to interfere, damage or misuse any items provided for their health, safety and welfare;
4. take reasonable measures to ensure the health and safety of others affected by their activities;
5. use items of work equipment, plant and PPE in accordance with information, training and instruction.

ACTIVITIES

Activity 2 – Safe handling of materials and equipment

Read through the following questions and answer them as fully as you can to help you develop your underpinning knowledge of this subject area.

1. What is the maximum weight a male and female can lift manually?
2. List the duties of employees under the Manual Handling Operations Regulations 2002.
3. Explain the safe process of manually lifting an object from the floor.
4. List four lifting aids.
5. What are the main causes of manual handling accidents and injuries?

SOURCES OF HEALTH AND SAFETY INFORMATION

CONSTRUCTIONSKILLS

ConstructionSkills was formally known as the Construction Industry Training Board (CITB). It is an employer-led organisation that works with partners in the government responsible for improving the skills and productivity in the sector. In recent years ConstructionSkills has reported a shortage of skilled operatives in the industry and a shortfall of new recruits for degree-level management courses. ConstructionSkills covers all areas of the construction industry from 'professionals', such as architects, to 'craft' workers, such as carpenters and joiners. Its aim is to attract people into the industry to fill significant skill gaps, provide training and help them to achieve their qualifications.

FREQUENTLY ASKED QUESTIONS

What does the term 'professional' mean?

Although, technically, any person that is paid for their services would be considered to be a professional, in the construction industry 'professionals' are people that have completed a university-level degree – for example, architects, structural engineers, civil engineers.

ConstructionSkills supports work-based learning construction apprenticeship frameworks, generally referred to as 'apprenticeships'. All trainees wishing to complete an apprenticeship must have a 'training provider' to support them through their programme. Some training establishments or colleges may provide their own 'work-based learning training provider' to fund trainees through their apprenticeships.

Qualifying construction companies are liable to pay the Sector Skills Council annually sums of money that reflect their turnover and the size of their firm; this is known as a 'levy'. The levy contributes to the funding needed to support apprentices while they are training. Companies that employ apprentices are entitled to several 'grants' in return for the experience and on-site training they provide to trainees.

APPRENTICESHIPS

An apprenticeship qualification is made up of a framework of different components to provide the apprentice with a rounded period of training to prepare them to meet the needs of the industry.

FIGURE 1.10 National Vocational Qualification flowchart

FREQUENTLY ASKED QUESTIONS

What is ERR?

The Secretary of State requires all apprenticeship frameworks to include 'Employment Responsibilities and Rights'. ERR is a compulsory short course of training that covers nine target areas:

1. employer and employee statutory rights under employment law and other legislation;
2. procedures and documentation;
3. sources of information on employment rights and responsibilities;
4. occupational roles within the industry;
5. occupational and career pathways;
6. roles and responsibilities of representative bodies;
7. sources of information and advice on the industry, occupation, training and career;
8. codes of practice and National Occupational Standards;
9. recognising and forming views of public concern that affect the industry.

CONSTRUCTIONSKILLS CERTIFICATION SCHEME (CSCS)

The ConstructionSkills Certification Scheme was introduced to the construction industry to improve health and safety awareness and standards on site, improve quality and help drive 'cowboys' out. Most construction site managers, employers and clients demand to see evidence that personnel have proven their understanding of and competence in basic health and safety in the industry before being allowed on a site. Enforcement of a card scheme helps employers to comply with current health and safety legislation.

Compliant employers require all persons entering or working on their building sites to have undertaken a theoretical CSCS health and safety test, and produce a certified card as evidence that they have successfully met the minimum requirement. The tests have to be booked through the CSCS card scheme and sat at a local testing centre. The test must be completed within 45 minutes and consists of 40 multiple-choice questions covering a range of various aspects of health and safety in the industry. Candidates must answer a minimum of 34 questions correctly to complete the test successfully, although no feedback is given other than 'pass' or 'fail'.

If candidates are unsuccessful in reaching the required standard, they are able to retake the test on another occasion. The test is selected at random from a bank of questions available from ConstructionSkills; these can be purchased either as a book or on CD-ROM.

TYPES OF CSCS SKILL CARD

There are various types of CSCS card available; each one is awarded to suit each individual and their level of qualifications in the industry:

- Red card – Trainee;
- Green card – Construction site operatives;
- Blue card – Experienced worker/craft;
- Gold card – Advanced craft/supervisory;
- Platinum card – Management;
- Black card – Senior management;
- White/yellow card – Professionally qualified person;
- Yellow card – Visitor with no construction skills;
- White card – Construction-related occupation.

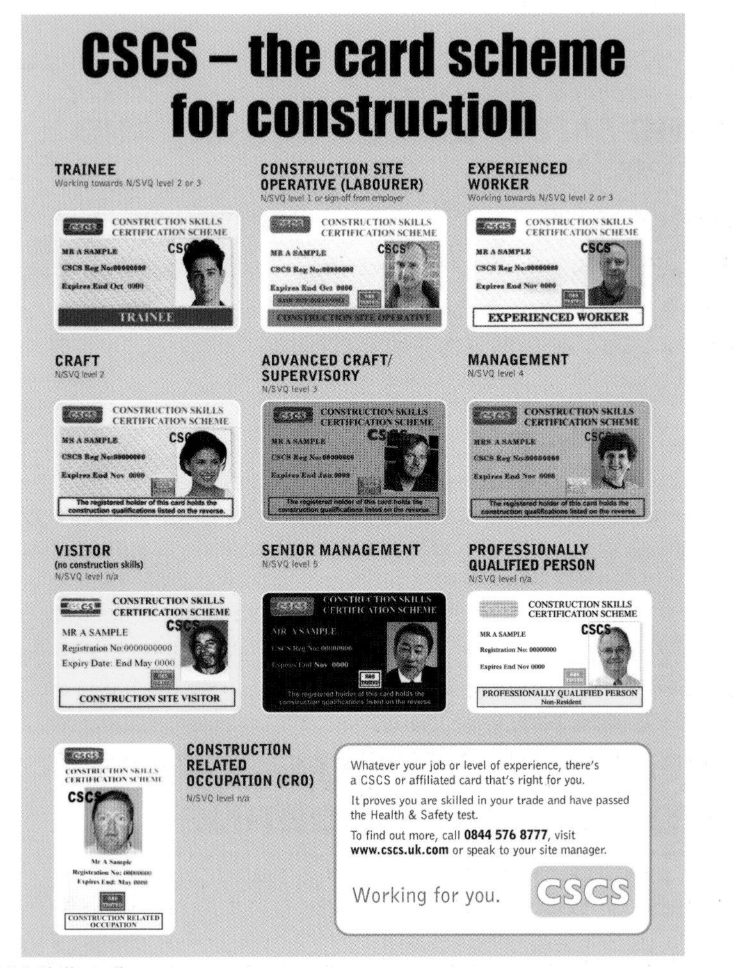

FIGURE 1.11 CSCS Skill cards

THE ROYAL SOCIETY FOR THE PREVENTION OF ACCIDENTS (ROSPA)

The RoSPA is a charity organisation that has been established for over 90 years. Its purpose is to provide sources of information and guidance to help reduce all types of domestic and industrial accidents, and help save lives. The RoSPA has reported the following astonishing statistics, occurring every year in the United Kingdom:

- approximately 350 fatalities to workers and members of the public due to reportable accidents at *work*;
- approximately 1000 deaths in *work*-related road crashes;
- approximately 12,000 early deaths due to past exposure to hazardous agents – for example, asbestos;
- over 36 million working days lost due to *work*-related accidents and ill health.

The RoSPA provides various sources of practical and technical information, guidance and support including:

- advice;
- conferences and events;
- links to further guidance and support;
- press releases;
- resources;
- safety groups;
- training courses;
- videos.

HEALTH AND SAFETY EXECUTIVE (HSE) AND COMMISSION (HSC)

As mentioned previously, part of the HSE and HSC role is to provide information and guidance to allow the duty holders, including employers, to comply with the law. The HSE produces many documents known as 'codes of safe practice'; these give practical advice to follow to comply with current legislation. Further guidance is published on 'information sheets'. These sheets provide specific requirements and details for the safe use of equipment and machinery, and can easily be obtained via the internet.

TOOLBOX TALKS

Developments and changes made on building sites require health and safety arrangements to be reviewed periodically. Consideration of the relevance of previous safety measures in the workplace, and new dangerous situations and risks arising as building work develops, should be highlighted to site personnel. This information is normally communicated to the workers through short verbal discussions known as 'toolbox talks'. Toolbox talks are commonly conducted by the site manager or safety officer during lunch breaks, or as personnel first enter the site at the beginning of the day.

ACCIDENT, FIRST AID, EMERGENCY PROCEDURES AND REPORTING

THE ACCIDENT BOOK (BI 510)

The accident book is a document used to record serious and minor injuries at work. Under the Reporting of Injuries, Diseases and Dangerous Occurrence Regulations (RIDDOR), any notifiable injuries should be reported to the HSE as soon as possible after the event. The accident book is used to meet the requirements of the management of health and safety law by recording the details shown in Figure 1.13.

FREQUENTLY ASKED QUESTIONS

▶ Who should complete the accident book?

An accident should be recorded as soon as practically possible after the event by the injured person. If that is not possible, then a witness or someone present at the time should complete the details.

Changes in legislation have prevented the information contained in the accident book from being disclosed to third parties without the injured person's consent. This prevents older accident books being used after December 2003. Details of accidents should be stored securely and separately from the accident book to prevent an infringement of personal information under the Data Protection Act (DPA) and confidentiality law.

If an accident has occurred at work, the area should be made safe to prevent further injuries, or fenced off in the event of a serious injury to allow an investigation to take place.

Employers and senior managers should review accident books regularly to investigate all recorded incidents, regardless of whether they were reportable under RIDDOR. This will allow improvements to be made in health and safety in the workplace, by preventing recurrences.

THE FIRST AID BOX

It is a known fact that failure to administer first aid during the early stages of an accident could result in minor injuries becoming major ones or even leading to death. All employers must provide suitable first aid equipment and competent first aiders for all workplaces and the

FIGURE 1.12 First aid box

ACCIDENT REPORT

Person affected/injured

Name: ....

Address: ....

.... Postcode: ....

Occupation: ....

Person reporting the incident

Name: ....

Address: ....

.... Postcode: ....

Occupation: ....

Details of accident

Date: .... Time: ....

Description of incident: ....

....

....

First aid administered: ....

....

....

Notifiable accident

Complete this box if the accident is reportable under the Reporting of Injuries, Diseases and Dangerous Occurrences Regulations 1995 (RIDDOR)

How was it reported? ....

Date reported: .... Signature: ....

Date form was completed: ....

Signature of injured person to disclose personal information to safety representatives: ....

FIGURE 1.13 Accident report form

self-employed, regardless of their size. The Health and Safety (First Aid) Regulations 1981 require employers to assess the needs for this provision to provide adequate and appropriate equipment, facilities and personnel to administer first aid to the injured or ill. The exact contents of a first aid box will depend on the number of people it is intended to serve and the type of work undertaken. In general, medication and tablets should not be kept in the first aid box or administered by a first aider. The box should be checked regularly to ensure that it is well stocked and the contents are not out of date. There is no definitive content list for first aid boxes, but the following items would normally be included in a place of work with no special requirements:

- disposable gloves;
- individually wrapped triangular bandages;
- information sheets on administering basic first aid;
- safety pins;
- sterile eye pads,
- various sizes of individually wrapped sterile wound dressings.

The HSE recommends that adequate notices are displayed in the workplace with the names of the nominated first aiders and the location of the first aid rooms and boxes.

Further details can be obtained from the relevant code of practice: The Health and Safety (First Aid) Regulations 1981 L74.

ACTIVITIES

Activity 3 – Accident, first aid, emergency procedures and reporting

Read through the following questions and answer them as fully as you can to help you develop your underpinning knowledge of this subject area.

1. Name two items that should not be included in a first aid box.
2. Who has the authorisation to complete the details in an accident book?
3. Where should an accident book be stored?
4. List five important details that should be recorded in an accident book after an accident on site.
5. What do the initials RIDDOR stand for?

ACTIVITIES

Activity 4 – Identifying hazards on construction sites

Read through the following questions and answer them as fully as you can to help you develop your underpinning knowledge of this subject area.

1. Who has a duty to carry out 'risk assessments' in the construction industry?
2. List five methods of good housekeeping on site.
3. What is a method statement?
4. Who should be informed of a 'near miss' on site?
5. If a hazard is identified on a construction site, what action should be taken?

HEALTH AND HYGIENE

WELFARE FACILITIES

All fixed construction sites must have adequate welfare facilities for site personnel, to meet the minimum requirements of the Construction (Design and Management) Regulations 2007. Employers must make suitable arrangements to provide welfare facilities for employees before any construction work starts. Larger sites may require more than one welfare facility to meet the demand of the increased number of workers on site, and to allow convenient access. Employers have a legal duty to provide the following facilities as a minimum requirement:

- hot and cold running water wash facilities, with soap and clean towels or driers;
- drinking water;
- toilets (*note* – men and women can share the same facility provided they are partitioned off from the urinals);
- drying rooms;
- canteen facilities with tables, chairs, a means of boiling water and heating food;
- secure storerooms for work clothing, with separate lockers.

Employers must also ensure that the facilities are clean and tidy (good housekeeping), and adequately lit and heated, with a good source of ventilation. All welfare arrangements should be clearly identified in the site 'Health and Safety Plan' as part of the Construction (Design and Management) Regulations (CDM).

NOISE AT WORK

Exposure to loud noise for long or short periods of time can cause temporary deafness or permanent hearing loss. The Control of Noise at Work Regulations 2005 requires employers to take the following measures to ensure the health and safety of their employees:

- assess the risks in the workplace (risk assessment);
- monitor noise levels in the workplace (these must not exceed the safe legal noise limits);
- provide adequate hearing protection (PPE);
- provide information, instruction and training (signage, safe use of safety devices, etc.);
- reduce noise exposure (length of time exposed to damaging noise levels, etc.);
- reduce the risks as far as practically possible.

Working in an area that requires you to raise your voice to be heard by a person approximately 2 metres away suggests that the noise levels in that area may be too high. Although you may consider the exposure to noise to be only a minor risk, the damage to the inner ear may increase gradually until it becomes noticeable. An early sign of damage to the nerve endings in the inner ear may be the inability to hear normal levels of conversation with background noise. In most cases this type of damage to the inner ear and nerve endings is irreversible and may result in 'tinnitus', permanent ringing in the ear.

FIGURE 1.14 Noise levels are measured and recorded in decibles (dB) with a sound meter

EXAMPLES OF SOUND LEVELS:

- television 20 dB;
- primary classroom 70 dB;
- power drill 90 dB;
- a busy bar or nightclub 100 dB;
- circular saw bench 102 dB.

Employers have a responsibility to make employees aware of the hazards and risks in their place of work by carrying out thorough
risk assessments and providing method statements. They should also display the correct warning signs and create zones where hearing protection should be worn at all times in areas of high risk. The HSE recommend the following safe guidelines:

- 80 decibels (dB) – employers should assess the risk to health;
- 85 decibels (dB) – hearing protection zones, personal protective equipment (PPE) should be worn (mandatory);
- 87 decibels (dB) – maximum exposure limit, taking into account the reduction provided by PPE.

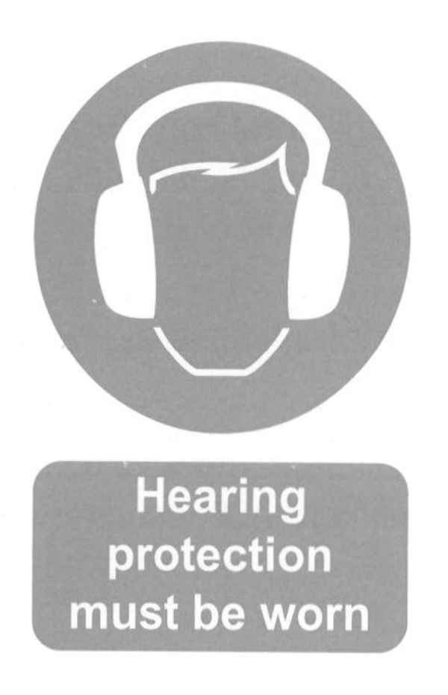

FIGURE 1.15

FIGURE 1.16 Hearing protection PPE

TRADE SECRETS

Hearing protection should be worn as recommended on the manufacturer's instructions. Failure to wear the equipment correctly may fail to provide adequate levels of protection or may even damage your hearing; for example, ear plugs inserted too deep into the ear canal may injure the inner ear. Disposable PPE should be worn only once by each user and correctly disposed of at the end of each use. Hearing protection that fits into the ear canal should never be shared with others, as this may cause possible infection to be passed from one person to another.

Hearing protection should reduce noise to a safe level but not eliminate it all together, as this could pose a potential risk if a person is operating machinery or equipment, or if an evacuation alarm is sounded.

ACTIVITIES

Activity 5 – Health and hygiene

Read through the following questions and answer them as fully as you can to help you develop your underpinning knowledge of this subject area.

1. List five essential welfare facilities that should be available on all construction sites.
2. How can you identify a dangerous noise level on site?
3. What measures can be taken to reduce the risk of potentially damaging noises at work?
4. Why is personal hygiene important on construction sites?
5. Men and women are permitted by law to share toilet facilities on site: true or false?

BASIC WORKING PLATFORMS

DANGERS OF WORKING AT HEIGHT

Several thousand major injuries are reported to the Health and Safety Executive every year as a result of falls or falling objects from height causing harm. The Working at Height Regulations 2005 (WAHR) place legal duties on everybody responsible for the welfare of personnel, including employers, the self-employed and managers. In general, duty holders are responsible for ensuring, as far as practically possible, that falls from height are prevented using the following measures:

- all work at height is properly planned and organised;
- all work at height takes account of weather conditions that could endanger personnel;
- assess the risks as a result of working at height;
- consider alternative methods of completing the work;
- ensure equipment for work at height is appropriately inspected;
- health and safety;
- produce method statements for the safe use of the access equipment;
- ensure the place where work at height is done is safe;
- the risks from falling objects are properly controlled;
- those involved in work at height are trained and competent.

Employees also have responsibilities to follow the training and instruction provided by their employers or safety representatives. In addition, they should also use the equipment and safety devices provided properly, and report any safety hazards encountered while carrying out their duties.

SELECTING EQUIPMENT FOR USE

Careful consideration should be given to the type of access equipment used for a particular task, to ensure the minimum amount of risk to the user and anyone in the immediate area. There are several other factors to consider that enable work to take place without personal risk of injury; these include:

- length of time required for use;
- maximum height to be accessed;
- frequency of use;
- static or mobile;
- type of work being carried out;
- amount of people required to use the equipment;
- ground conditions (level or uneven?/firm or soft?/slippery?)
- weather conditions and lighting;
- consider the position of the access equipment; will it pose a threat to the general public or people working underneath?

TYPES OF ACCESS EQUIPMENT AND BASIC WORKING PLATFORMS

Access equipment is interpreted as any item of equipment that allows personnel, tools and materials to gain safe entry or exit to one or more different levels. There are various types of access equipment available for use; these include:

- hop-ups;
- leaning ladders;
- stepladders;
- trestle platforms;
- mobile tower scaffolds;
- independent scaffolds (Figure 1.17).

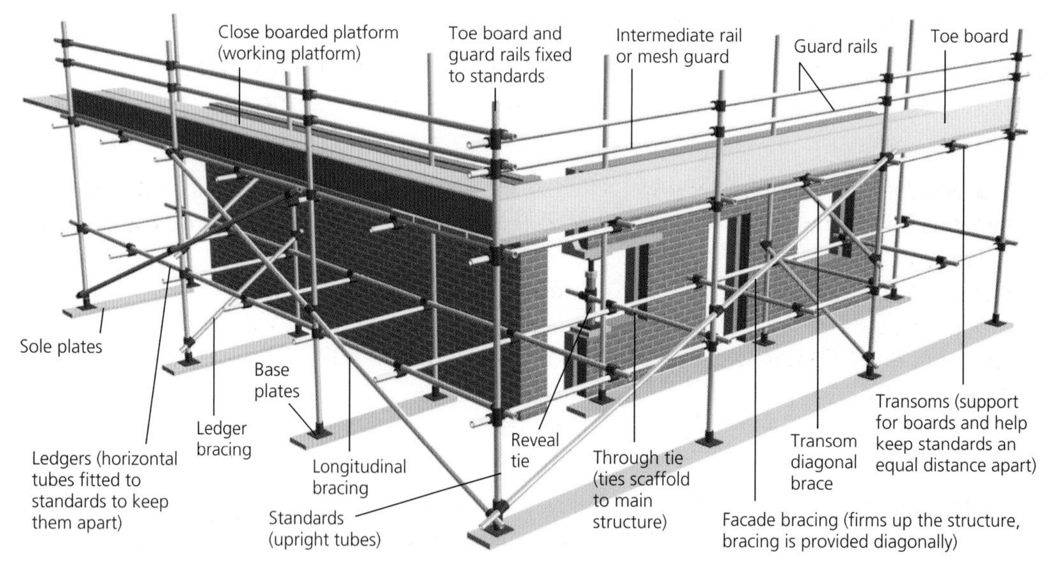

FIGURE 1.17 A section through an independent scaffold

Each item of access equipment requires the user to be suitably trained and informed about its safe use, maintenance and inspection. However, independent scaffolds should only be erected and inspected by fully trained scaffolders, known as 'card holders'.

HOP-UPS

Hop-ups are the smallest items of access equipment, with a maximum of two to three steps giving access to lower levels. This type of equipment should be used only for short periods of time because of its lack of handrail and platform space.

LADDERS

The term 'ladders' refers to several different types of access equipment. These include:

1. leaning ladders;
2. stepladders.

Leaning ladders

Single, double and triple extension ladders are all styles of 'leaning ladders'. Once fully erected they should be positioned at a safe working angle of 75°, or a ratio of 1:4 (one in four units). Wherever possible, leaning ladders should be securely tied at the top and secured to a temporary stake in the ground at the lower level. Alternatively, if the ladder is going to be used only for short periods of time it should be supported by an additional person 'footing' the bottom rungs. Ladders used to access a working platform should extend above the stepping-off point by 1 metre; this allows the top rungs to be used as handrails and reduces the risk of the ladder slipping off the platform.

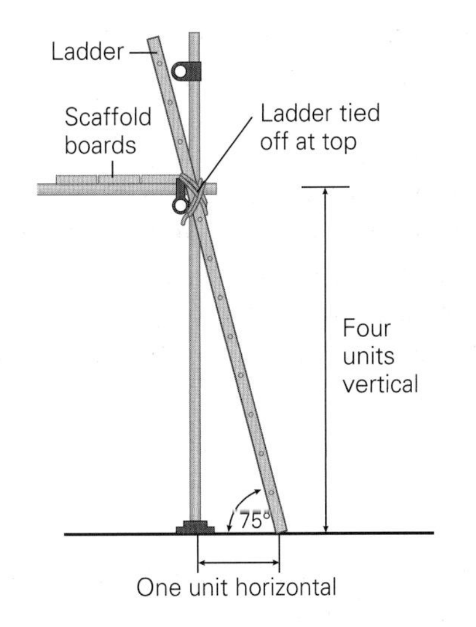

FIGURE 1.18 Safe use of a leaning ladder

It is good practice to use a single leaning ladder to access a working platform, especially if it is required for a long period of time. Wooden or steel 'pole' ladders are normally used for this purpose, because they do not require overlapping mid-span and therefore have no weak spot. Pole ladders commonly range in size from 3 metres up to 10 metres.

Ladders should be thoroughly inspected at the beginning of the working day by the person intending to use the equipment. The following areas should be considered as part of these daily inspections:

- wooden components should be checked for splits, cracks and significant signs of damage;
- avoid using and report any painted items of access equipment as this may be concealing defects;
- check the rungs are in good condition, not loose, bent, split or missing;
- check the stiles are not bent as this may lead to the ladder collapsing;
- check the non-slip feet of the stiles are not missing or excessively worn.

Stepladders

Traditionally, wooden stepladders were used in the construction industry, although they are rarely used nowadays because of their ability to become unstable through continued use and wear. These types of stepladder are also considerably heavier than their modern aluminium and fibreglass equivalents. Fibreglass ladders are commonly used by electricians because they are safer, due to their inability to conduct electricity in the event of an emergency.

The following inspections should be carried out on stepladders before use:

FIGURE 1.19 A stepladder

- check the locking bars are not bent or missing;
- ensure the stepladder is set up on firm, level ground;
- ensure the stepladder is fully open, with the locking bars fully engaged;
- check the feet of the stiles are not missing and they are in good condition;
- check the platform is not buckled or split;
- check the steps are free of mud and dirt, to avoid slipping;
- check the steps are secure;
- check the stiles are not bent or damaged;
- check the ropes are in good condition and taut when the steps are erected (if applicable).

TRESTLE PLATFORMS

Adjustable steel trestles are normally used used by bricklayers, and suitably heightened as the building work progresses. Alternatively, 'A' frames can be positioned, at either end of trestle boards to create and support a stronger working platform. Trestle platforms (or scaffolds) should be fitted with handrails in positions where the risk assessment has highlighted the need for edge protection. In some cases it is not always practical to fit handrails on both sides of the platform because the type of work would be restricted. The distance and consequences of falling from trestle scaffolds should always be minimised with 'fall arrest equipment' or restricting the maximum height. The duration of work on trestle scaffolds should always be keep to a minimum to reduce the risk of falls from height; alternative methods should be sourced for longer periods.

Trestle scaffolds should always be set up on firm, level ground, with a minimum platform width of 450 mm. The overhang of the working platform on either end of the trestles should be a minimum of 50 mm and maximum of four times the thickness of the boards. These guidelines prevent the boards slipping off the edge of the trestles and flipping up in the event of somebody stepping on their end. Working platforms should also never be positioned higher than two-thirds the overall height of the trestles.

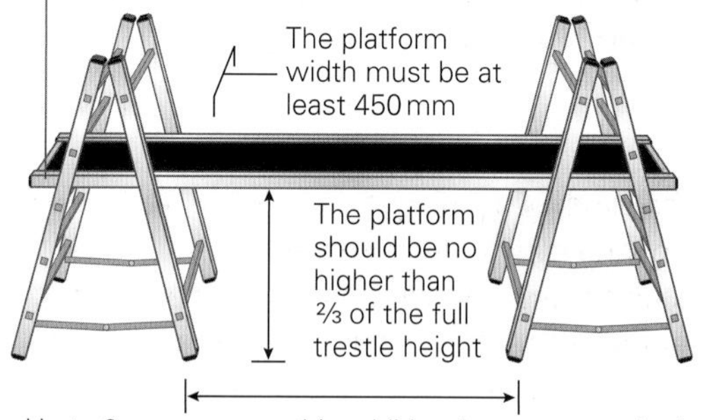

FIGURE 1.20 Safe use of a trestle platform

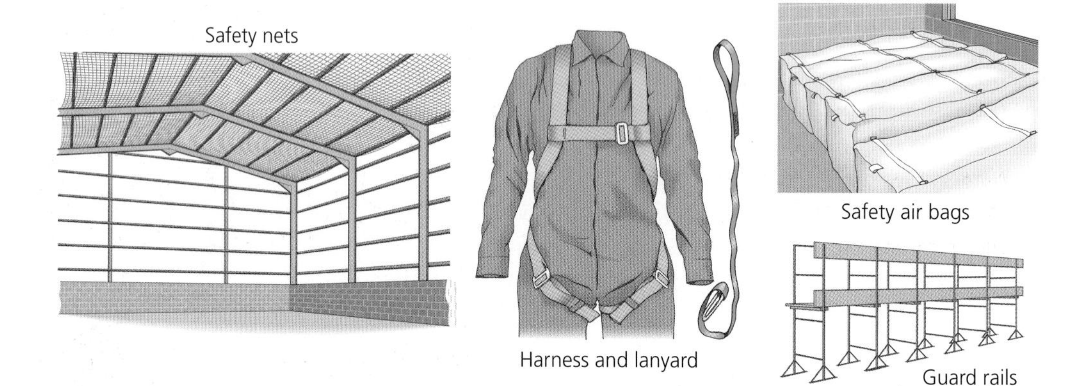

FIGURE 1.21 Safety devices used at height

MOBILE TOWER SCAFFOLDS

Mobile tower scaffolds are used for a variety of maintenance, inspection and building work. The loose component parts of the tower allow them to be adjusted to several different heights, with working platforms at intermediate positions in between. Mobile towers should be erected only by trained and competent people authorised to do so, following the manufacturer's instructions.

In general, mobile tower scaffolds should comply with the following recommendations for safe use:

- use only suitable component parts;
- never assemble the tower with damaged or broken parts;
- ensure all handrails are fitted at a minimum height of 950 mm;
- ensure all intermediate rails are fitted below the handrails, leaving a maximum gap of 470 mm between;
- never overload mobile towers with tools, equipment or personnel;
- never move towers with people on board;
- never use in high winds or adverse weather conditions;
- never climb the outside of towers (mountaineering);
- tower should not exceed 4 metres when it is being moved;
- ensure the brakes are engaged before use;
- consider overhead wires, electricity cables and obstructions;
- ensure the ground conditions are suitable for use (e.g. flat, level, firm);
- always move towers manually and never by mechanical means (e.g. vehicles);
- isolate the working areas around mobile towers and use suitable signage to inform people of the hazards.

INSPECTION

The Working at Height Regulations 2005 (WAHR) state that any item of access equipment or working platform where there is a risk of falling 2 metres and over (except mobile towers) should be inspected no more than seven days before use. Further inspections should be carried out at regular intervals to be determined by the person carrying out the method statement. In general, the period of time between re-inspections should not compromise the health and safety of the users or people affected by its use. Static scaffolds are normally inspected by a competent

TRADE SECRETS

While working at height, materials may have to be removed to a lower level. These items should never be thrown or 'bombed' over the edge of the access equipment, or from any height, because of the risks of personal injury to people below. Alternatively, they could be sent safely down a 'rubbish chute' into a covered skip below.

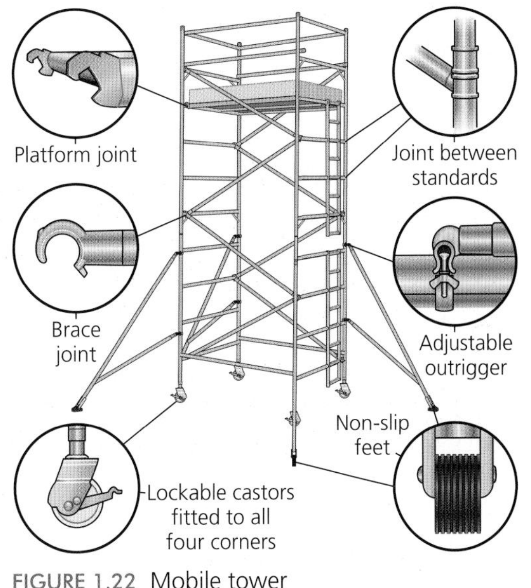

FIGURE 1.22 Mobile tower

person once they have been erected, and at least every seven days thereafter. During adverse weather conditions, such as frost, snow, high winds and heavy rainfall, further inspections should take place. All inspections carried out should be recorded within 24 hours and kept on the site with the health and safety file. Once the project has been completed, the inspection reports should be stored for a further three months at the company's head office.

ACTIVITIES

Activity 6 – Basic working platforms

Read through the following questions and answer them as fully as you can to help you develop your underpinning knowledge of this subject area.

1. List three precautions that could be used to prevent falls from height.
2. What is the difference between a 'rung' and a 'step'?
3. At what height should handrails be used on trestle scaffolds?
4. What is the maximum height a mobile tower scaffold should not exceed when it is to be moved?
5. How often should scaffolding be inspected?

WORKING WITH ELECTRICITY

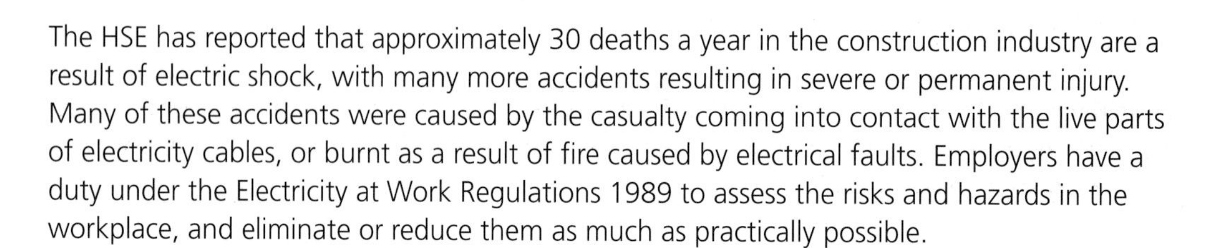

The HSE has reported that approximately 30 deaths a year in the construction industry are a result of electric shock, with many more accidents resulting in severe or permanent injury. Many of these accidents were caused by the casualty coming into contact with the live parts of electricity cables, or burnt as a result of fire caused by electrical faults. Employers have a duty under the Electricity at Work Regulations 1989 to assess the risks and hazards in the workplace, and eliminate or reduce them as much as practically possible.

FREQUENTLY ASKED QUESTIONS

What is the difference between a 'risk' and a 'hazard'?

The term 'risk' means that there is a chance of injury.

The term 'hazard' means that something *will* cause harm or personal injury.

Domestic mains supply electricity is 230 volts; this level is powerful enough to kill anyone who comes into direct contact with it. To reduce the risk of death on site as a result of electrocution, it is recommended that the supply is reduced to a safe level. Most construction sites now use 110 volts by reducing the mains supply with a 'transformer'; 110 volts electricity cables and transformers are easily identified by their yellow casing and round three-pin plugs and sockets.

The risk of serious electric shock from mains supplies can be prevented by plugging a residual current device (RCD) into the power source. RCDs are small, sensitive devices that have the ability to rapidly cut the supply of electricity, making it safe. If an RCD does 'trip out' it is probably because of an electrical fault and should be investigated further before continued use.

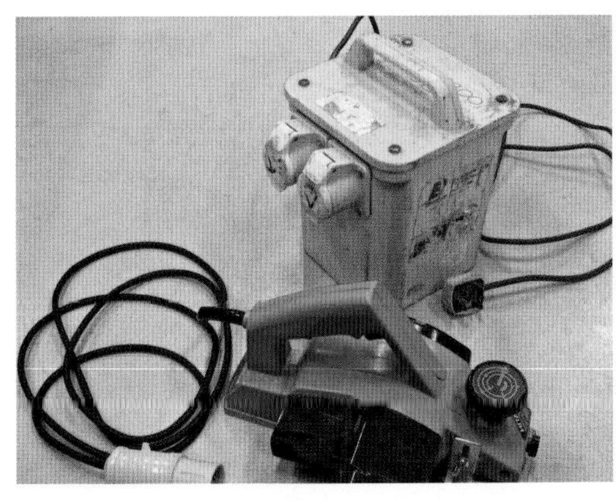

FIGURE 1.23 Transformers and 110 volt power tools

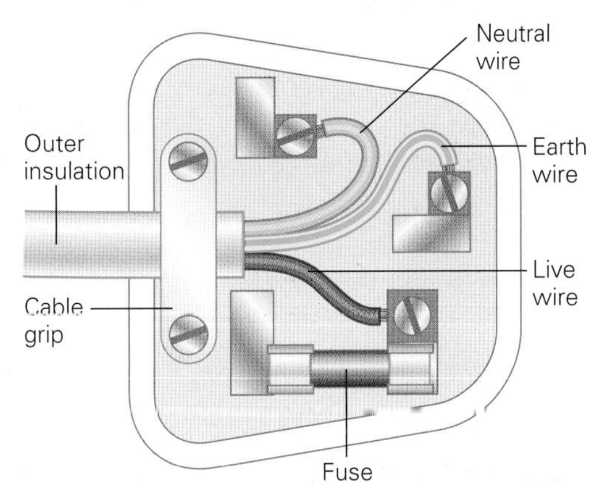

FIGURE 1.24 Wiring a domestic three-pin plug

FREQUENTLY ASKED QUESTIONS

If I use 110-volt power tools or an RCD with mains power, does this mean the risk of electric shock is reduced?

No, these methods will not reduce the risk of being electrocuted. They will only reduce the risk of death as a result of electric shock.

In addition to 110 volts and 230 volts, 415 volts (three phase) must be used to operate large items of equipment and machinery. Sockets, plugs and leads with this power supply are easily recognisable by their blue casing and sheaths.

SAFETY PRECAUTIONS TO BE TAKEN WHEN WORKING WITH ELECTRICITY

The possible risks of electrocution on site apply not only to electricians working directly with wires, cables, etc., they potentially affect everyone on site. The following guidelines should be followed by all personnel.

- If possible, avoid using 230 volts on construction sites.
- Check that plugs and cords on power tools and extension leads are not cracked, split or damaged.
- Check wires are not exposed around plugs and sockets.
- Use another source of power in wet or damp conditions, such as 'compressed air' or 'battery' powered tools.

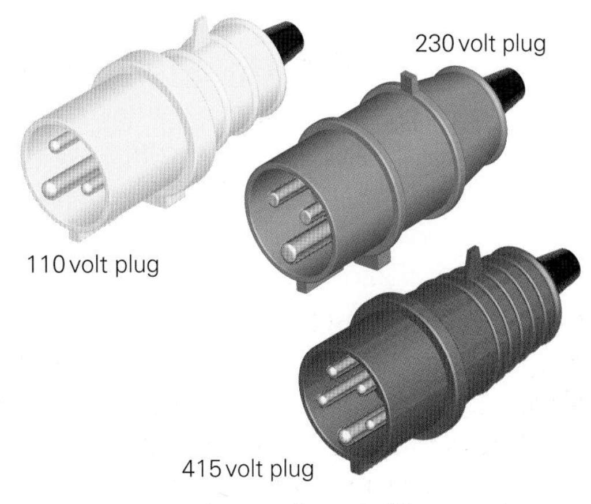

FIGURE 1.25 Colour coding of different voltages

- Avoid overloading plug sockets.
- Keep leads and power cables above head height if possible to avoid trip hazards, etc.
- Never use damaged items of electrical equipment.
- Follow the site rules and the employer's method statements.
- Always isolate power tools before leaving them on site.
- Keep walkways free of electric power tools and electrical cables.
- Use only double-insulated power tools.
- Ensure all electrical items have a current PAT testing certificate, and that they are regularly maintained.

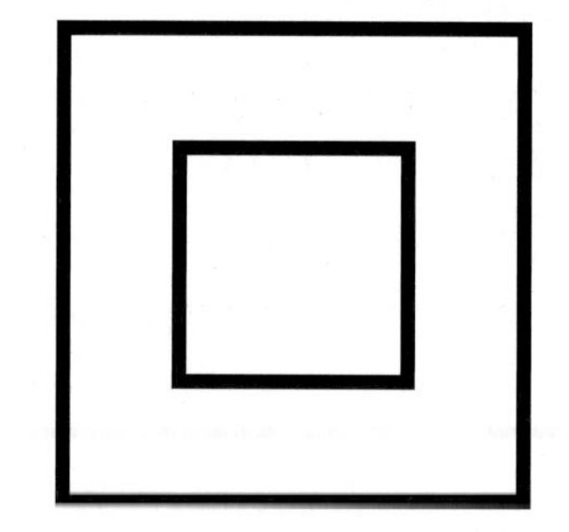

FIGURE 1.26 Double insulation symbol

FREQUENTLY ASKED QUESTIONS

What is 'PAT' testing?

'PAT' is an abbreviation for 'portable appliance testing'. The Electricity at Work Regulations require all portable electrical appliances to be tested regularly to ensure they are safe for use. There are no recommended guidelines stating the period of time between inspections; this generally depends on the frequency of use.

EFFECTS OF ELECTRIC SHOCK

If high levels of electricity run through the human body the current normally heats up the tissue and results in deep burns that are likely to require surgery. The victim may also suffer muscular spasms, cardiac arrest and may stop breathing. All these injuries are potentially life threatening, although the amount of electricity and length of time the victim came into contact with it will have a bearing.

DEALING WITH AN INCIDENT INVOLVING ELECTRIC SHOCK

1. Raise the alarm, without leaving the victim.
2. Do not touch the casualty, as they may still be in contact with the electricity supply.
3. Isolate the power at the source. If this is not possible and the victim is still in contact with the supply, use a non-conductible item, such as a wooden broom, to move the hazard away.
4. Check for signs of a pulse and breathing. If none is found, start mouth-to-mouth resuscitation and chest compressions (consult a trained first aider before attempting).
5. Check for signs of shock.
6. Treat the victim's burns with sterile dressings.

Note: these steps should only be used as a guide. Professional medical advice and training should be sought before treating burn victims.

ACTIVITIES

Activity 7 – Working with electricity

Read through the following questions and answer them as fully as you can to help you develop your underpinning knowledge of this subject area.

1. What colour extension cables should be used with 110-volt portable power tools?
2. What is a transformer?
3. In a 240-volt three-pin plug, what colour sheathing should be used on the neutral wire?
4. What do the initials RCD stand for?
5. What is the purpose of 'PAT' testing?

PERSONAL PROTECTIVE EQUIPMENT (PPE)

Personal protective equipment (PPE) should be worn only as a last resort, after eliminating or reducing all other possible risks and hazards from the working area. Some items of PPE are compulsory on construction sites due to the nature of the project; these items may include:

- safety footwear (boots/shoes/riggers);
- high-visibility clothing (commonly referred to as a 'hi-vis');
- safety helmets;
- builders' gloves.

FIGURE 1.27 A construction site notice

The minimum requirements for PPE on a construction site will be displayed on information notices and posters on the boundary fence at the entrance to the site. Site personnel will also be informed of the site manager's requirements during their induction onto the site. In addition to these requirements other items of PPE should worn if there is a risk of personal injury.

The Personal Protective Equipment at Work Regulations 1992 requires all employers to provide employees and agency workers with all necessary PPE free of charge to allow them to carry out their job safely. PPE is considered to be any item of protective equipment, including weatherproof clothing.

EXAMPLES OF PPE TO PROTECT VARIOUS PARTS OF THE BODY

EYE PROTECTION

Safety glasses/safety goggles/face shields/visors. *Safeguards against* – chemical splashes, dust and dirt, flying objects, gas and vapour, ultraviolet radiation (sun rays).

HEAD PROTECTION

Hard helmets/bump caps. *Safeguards against* – risk of bumping head, impact from falling or flying objects.

FREQUENTLY ASKED QUESTIONS

What is a 'bump cap'?

A bump cap looks very similar to a baseball cap with built-in head protection. Bump caps are usually lighter and more comfortable to wear than hard hats, and are usually used in areas where the risks of head injury are reduced.

FIGURE 1.28 Eye protection

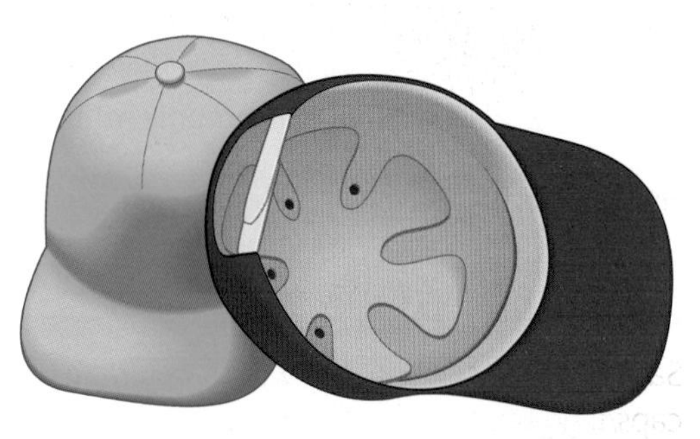

FIGURE 1.29 Bump caps

RESPIRATORY PROTECTION

Disposable moulded dust masks/dust mask with replaceable filters/half- or full-face respirators/air-fed helmets. *Safeguards against* – dust, dangerous gases and vapour.

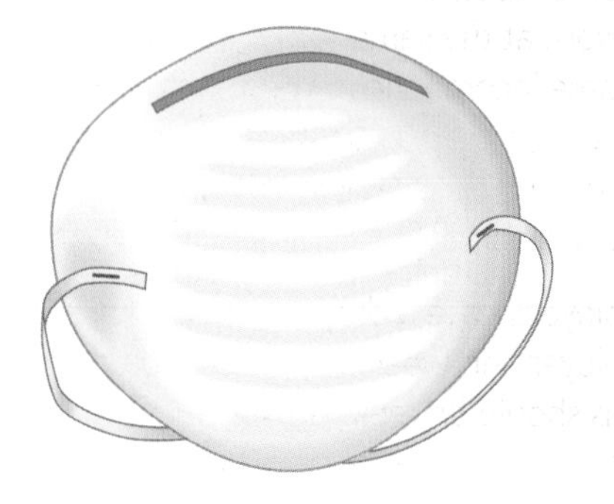

FIGURE 1.30 Dust mask

BODY PROTECTION

Boiler suits/disposable overalls/aprons/high-visibility vests and coats. *Safeguards against* – entanglement of own clothing, adverse weather, chemical splashes, spray from pressure leaks or spray guns, contaminated dust.

HAND/ARM PROTECTION

Gloves/gauntlets/armlets/wrist cuffs. *Safeguards against* – abrasions, cuts, extreme temperatures, chemicals, skin infection.

FIGURE 1.31 High-visability jacket

FIGURE 1.32 Gauntlets

FIGURE 1.33 Safety boots

FOOT/LEG PROTECTION

Safety boots, shoes and riggers with steel toe caps/leggings/gaiters. *Safeguards against* – slipping, cuts, falling objects, chemical spills, abrasion, wet conditions.

In some cases, several items of PPE may need to be worn at the same time, causing them to become incompatible with each other. This can be overcome by using specially designed equipment; for example, mounted ear defenders and visors can be added to hard hats.

FIGURE 1.34 Hard hat with ear defenders and visor

Employees have a duty to request PPE from their employer and wear it correctly. Any damaged items should be replaced immediately to ensure that the equipment functions properly. Employees should also look after the equipment they have been provided with by storing it correctly and taking reasonable measures to make sure that it is not damaged through neglect.

ACTIVITIES

Activity 8 – Using appropriate personal protective equipment (PPE)

Read through the following questions and answer them as fully as you can to help you develop your underpinning knowledge of this subject area.

1. List three mandatory items of PPE to be worn on all sites.
2. Employees should provide their own safety boots in the construction industry: true or false?
3. On what part of the body are gauntlets worn?
4. How often should PPE be replaced?
5. If ear defenders are supplied by your employer and you are wearing a hard hat, you are permitted to remove it for a short period of time on site: true or false?

FIRE AND EMERGENCY PROCEDURES

THE FIRE TRIANGLE

Fuel, heat and oxygen are the three elements needed to allow fire to ignite and burn; collectively these are known as the 'fire triangle'. If one of these elements was to be removed the fire will cease to continue burning. These principles are applied to tackle and extinguish fires in the event of an emergency, although the exact method needed will depend on the type of fire. For example, a wood or paper fire could be extinguished with water; however, if the same method was used for an electrical fire there would be a risk of electric shock.

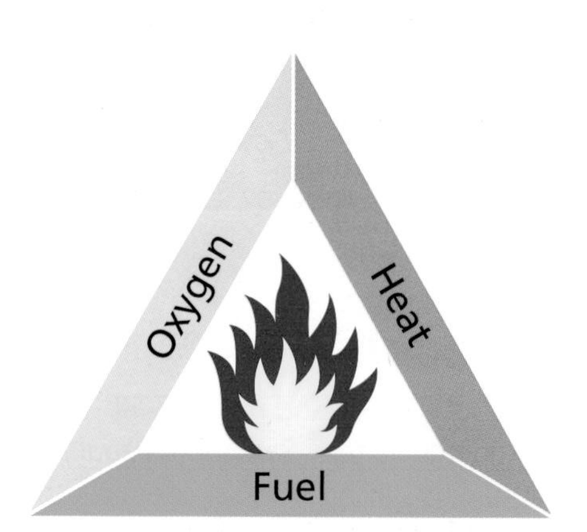

FIGURE 1.35 The fire triangle

There are many different types of fire-fighting equipment and fire extinguishers to deal with various fires. In the past the contents of a fire extinguisher were identified by the colour of its container. Changes in legislation now require all types of fire-fighting equipment and fire extinguishers to be coloured red, and identified by their coloured label and the information contained on it.

CLASSIFICATIONS OF FIRE

British Standards BS EN:2 1992 has categorised fires into five different classes so that the type of equipment can quickly be identified and selected for use.

1. Class A – fires involving solid materials, usually of an organic nature; 1
2. Class B – fires involving liquids or liquefiable solids;

3. Class C – fires involving gases;
4. Class D – fires involving metals;
5. Class E – fires involving electrical equipment.

TYPES OF FIRE-FIGHTING EQUIPMENT

Type of Extinguisher	Class of fire to be treated
Water extinguisher 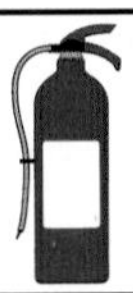	A
Foam extinguisher 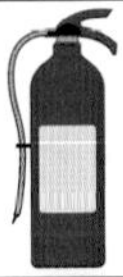	A, B
Carbon dioxide extinguisher 	B, C, E
Dry powder extinguisher 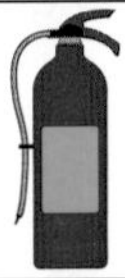	A, B, C, D, E
Fire blanket	A, B, C, D, E (Will smother all fires. Particularly useful for when clothing is alight as there is no risk to skin or breathing. Other extinguishers may be harmful.)

FIGURE 1.36 Fire-fighting equipment

METHODS OF FIRE PREVENTION

Employers should conduct a thorough risk assessment of their place of work to establish the equipment and materials needed in the event of an emergency. They should also provide information and training to employees to eliminate or reduce the risks at their establishment.

During routine risk assessments, employers and people responsible for the health and safety of others should consider the following hazards:

- storage of waste materials on site and good housekeeping;
- storage and quantity of dangerous substances used on site, including petrol, liquid petroleum gas (LPG), paints and varnishes, solvents and dust;
- dust, vapour, gas and mist pose a high risk of fire or explosion if mixed with air.

Employers also have a duty to provide suitable warning and safe conditions signs as well as visual and audible alarm systems. All personnel should be made aware of the escape routes to the nearest assembly points on site, so that they can be accounted for in the event of an emergency evacuation. In preparation for such an event, regular practice drills should be conducted by employers to eliminate any confusion and reduce the length of time taken between raising the alarm and assembly.

FREQUENTLY ASKED QUESTIONS

Should I return to a burning building to use the equipment provided to fight the fire if I have received training?

No, even if you have received training in the use of the equipment supplied. These items are provided to gain a safe route from a burning building in the event of a fire. Remember to follow your employer's evacuation plan in the event of an emergency and remain at the assembly point until instructed to do otherwise.

ACTIVITIES

Activity 9 – Fire and emergency procedures

Read through the following questions and answer them as fully as you can to help you develop your underpinning knowledge of this subject area.

1. What element is missing from the following fire triangle: heat, fuel and … ?
2. What class of fire involves electrical equipment?
3. Other than fire extinguishers, name an item of fire-fighting equipment.
4. Extinguishers with black labels contain what substance?
5. What type of extinguisher should be used on gas fires?

MULTIPLE-CHOICE QUESTIONS

1 Accidents at work must be reported in accordance with which one of the following regulations?
 a MHOR
 b PUWER
 c RIDDOR
 d COSHH

2 Under the Health and Safety at Work Act, which one of the following is a responsibility of an employee?
 a To wash their hands prior to eating
 b Not to interfere or misuse PPE
 c To examine or investigate accidents
 d Not to report the lateness of workers

3 Which one of the following areas of is covered by PUWER?
 a Personal protective equipment
 b Lifting and handling procedures
 c Safe working at height
 d Use of work equipment

4 How many days' absence from work as a result of an injury can be taken before it must be reported to the Health and Safety Executive?
 a 2
 b 3
 c 4
 d 5

5 A prohibition notice is served to stop:
 a all work on site continuing
 b lorries from entering the site
 c people drinking alcohol on site
 d dangerous work on site continuing

6 The **main** reason for using correct lifting techniques is to:
 a work quicker
 b prevent injuries
 c save time when loading/unloading
 d allow the load to be carried further

7 The recommended voltage supply used on site is:
 a 24 V
 b 55 V
 c 110 V
 d 230 V

8 Which one of the following is used to reduce the voltage from 230 v to 110 v on site?

a Inducer
b Reducer
c Transformer
d Residual current device

9 The correct pitch for the safe use of a ladder is

a 1:2
b 1:4
c 1:6
d 1:8

10 Which one of the following acts would legislate on the control of dust?

a HASAWA
b RIDDOR
c COSHH
d MASK

CHAPTER 2

INFORMATION, QUANTITIES AND COMMUNICATING WITH OTHERS 2

LEARNING OUTCOMES

By the end of this chapter you should have developed a knowledge and understanding of:

- interpreting and producing building information;
- estimating quantities of resources;
- communicating workplace requirements efficiently.

INTRODUCTION

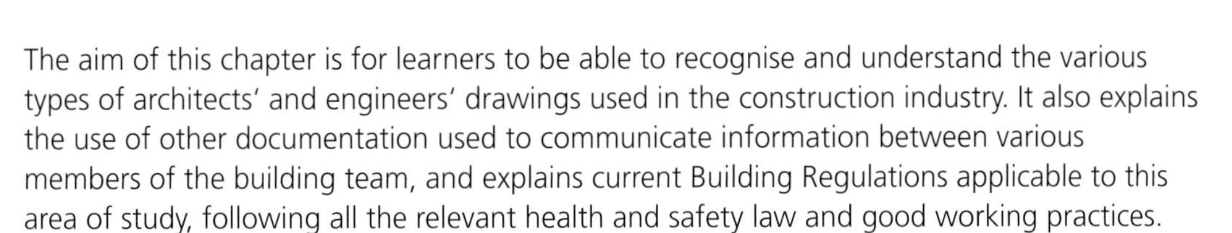

The aim of this chapter is for learners to be able to recognise and understand the various types of architects' and engineers' drawings used in the construction industry. It also explains the use of other documentation used to communicate information between various members of the building team, and explains current Building Regulations applicable to this area of study, following all the relevant health and safety law and good working practices.

SOURCES OF INFORMATION

Effective lines of communication in the building industry are vital to ensure that projects run smoothly, within budget and are completed within an agreed timescale. Communication begins on a building project with the first contact between the client and the architect to discuss a concept or idea, through to the site manager inducting tradesmen/women on to the site. Lack of communication between the client, the main contractor, designers, architects and different trades can result in costly mistakes and delays on a project. Effective communication between members of the building team is vital if changes are made to the contract documentation, or matters of health and safety arise on site. The following problems commonly occur in the construction industry as a result of poor communication:

- labour/tradespeople not on site and available at the right time and in the right sequence;
- materials not ordered in advance, resulting in delays on site; for example, carpenters waiting for the kitchen units to arrive so that the plumber can 'plumb in' the sink etc. to complete his second fixing;
- skips not ordered or emptied, resulting in stoppages;
- visits by the Building Control Officer are not requested in advance, delaying further progress on the build;
- changes to build details and contract documents, resulting in several sets of information being used by different members of the building team.

A line of communication can be as simple as a conversation between two parties, a memorandum (memo) on a notice board or even an employee's pay slip. Verbal dialect, hand signals and even body movement, posture and facial expressions are all methods of communication. Although we use these methods every day while undertaking work activities, there are several disadvantages. These include:

- they can often be misinterpreted;
- they are often misunderstood;
- there is no written evidence that can be referred back to at a later date.

FIGURE 2.1 Methods of communication

The term 'written information' refers to all forms of documentation, including architects' drawings, sketches, programmes of work, etc., as well as text. Problems and queries arising throughout a construction project are usually passed through a hierarchy of people until the issue is resolved.

FREQUENTLY ASKED QUESTIONS

What is a 'hierarchy'?

A 'hierarchy' is term used to refer to the chain of command or pecking order on a construction site. The client is the most important member of the building team because they are usually funding the project; they are therefore always at the top of the ladder (see Figure 2.2).

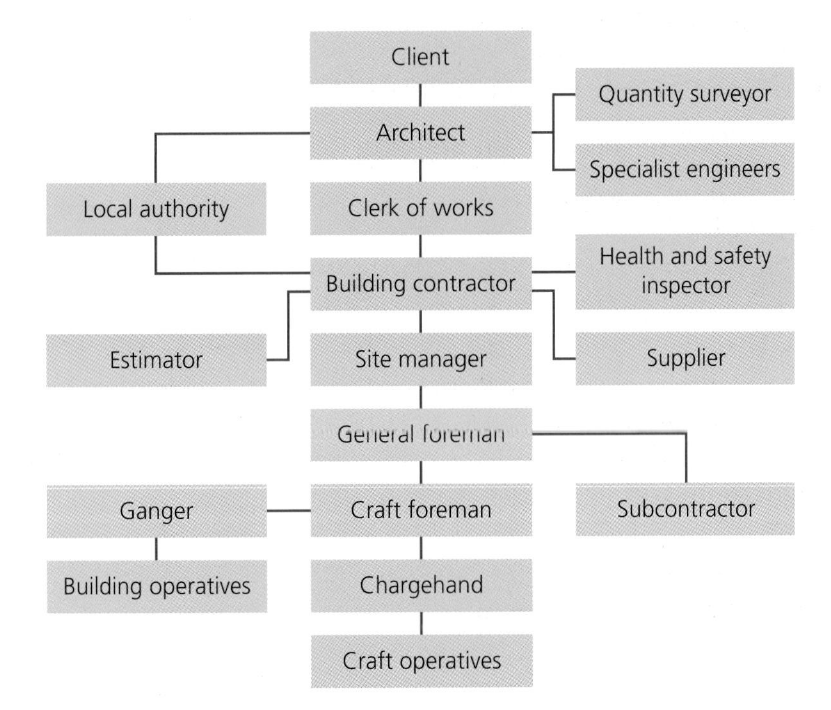

FIGURE 2.2 The hierarchy of a building team

The processes of controlling materials, labour and progress on a building site are usually down to one person – the 'site manager'. Although everyone on a construction site is usually working indirectly for the client, it is the site manager who will refer to the contract details to plan and organise the project. The most important sources of information a site manager will use are:

- architects' and structural engineers' drawings;
- bill of quantities;
- work programmes/schedules;
- specifications;
- manufacturers' information;
- conditions of the contract between the main contractor and the client.

These sources of information are compiled between the architects, the structural/services engineer and the quantity surveyor for the approval of the client before a project begins. When 'planning permission' and 'Building Regulations approval' have been granted by the local authority, the client will then pass these documents to the nominated main contractor. When an agreement has been signed between both parties, these papers are then referred to as the 'contract documents'. Any further changes or developments throughout the project must be in addition to the original agreement and are therefore an amendment. These changes to the original contract must be made in writing on a 'variation order' to prevent disputes when extra costs are claimed by the main contractor at the end of the project.

FREQUENTLY ASKED QUESTIONS

Where can I obtain Building Regulations information?

The local planning office will answer any questions you have regarding current Building Regulations. In most cases the architect and structural engineer will design the project, following the guidelines laid out in the approved documents below:

- Part A – Structure;
- Part B – Fire;
- Part C – Site preparation and resistance to moisture;
- Part D – Toxic substances;
- Part E – Resistance to the passage of sound;
- Part F – Ventilation;
- Part G – Hygiene;
- Part H – Drainage and waste disposal;
- Part J – Heat-producing appliances;
- Part K – Stairways, ramps and guards;
- Part L – Conservation, fuel and power.

When a project has been granted planning permission, Building Regulations approval must be sought to commence the build. Alternatively, the building work may start on a 'Building Notice' issued by the local authority, providing any work carried out complies with current regulations. At several stages throughout the building work, the Building Inspector will visit the project to ensure that the progress made satisfies the requirements of Building Regulations.

INTERPRETING INFORMATION

ARCHITECTS' DRAWINGS

BLOCK PLANS

Block plans identify the position of the building plot in relation to the surrounding area. They are normally drawn in scaled-down proportion of the real size, but rarely contain measurements. Block plans are used to submit information to the Planning Office in order to obtain planning permission.

SITE PLANS

Site plans are drawn to a bigger scale to illustrate the position of the proposed building on the plot. They usually contain basic information, including site dimensions, tree positions, drainage and main roads, etc.

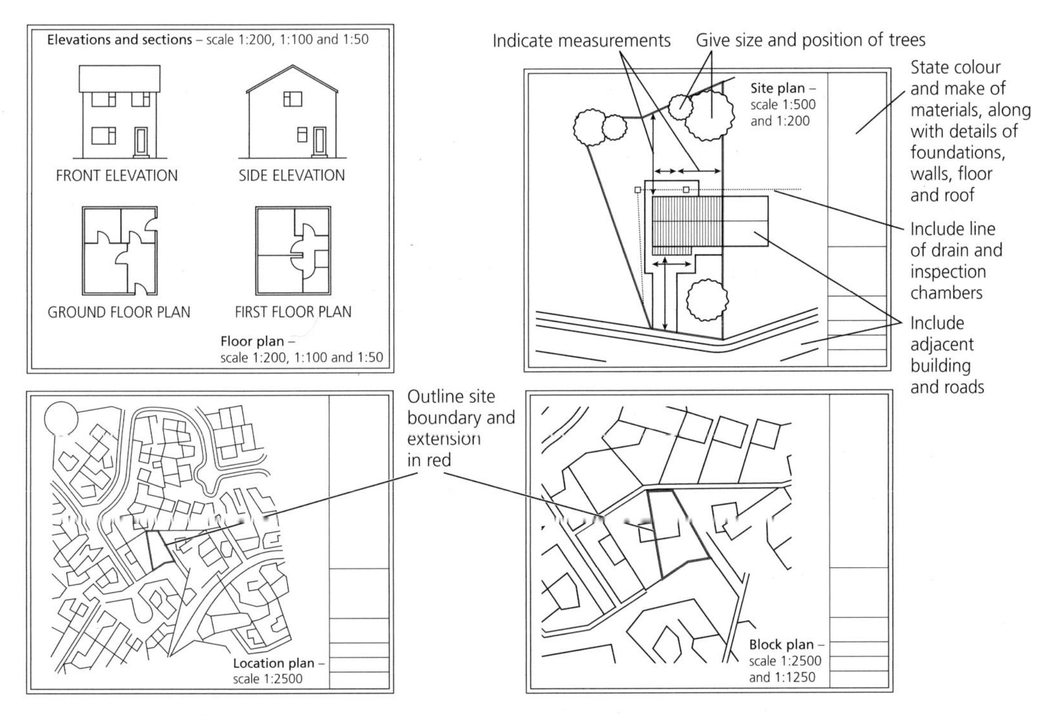

FIGURE 2.3 Site plan

ELEVATIONS

Elevations illustrate views of the building from all four sides to give a visual impression of the aspects. These are not working drawings and should not be used to construct the building.

SECTIONS

Sections through a building will be drawn by the architect to demonstrate the general construction of the floors, walls, ceiling and roof. Simple dimensions, such as room heights, and general notes are also contained within the drawings.

FLOOR PLANS

Floor plans are a section through a building at approximately 1 metre above floor level. Their purpose is to indicate the layout of all the internal walls, internal and external doorways, windows and staircases within the proposed building.

DETAILED/TECHNICAL DRAWINGS

These are drawn by architects and structural engineers illustrating the information and calculations required to complete particular areas of the build. These may include:

- foundations;
- wall construction (e.g. brick and block cavity wall, timber framed);
- ground and upper floor construction;
- positions of any steel beams or load-bearing walls;

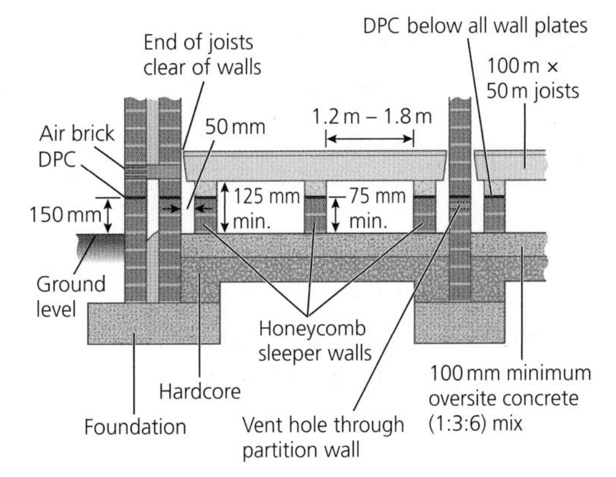

FIGURE 2.4 A technical drawing

- roof construction;
- eaves finishing detail.

RANGE DRAWINGS

Range drawings are used as references to detail the full 'range' of particular components; for example, a 'door range drawing' will illustrate the front views of all the internal and external doors, with basic information such as the overall dimensions (see Figure 2.5). They would also contain a simple referencing system to help identify the doors and their location on the floor plans. Range drawings are particularly useful when 'tendering' for a contract or placing purchase orders through the company 'buyer'.

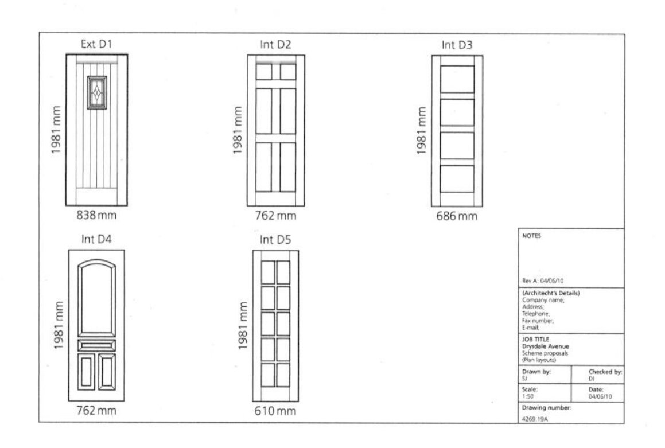

FIGURE 2.5 Example of a door range drawing

FREQUENTLY ASKED QUESTIONS

What does the term 'tendering' mean?

'Tendering' is a process conducted by the client or quantity surveyor on larger builds to determine the main contractor for a building project. A list of work required to complete a project is written by the quantity surveyor and forwarded to a number of building companies. This list is known as a 'bill of quantities'. Each company will itemise the costs for each phase of the build on the bill of quantities and return it for the client's approval.

TITLE PANELS

Title panels are sources of written information that should be contained on all forms of architectural and technical drawings. They are usually contained on the right-hand side of the main drawing and provide useful information, including:

- the name, address, telephone/fax and email address of the architect's or structural engineer's practice;
- job title – for example, the name and address of the client or project;
- drawing title – an explanation of the details drawn; for example, 'Building Regulations First Floor Plan';

- name of the author (the person who developed the drawing);
- name of the person who checked the drawing;
- the scale and size of the paper containing the drawing;
- date first drawn;
- drawing number;
- revision number or date.

'SCALED' MEASUREMENT

A scaled measurement is used to reduce the full size of an original item to a small enough measurement so that it will fit onto a single piece of drawing paper. This allows buildings or projects of any size to be drawn to a true representation of the original proportion. Scaled drawings are traditionally produced by hand on a drawing board with a range of drawing tools, which include a 'scale rule'.

NOTES Rev A: 04/06/10	
(Architecht's Details) Company name; Address; Telephone; Fax number; E-mail;	
JOB TITLE Drysdale Avenue Scheme proposals (Plan layouts)	
Drawn by: SJ	**Checked by:** DJ
Scale: 1:50	**Date:** 04/06/10
Drawing number: 4269.19A	

FIGURE 2.6 Example of a title panel

The disadvantages of this method are that these drawings are slower to produce, they are not as accurate as electronic versions, and the paper will shrink and expand with changes in the climate. Nowadays they are regularly produced by architects, engineers and draughtsmen on 'AutoCAD®' computer programs. There are several advantages of having the drawings electronically; these include:

- images can be distributed quickly between members of the building team;
- the images can easily be magnified, manipulated and amended;
- AutoCAD® (computer-aided design) is 100 per cent accurate.

FIGURE 2.7 An example of an architect's drawing completed with a computer program

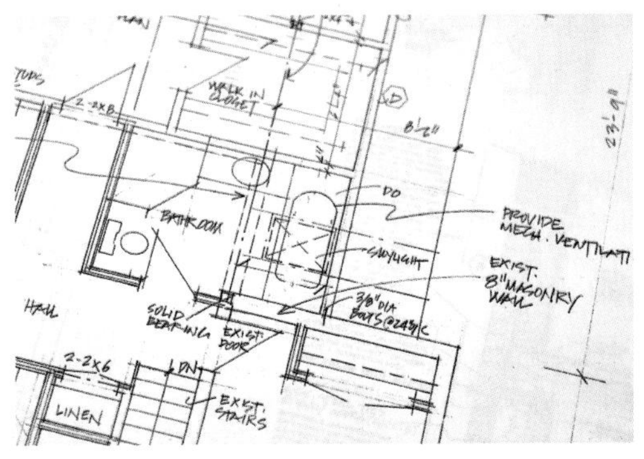

FIGURE 2.8 An example of an architect's drawing completed by hand

CONVERTING SCALE

Only essential information and measurements are usually contained on scaled drawings, other sizes are normally scaled directly from the images. This is simply achieved by referring to the title box to first establish the scale used to draw the pictures. Measurements can then be read from the drawing using the same conversion on a scale rule; alternatively, a standard metric rule can be used. In this case the size measured from the drawing must be multiplied by the scale. Here are some examples.

Example 1

A wall measures 40 mm on a drawing with a scale of 1:100

$40 \times 100 = 4000$ mm

= 4 metres (this is the actual size of the wall)

Example 2

A window opening measures 7 mm on a drawing with scale of 1:50

$7 \times 50 = 350$ mm

= 0.350 metres (this is the actual size of the window)

Example 3

If a floor joist measures 3.8 metres in length and has to be drawn to a scale of 1:50, the actual size it will be drawn will be 76 mm. This is simply calculated by breaking the 3.8 metres down into millimetres and dividing it by the scale size.

3.8 metres = 3800 mm $\div$ 50 = 76 mm (this is the actual size of the floor joist on the drawing)

Whenever possible, any measurements specified on working drawings should always take priority, and scaling directly from them should be the last resort.

TRADE SECRETS

Do not assume that all drawings have been completed by the architect. Always check the title panel before scaling from the drawings, to ensure that the architect has not made a comment. In some cases the architect will provide instructions not to scale from the drawing, and all measurements are to be checked on site.

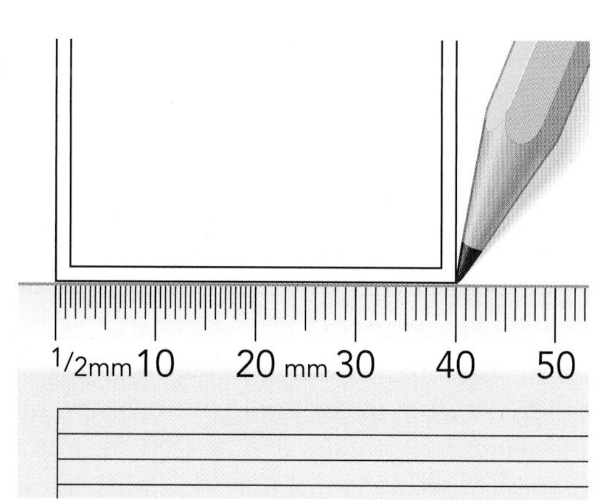

FIGURE 2.9 Example 1

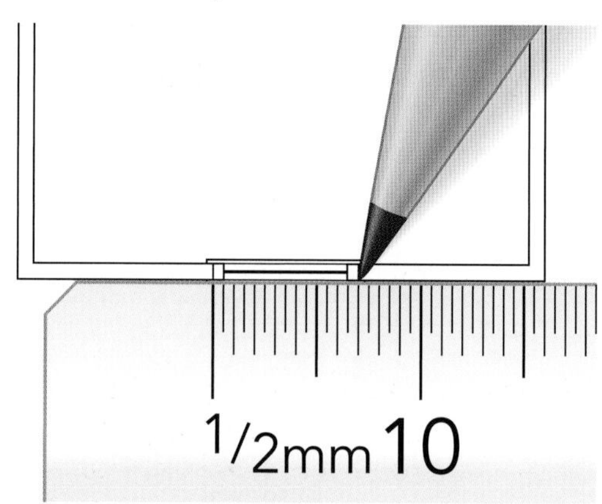

FIGURE 2.10 Example 2

DRAWING SYMBOLS AND ABBREVIATIONS

Drawing symbols and abbreviations are commonly used on architects' and engineers' drawings to prevent full explanations cluttering up the details. The symbols and abbreviations used by architects and engineers have been standardised by the British Standards Institute (BS 1192) to prevent confusion (see Figure 2.11).

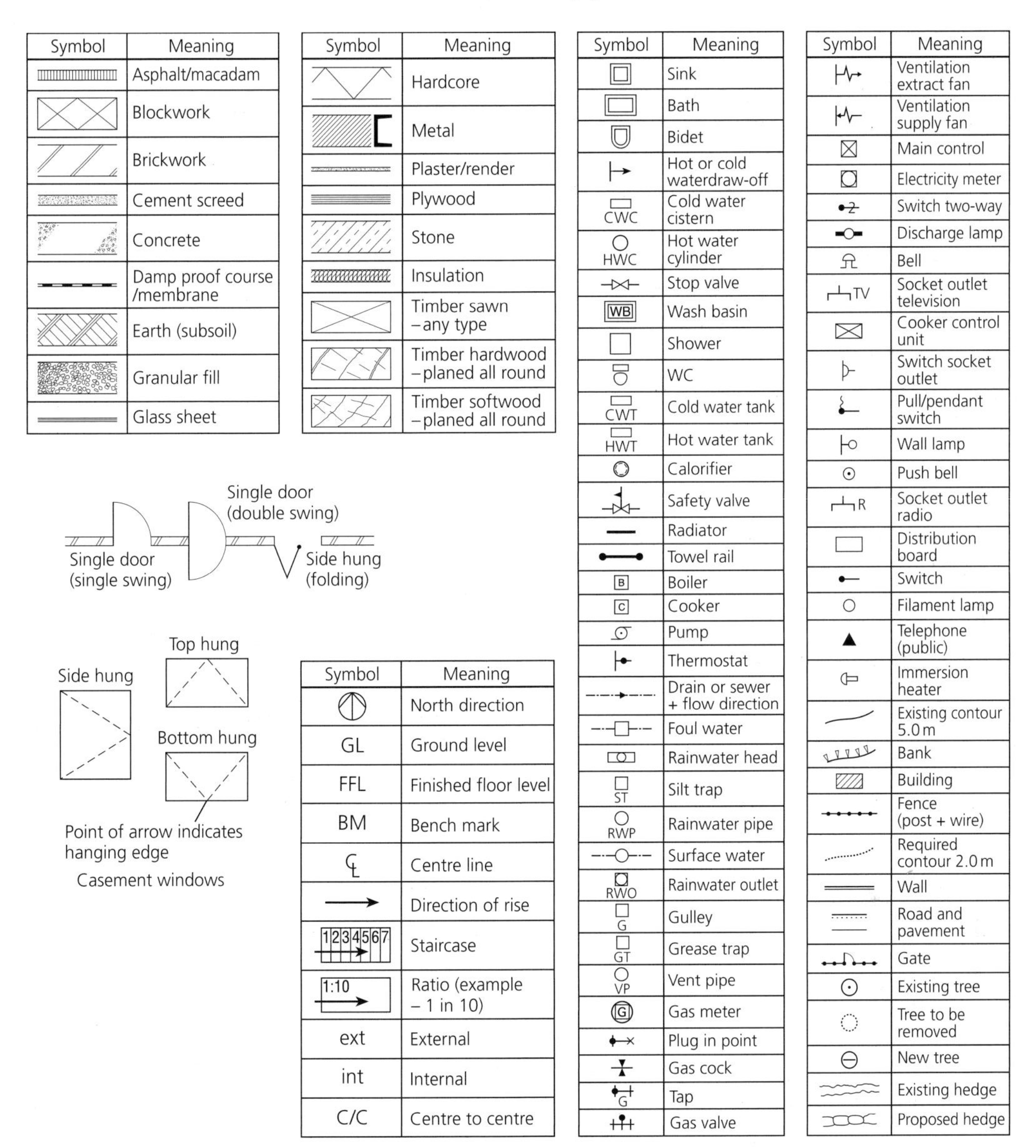

FIGURE 2.11 Drawing symbols and abbreviations

BILL OF QUANTITIES

A bill of quantities is a list of itemised materials, labour and parts required for a building project. The list is completed either by a 'quantity surveyor' or by an 'estimator', using the information and details contained in the architect's and engineer's drawings, the specifications and schedules. The bill of quantities is sent to a number of different companies at the tender stage to get a breakdown of estimated costs for the project. The breakdown of costs allows the client and project manager to make variations to the original contract documents throughout the project without having to re-tender.

WORK PROGRAMMES

Once all the tenders have been received, the client and quantity surveyor will review the costs submitted by the various companies and make an informed decision to award the contact. The company with the cheapest tender estimate is not necessarily the one that is chosen to complete the building work. There are other factors to consider; these include the following.

- Start date – the client's start date may not always suit the contractor's schedule. If possible the client may have to be flexible if they expect to use some larger, well-established companies.
- Quality of work – before awarding a tender to a company you should first carry out an investigation into the quality of previously completed projects and obtain references from other clients.
- Completion date – the client may instruct the quantity surveyor to request that a programme of work is submitted with each tender; this will provide details of the expected start and completion dates; in some cases the project may have to be finished within a designated time period.

TRADE SECRETS

If a company is short of work or able to start the project straight away, it is probably for a good reason. This may include:

- they are too expensive to win any other tenders;
- poor-quality workmanship.

Remember – it can take years to build a good reputation but only five minutes to lose one.

Item	Description	Quantity	Unit	Rate	Amount £	
	Preliminaries Name of parties Client Mr J Crouch Quality Printing Cornwallis House Devon					
	Architect RDJ Contractors					
	Preambles					
	Woodwork Supply and fit new impregnated sawn softwood joists to replace existing damaged joists					
A	Impregnated timber has been pressure impregnated with an approved preservative. Any timber cut on site must have a brush application of the same preservative in accordance with the manufacturer's instructions.					
B	50 mm x 200 mm joist					
C	75 mm x 200 mm joist					
D	Provide the P.C. sum of *Five hundred and eighty pounds* £580 for the supply of impregnated sawn softwood joists.				580	00
E	Add for expenses and profit			%		
F	Include the provisional sum of *two hundred pounds* £200 for contingencies.				200	00
				£		

FIGURE 2.12 Sample bill of quantities

- Payment terms – in general, the contractor will stipulate their payment terms as part of their tender. Terms and conditions will vary between companies, although bigger projects are usually settled with 'interim' or 'stage' payments over the course of the build.
- Retention – a period of between three and six months is usually agreed between the contractor and the client to retain approximately 5 per cent of the final project cost. The 'retention' of money allows the client to inspect the work completed and ensure that any 'snagging' is completed to a satisfactory standard. If the contractor fails to complete any snagging, then the retention money is used to employ other contractors. On satisfactory completion, any outstanding payments are made to the main contractor.
- Penalty clauses – as previously mentioned, some projects will have to be completed within a certain time period before the client starts to incur losses. To prevent the contractor running over the expected completion time, a penalty clause is usually built in to the contract between both parties. Penalty clauses can amount to huge losses for the main contractor, who may pass on these costs to subcontractors. If the project looks likely to overrun, the employer usually offers incentives to the workforce, increases the working hours on site or employs further labour to complete the project on time.

  To avoid costly penalty clauses, the main contractor will plan a programme of work at the start of the contract to track progress made through to the end of the build. There are several different types of programmes of work commonly used in the building industry; these include:

  - Gantt charts;
  - critical path analyses;
  - bar charts;
  - time lines.

W/C	2 July	9 July	16 July	23 July	30 July	6 August	13 August	20 August	27 August	3 September	10 September	17 September	24 September	1 October
Site preparation														
Setting out														
Excavate foundations and drains														
Concrete foundations and drains														
Brickwork to DPC														
Concrete to ground floor														
Brickwork to first floor														
First-floor joists														
Brickwork to eaves														
Roof structure														
Roof tile														
Internal blockwork partions														
Carpentry and joinery					1st fix			2nd fix						
Plumbing						1st fix		2nd fix						
Electrical							1st fix	2nd fix						
Services														
Plastering														
Decoration and glazing														
Internal finishing														
External finishing														
Labour requirements	3	2	3	1	1	1	1	1	1	2	2			

NOTES
target
actual

RDJ CONTRACTORS

JOB TITLE SEAVIEW DEVELOPMENT

DRAWING TITLE Programme of work

JOB NO.		DRAWING
SCALE	DATE	DRAWING
	2009	Nº65A

FIGURE 2.13 Example of a Gantt chart

Gantt charts

In simple terms, a Gantt chart is a programme of work designed by the site manager to complete the project in a timely manner. It consists of the starting and finishing dates along the top of the chart, with contingency weeks either end to allow for the project overrunning. These dates are broken down into the number of weeks the project is expected to last. Adjacent to this is a list of the different trades and phases of the build in the order they are expected to be introduced into the project. The projected target times are plotted on the chart for each trade, items of plant and equipment, as well as the various stages of the construction process. As progress is made, amendments may have to be plotted on the Gantt chart to indicate the actual stages of the building project. Adjusting the programme throughout the building work will allow the project manager to rectify delays caused at an earlier stage to complete the project on time.

SCHEDULES

Building drawings alone are unable to contain all the information needed for second-fix carpentry, and electrical and plumbing fitments; for example, internal doors will be shown on a plan and elevation drawings but rarely contain details regarding their design and use of ironmongery. To overcome this issue, 'schedules' are produced with a referencing system to the architect's drawings. In this case a door schedule may contain the following information:

- material used to construct the door (hardwood/softwood);
- type of glazing (if applicable);
- overall dimensions;
- type of door (flush, panelled, fully glazed, etc.);
- finish (painted, stained, etc.);
- types of ironmongery used (mortise lock, number of hinges, etc.).

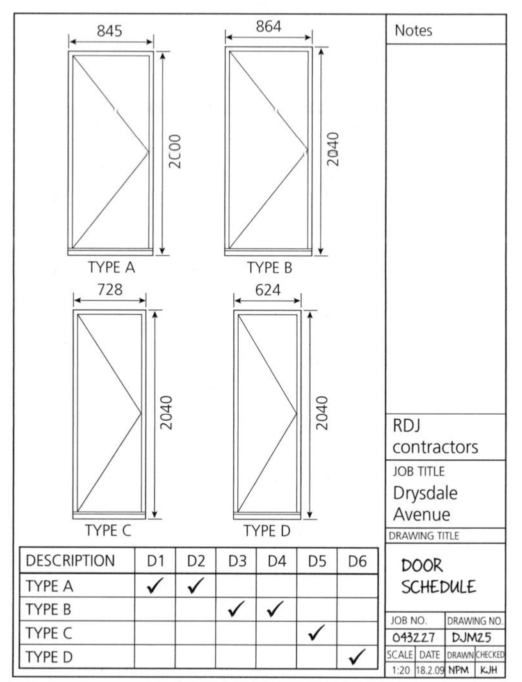

DESCRIPTION	D1	D2	D3	D4	D5	D6
TYPE A	✓	✓				
TYPE B			✓	✓		
TYPE C					✓	
TYPE D						✓

FIGURE 2.14 Example of a schedule

Schedules are produced by quantity surveyors to form a part of the tendering package used to estimate material and labour costs. They are also used throughout the building process to measure and order materials, check deliveries and help to identify the location of each item.

SPECIFICATIONS

As mentioned previously, it is impractical for an architect to be able to contain all the information needed to complete a building project on a set of drawings, so usually a document known as a 'specification' is used in conjunction. A specification is a written document that includes essential information that cannot be contained on the architect's drawings; for example:

- material information (e.g. dimensions, shape);
- finishes and quality;
- fixing methods;
- removal of materials from site;
- details of services (e.g. water, gas, electricity, telephone);
- recommended specialist suppliers;
- etc.

SPECIFICATION

(Architect's Details)
Company name;
Address;
Telephone number;
Fax number;
Email;

Site Address:

Job No:

1.0) INTERNAL PARTITION WALLS
Stud walls to comprise of 100 mm x 50 mm treated softwood studs built off 100 mm x 50 mm sole plate and complete with 100 mm x 50 mm head plate. Provide noggins mid height. Face both sides with 12.5 mm plasterboard, joints taped and skimmed to form a smooth, level finish. All partitions separating rooms to be filled with sound-deadening insulation quilt. All timber to be preserved and pressure treated.

2.0) STAIRCASE
Oak risers and treads between floors. Total rise as indicated on the stair details comprising equal risers and treads (minimum going of 225 mm) at a max pitch of 42°. Construction to comprise ex 32 mm strings, ex 30 mm treads, ex 12 mm risers and 94 mm x 94 mm straight stop chamfered newels. Handrails to be 900 mm above the pitch line. 25 mm x 25 mm oak vertical balusters at 100 mm centres to balustrades and flights where necessary. Landing to be provided with 1100 mm high balustrade comprising 25 mm x 25 mm oak vertical balusters at 100 mm centres.

3.0) ROOF CONSTRUCTION
For roof structure layout see layout drawings. Generally prefabricated trussed rafters at 600 mm centres, designed, manufactured and installed in accordance with BS.5268. Roof bracing to be carried out in accordance with BS.5268 Part 3: 1985 and truss manufacturer's recommendations with a 38° pitch.

4.0) LINTELS
Lintels to suit the size of opening and engineer's recommendations with minimum 150 mm bearing on either end. Paint underside of steel lintels with white paint over min. 2 coats metal primer and rust inhibiter.

FIGURE 2.15 Example of a specification

ACTIVITIES

Activity 10 – Interpreting and producing building information

Read through the following questions and answer them as fully as you can to help you develop your underpinning knowledge of this subject area.

1. What is a 'specification'?
2. List three different types of architect's drawings.
3. Explain the term 'tendering'.
4. What document itemises all the materials, labour and parts required for a building project?
5. Explain the terms 'retention' and 'penalty clause'.

MANUFACTURERS' INFORMATION

Manufacturers' information is usually distributed through catalogues and the internet to builders and the general public. Material specifications and design are regularly developing and updated in the industry, so it is important to understand these changes as they happen. Suppliers usually have staff members that are technically competent with specific materials, and in some cases their knowledge and advice can prove useful, saving time and money.

METHODS USED TO ESTIMATE QUANTITIES OF MATERIALS

ESTIMATING QUANTITIES

Tradesmen/women will be expected to calculate the quantity and cost of materials and labour needed to complete certain aspects of their job. This may range from a painter and decorator needing to work out the amount of wallpaper to cover a dining room, to a site carpenter needing to estimate the total length of skirting boards required for several rooms. In general terms these quantities and many more can be calculated using the following basic methods of mathematics (see Figures 2.16–2.19):

- size;
- area;
- perimeter;
- volume;
- percentage.

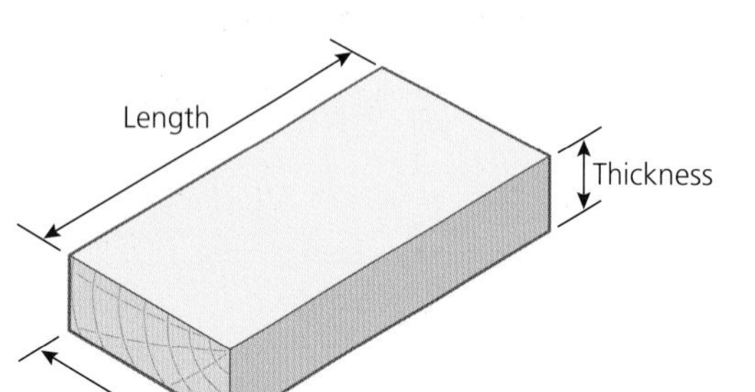

FIGURE 2.16 Size

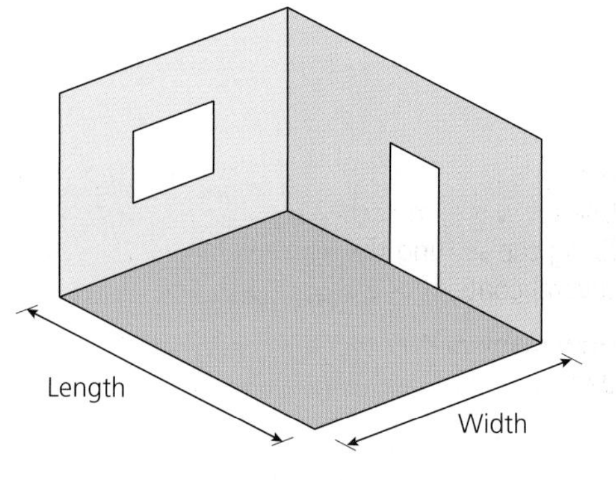

FIGURE 2.17 Area

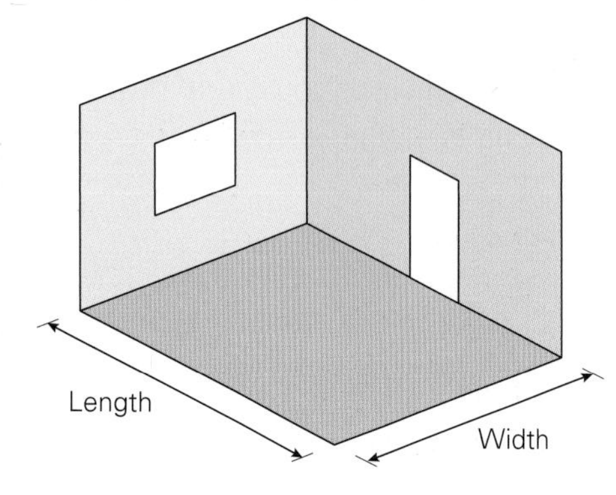

FIGURE 2.18 Perimeter

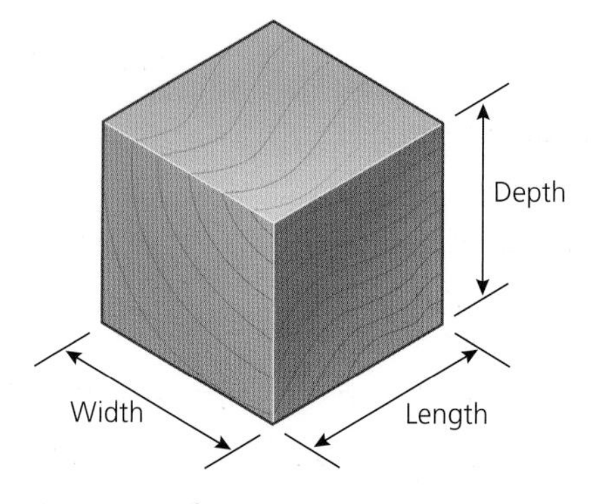

FIGURE 2.19 Volume

TRADE SECRETS

- *Careful estimating of materials, plant, equipment and labour costs is essential to prevent over- or under-costing; for example, one decimal point in the wrong position on a tender could result in overestimating and losing the work. On the other hand, if the decimal point is in the wrong position in the opposite direction, this could result in underestimating, and monetary losses.*
- *It is common practice to add between 5 and 10 per cent on the quantity of timber-based materials to allow for saw cuts and waste materials when ordering.*

EXAMPLE

A carpenter requires 625 metres of skirting plus 5% for waste to complete a project. Prior to fixing the skirting the decorator needs to apply several coats of oil to seal the timber.

How many metres of skirting should the decorator oil?

Percentage %

Length of skirting — Answer equals 1% of the total length of skirting

625 ÷ 100 = 6.25 metres

Percentage of waste

1% of the total length of skirting — Total length of waste

6.25 × 5 = 31.25 metres

31.25 + 625 = <u>656.25 metres</u>

Total length of waste — Length of skirting — Total length of skirting required

FIGURE 2.20 Percentage

ACTIVITIES

Activity 11 – Estimating quantities of resources

Read through the following questions and answer them as fully as you can to help you develop your underpinning knowledge of this subject area.

A room measures 3.5 metres × 4.3 metres × 2.4 metres high.

1. Tongue and groove boards are required to cover the entire floor area. Calculate the quantity required.
2. Skirting boards are also needed. Calculate the quantity required.
3. Calculate an additional 5 per cent of skirting board for waste.
4. Calculate the volume of the room.

COMMUNICATING WORKPLACE REQUIREMENTS EFFICIENTLY

THE BUILDING TEAM

Throughout this book and while working in the construction industry you will often hear the term the 'building team' referred to. The building team is made up of all the key members of the design, planning and building stages of a project. The most important member of the team is the client, without whom the project would not be funded. Other members of the building team include those listed below.

ARCHITECT

The architect works closely with the client, the local authority and specialist engineers to design and plan buildings to the client's specifications. Once drawn, the plans are submitted with a planning application to the local planning office for 'planning permission'. Once planning permission has been obtained, the architect may then liaise with specialist engineers and manufacturers to design detailed drawings and a specification detailing the method of construction. These details are then submitted again to the local planning office to obtain 'Building Regulations approval'.

BUILDING CONTRACTOR

The main contractor is responsible for the day-to-day running and progress made on site. They are normally appointed by the client after submitting a tender for the intended works, and signing the contract documents. The building contractor regularly liaises with the architect and engineers to discuss issues encountered throughout the build.

BUILDING OPERATIVES

These are generally referred to as 'labourers' in the industry because of the range of work that they carry out. They usually support craft workers by lifting, moving and mixing a variety of building materials such as plaster and mortar. They may also dig trenches, lay paths, and operate machinery such as cement mixers, drills and pumps.

CHARGEHAND

These are general workers ranking just below the foreman with supervisory responsibilities on larger construction sites.

CLERK OF WORKS (COW)

The architect or client usually appoints this person as the client's representative on site. Their primary role is to ensure the quality of the materials used on the building work, and that the workmanship is in accordance with the specification and engineer's details.

CLIENT

The most important person involved in the building team, and usually the source of funding for the building works.

CRAFT FOREMAN

The craft foreman supervises skilled and unskilled labour on site, reporting to the general foreman. These types of foreman are normally very experienced and responsible only for their own craft workers.

CRAFT OPERATIVES/SKILLED WORKERS

The majority of the workforce on a construction site at any one time will be made up of skilled tradesmen/women. These workers have completed a period of several years' training to assess their competence as skilled workers before qualifying in their areas of expertise. Skilled workers are usually responsible to the trade foreman or the general foreman on smaller sites.

ESTIMATOR

Working directly for the main contractor, the estimator calculates the quantities of materials, plant and labour needed to complete a building project. They usually produce a list of these items on a document known as a 'bill of quantities' (see Figure 2.12). The bill of quantities may be sent to several subcontractors to complete the total cost for the building work.

FREQUENTLY ASKED QUESTIONS

What does the term 'plant' mean?

The term 'plant' refers to the heavy machinery used during the construction process. Typical items of plant include:

- dumper;
- cement mixer;
- mini digger.

GANGER

This term is used for a person responsible for a group, or 'gang', of construction workers.

GENERAL FOREMAN

These operatives are responsible for coordinating a large team of construction workers, including the craft foremen and subcontractors. The general foreman reports directly to the site/general manager, and is partly responsible for hiring and firing of the labour on site.

FIGURE 2.21 Plant

HEALTH AND SAFETY INSPECTOR

The Health and Safety Inspector works for the Health and Safety Executive (HSE) to enforce the Health and Safety at Work Act 1974 (HASAWA). They may enter a site at any time throughout the building work to ensure that matters of health and safety are meeting with current laws and legislation. The Health and Safety Inspector has the right and power to:

- enter a premises without invitation;
- investigate the causes of incidents or accidents that have occurred on site;
- establish whether there has been a breach of health and safety law.

In addition to this, they can:

- take measurements, samples and photographs as evidence;
- record statements from employees;
- seize and make safe work equipment;
- dismantle or destroy dangerous machinery and equipment;
- inspect and copy documentation.

Any breaches of health and safety law can result in the following actions being taken by the Health and Safety Inspector:

- issue an *informal warning*, either written or verbally;
- issue an *improvement (prohibition) notice* giving an employer the opportunity to rectify, change or destroy dangerous items of equipment or machinery or systems of work;
- *prosecute* companies or individuals responsible for endangering the lives of employees and the general public.

LOCAL AUTHORITY

The local authority is responsible for issuing planning consent and Building Regulations approval for a given project. At several stages throughout the building work a representative of the local authority, known as the 'building control officer', will visit the site to inspect the foundations and brickwork to damp-proof course level and wall plate level. Further visits will be made to the site by the building control officer as and when required. The purpose of each visit is to ensure that the building work is carried out as per the details submitted for approval at the planning stage. Any substandard work must be taken down and rebuilt to satisfy the building control officer and current Building Regulations.

QUANTITY SURVEYOR

The quantity surveyor works closely with the client to manage and control costs resulting from building work. S/he is also normally involved in the production of the bill of quantities and the appointment of the main building contractor.

SITE AGENT

The site agent, also known as the site manager, is appointed by the main contractor to manage the day-to-day running of a building project. S/he is also responsible for ensuring the health and safety of all site workers, general appointment of subcontractors and maintaining strict quality control. Site agents will also produce a programme of work to ensure the building project is completed within the terms of the contract agreement.

SPECIALIST ENGINEERS

These professionals will work alongside the architect to develop the plans and structural calculations needed for specialised non-conventional building work. Engineers may visit the site from time to time to ensure that any specially designed metalwork etc. is being installed as it was intended.

FIGURE 2.22 A steel structure designed by a specialist engineer

SUBCONTRACTORS

Subcontractors are commonly appointed by the main contractor to undertake part of the building work. In many cases, subcontractors are used by larger companies to reduce the risk of employing full-time staff for the duration of a construction project. The danger of employing skilled labour is the potential for the specialised work to dry up, resulting in the employer having to pay wages to tradesmen and tradeswomen without using them to their full potential.

SUPPLIERS

Suppliers play a vital role in the construction process. Ensuring that materials, equipment and plant are available at the right time to purchase and hire requires good communication with the site manager/buyer. Phasing deliveries to site prevents materials being stored on site for long periods of time, taking up valuable space and becoming potential obstructions.

METHODS OF COMMUNICATION IN THE WORKPLACE

VERBAL COMMUNICATION

Verbal communication is probably the most widely used method of passing information and instruction from one person to another on a building site. Although this method of communicating is fast, enabling the workforce to react quickly, it is often misunderstood and misinterpreted. Communication between members of the management team is usually conducted through drawings, letters, memos, emails and 'minutes' of meetings that have taken place. These methods of passing information between members of the building team provide documented material that can be referred back to at a later date. This is especially important if disputes occur either throughout the build or upon completion.

FREQUENTLY ASKED QUESTIONS

What does the term 'minutes' mean?

'Minutes' are a written overview of a meeting. They are usually recorded by an administrator as a meeting progresses, accounting for the people present and those absent. Generally the structure and topics to discuss are written on an 'agenda' before the meeting takes place. Any responses to each item discussed are then recorded and written up logically after the meeting has finished – these are known as the 'minutes'. Copies of the minutes are sent to those who were present and absent from the meeting, for their records.

FORMAL BUSINESS LETTERS

These are used regularly to communicate information between members of the building team. Letters should be written in a clear, legible format and contain all the information required without being longer than necessary. The following guidelines are generally used to create business letters (see also the example in Figure 2.23).

- Addresses – the name and address of the person, company or organisation sending the letter should be written in the top right-hand corner of the page. The receiver's details are usually written on the left-hand side, just below the ending point of the other address.
- Date – this is written on the line below the last line of the address on either the right- or left-hand side.
- Greeting – 'Dear Sir or Madam' is used if you are unsure of the person you are sending the letter to, or alternatively 'Dear Mr (Mrs, Miss, Ms, Dr, etc.) Jones' if you know the title of the person. Never introduce a formal letter with the person's full name.
- Content of letter – the first paragraph usually explains the purpose of the letter, before leading on to the main text containing the relevant information. The final paragraph should explain the action required (if any) from the recipient.
- Ending – if you introduced the letter with 'Dear Sir or Madam' you should finish it with 'Yours faithfully'; if you started with the person's name, you should finish it with 'Yours sincerely'. Complete the letter with your signature and printed name underneath.

Large companies usually have an administrator to write the details of a dictated letter in shorthand before writing it up fully and signing it on behalf of the sender. This type of ending is normally identified with the initials 'PP' either before or after the signature. Some other common abbreviations found on formal letters are listed below:

- ASAP – as soon as possible;
- CC – carbon copy; this lets the receiver know that the letter has been duplicated and sent to more than one person;
- ENC – enclosure; this references documents contained within the letter;
- PP – per procurationem; this is Latin to explain that a letter or document as been signed on behalf of somebody else;
- PTO – please turn over;
- RSVP – répondez s'il vous plaît; this is a French phrase that translates to 'reply, please'.

FAX (SHORT FOR FACSIMILE)

Faxes are used regularly on small and large construction sites because the system is simple to set up and use with a telephone point. Messages, pictures and sketches can be sent and received on site very quickly with a fax machine; the disadvantage is that the images are in black and white. Any messages received will have the time and date recorded on the top or bottom of the fax page, along with the number of pages and the telephone number of the sender. Outgoing faxes will be recorded with a receipt printed off after a successful transfer has taken place.

EMAILS

Emails are normally processed through the computers in the site manager's office. They are advantageous because correspondence is clear, easily stored and accessible, which prevents the need for various unnecessary paper-based or 'hard' copies on site. Digital photographs

Mrs Architect
35 North Road
Bristol
BS12 7KT

Mr and Mrs Client
164 Lodge Avenue
Cardiff
CF24 8QR

16 February 2009

Dear Mr and Mrs Client

Ref: Development at 37 Sea View Heights, Little Town

Many thanks for your recent instruction regarding the changes to the design brief for the staircase position at Sea View Heights. I have made the appropriate alterations to the initial drawings, and have the pleasure of providing the revised layout of the proposed development. Please let me have your comments as soon as conviently possible so that we can submit a full planning application.

I look forward to your comments.

Yours faithfully

Mrs Architect

Mrs Architect

FIGURE 2.23 Formal business letter

FIGURE 2.24 Communication by Fax

FIGURE 2.25 Communication by Email

may also be taken on site of defects or progress made, and quickly uploaded onto a computer, before sending as attachments via email. The disadvantages of this method are that sketched images are difficult and slower to send through a computer system, and smaller sites may not have an internet connection.

TELEPHONE

Generally, telephones, fax machines and answer phones are combined units to save valuable space on site. Other similar methods of communicating via a phone include:

- two-way radios – used to transmit messages quickly between personnel on site, over short distances;
- mobile phones – used as an alternative to the two-way radios on site, with the advantage of being able to communicate verbally or with the use of text messages if the person leaves the site.

MEMORANDUM (MEMO)

Memos are used to record and pass information between operatives internally on construction sites. They normally contain only brief messages recorded by the site administrator from telephone calls or from notes passed between site workers about general information. A memo should be recorded on headed paper with the following information:

- name of person recording the message;
- name of the intended recipient/receiver;
- date and time the message was recorded;
- the message.

Telephone message

Date: .... Time: ....

Message for: ....

From: ....

Tel: ....

Message: ....

....

....

....

Message taken by: ....

FIGURE 2.26 Memorandum

POSTERS

Information contained on site message boards, posters and signs displayed around a construction site are all sources of important written instructions. Generally, site information is displayed at the entrance to the site to inform new personnel entering the area, in the site manager's office/hutment (hut) and in communal areas such as the site canteen. Displaying posters is a simple method of communicating best practice, as well as the dos and don'ts while on or around the site.

FIGURE 2.27 An information board

TIME SHEETS

Employees complete time sheets on a daily basis, logging the hours worked on each task while on a project. The time sheets are normally completed for a whole week before employees and their line managers sign and date them before submitting to the site administrator. The information contained on the time sheets is usually input onto a computer database to monitor the progress made by the employees, and also account for expenditure and wages. They may also be referred to in the future to analyse the time taken on a particular project, and to plan upcoming programmes of work and contracts.

RDJ contractors
Weekly Time Sheet

35 North Road
Bristol
BS12 7KT

Employee Peter Smith

For period 06/04/09 to 10/04/09

Job name Drysdale Avenue

Date	Work carried out	Total hours
06/04/09	Brickwork to first floor	8
07/04/09	Brickwork to first floor	8
08/04/09	Roof structure	8
09/04/09	Roof structure	8
10/04/09	Roof tiles	8

Signature ____________

Standard hours ____________

Job overtime ____________

Total ____________

FIGURE 2.28 Time sheet

DAYWORK SHEETS

These documents (e.g. Figure 2.29) are normally completed by subcontractors to account for work completed on site that has not been estimated prior to starting; for example, additions or variations to the original contract details or specification.

SITE DIARY

Entries into the site diary may be made several times throughout a day by the site manager to record the activities on site. The information contained in the site diary is normally duplicated and forwarded to the head office, and may also be used as evidence in the case of disputes between the main contractor, subcontractors, suppliers, etc. A typical site diary would include the following details:

- date and time of entry;
- adverse weather conditions;
- site visitors;
- delays/stoppages;
- late starts or subcontractors not attending the site;
- telephone/fax messages.

ORDERS/REQUISITIONS

Larger companies usually employ a 'buyer' who is responsible for sourcing and purchasing materials, equipment and plant for a construction project. The buyer would normally go through a process of sourcing the items through several companies to establish the most competitive quotes/prices, before completing a purchase order. Every purchase made should be documented on a form known as a 'requisition'.

RDJ contractors
Daywork Sheet

35 North Road
Bristol
BS12 7KT

Sheet no. 35/08

For period 06/04/09 to 10/04/09

Job name Drysdale Avenue

Work carried out
Second fixing to plot 8

Labour	Name	Craft	Hours	Gross rate	Total	
35/08	P. Smith	C & J	3	£25	£75	00
				Total labour	£75	00
Materials		**Quantity**	**Rate**	**% Addition**		
New door		1	£80	20	£100	00
				Total materials	£100	00
Plant		**Hours**	**Rate**	**% Addition**		
				Total plant	00	00
				Sub total	£175	00
				VAT (where applicable) N/A %	—	—
				Total claim	£175	00

Note Gross labour rates include a percentage for overheads and profit as set out in the contract conditions

Site manager/foreman ______________________
Architect ______________________

FIGURE 2.29 Daywork sheet

DELIVERY NOTES/RECORDS

Phased deliveries are normally made to a site at various times throughout the construction of a building. Upon the delivery of materials the driver will present a 'delivery note' containing the following details:

- items delivered;
- quantity of the goods;
- delivery address;
- company address;
- the date delivered.

Some suppliers and couriers prefer not use paper-based documentation for deliveries, and use electronic handheld notebooks instead. These are simply operated by the driver entering the delivery address into the notebook to bring up details of the order prior to the receiver signing the digital screen.

Only authorised personnel should sign for deliveries on site after carrying out the following checks.

FIGURE 2.30 Hand-held electronic delivery notebook

1. Check the delivery address on the delivery note to ensure that the correct goods are being delivered to the correct site.
2. Check the delivery against the original requisition (order).
3. Check the quantity. Large quantities of goods can often be overlooked if some items are missing.
4. Check the quality. Any damaged goods should either be sent back to the supplier or accepted with a comment made on the delivery note with a signature from the delivery driver as a witness.
5. Check the delivery has been made on the correct day. Phased deliveries to site prevent unnecessary storage for long periods of time.

SAFETY SIGNS

There are numerous amounts of different types of signage used in the workplace to convey information and instruction quickly and clearly. The Health and Safety (Safety Signs and Signals) Regulations 1996 place a legal responsibility on employers to display and maintain signs wherever there is a significant risk. British Standards BS5378: Parts 1 and 3: 1980 Safety Signs and Colours, has standardised the use of signs in the workplace so that they are interpreted in the same way, and have the same meaning. New members of staff that are unfamiliar with the safety signs used should have them explained as part of their induction.

Safety signs are usually identified by:

- shape;
- colour;
- pictograms;
- written information.

Safety signs fall into six different categories.

FIGURE 2.31 Prohibition

1. Prohibition – identified by a round, red and white sign, together with a pictogram in the middle, they mean you 'must *not* do'. For example, a sign with a picture of a mobile phone in the centre would mean 'you must *not* use a mobile phone'.
2. Mandatory – identified with a white pictogram on a blue background, they mean 'you *must* do'. For example, a sign with a picture of a pair of boots would mean 'safety footwear *must* be worn'.

FIGURE 2.32 Mandatory

FIGURE 2.33 Warning

3. Warning – identified with a black pictogram on a yellow background, they indicate a 'warning'. For example, a sign with a skull and crossbones would 'warn of toxic material'.
4. Safe conditions – identified with a white pictogram and text on a green background, they indicate 'safe conditions' or route. For example, a sign with a running stick man, an arrow and 'Fire exit' written, would indicate 'the direction of the emergency escape route'.
5. Information – identified with text on a white background, they provide written information and instruction (e.g. Figure 2.35).
6. Fire fighting – identified with white text on a red background, they display the position of 'fire fighting equipment' – for example, fire blankets, fire extinguishers.

FIGURE 2.34 Safe conditions

FIGURE 2.35 Information

FIGURE 2.36 Fire fighting

SITE MEETINGS

Every year the construction industry repairs and rectifies billions of pounds worth of defects, some of which are the result of poor communication, poor detailed drawings, incorrect and delayed information. Many of these problems could have been avoided if good management systems were in place to ensure effective lines of communication between the client, main contractor, subcontractors, etc.

General site meetings offer a very effective method of communicating between members of the building team. They are normally held regularly to freely discuss problems arising (and possible solutions) and to update/inform the client of progress made with the project.

There are several different types of meetings that will occur on site throughout the duration of a building project; these may include the following.

- 'General site meetings' are normally well structured and involve the client, main contractor, architect, structural engineer and the subcontractors. These meetings are normally held on site at regular intervals to discuss progress made, problems or issues that have arisen, and changes to the specification, drawings or contract documentation.
- 'Domestic site meetings' are held weekly between the main contractor, employees and subcontractors to discuss general issues, delays and the programme of work.
- 'Informal meetings' are often arranged at short notice to resolve problems arising as the build progresses. These types of meetings are unofficial and are usually conducted without a 'chairperson' or an agenda. The disadvantages of informal meetings are that they are not recorded and often result in misinterpreted information, mistakes and conflict between site personnel.

FREQUENTLY ASKED QUESTIONS

What does a 'chairperson' do?

A 'chairperson' is the head of a meeting or committee. They are responsible for:

- arranging the date, start and finish times for the meeting;
- writing and distributing a programme for discussion, known as the 'agenda', to personnel invited to attend the meeting;
- appointing a person to record the points of discussion throughout the meeting; These are commonly referred to as the 'minutes';
- distributing a copy of the minutes to members that did not attend;
- making sure that the agenda of the meeting is followed in a controlled and orderly fashion.

STRUCTURED SITE MEETINGS

Well-planned meetings generally follow a similar structure, with specific points for discussion pre-planned and recorded on an 'agenda'. The agenda lists points for discussion and the order they will be addressed throughout the meeting. A typical agenda is structured as shown in the example that follows.

FIGURE 2.37 A site meeting

Agenda

1. *Members present* – names of the people actually attending the meeting.
2. *Apologies* – names of the people invited to the meeting but unable to attend. Minutes recorded at the meeting should be forwarded to these individuals to ensure they are kept up to date with the issues discussed.
3. *Matters arising from the previous meeting* – a copy of the minutes from the previous meeting should be available to review any action points previously raised and whether further action is required to resolve the issue.
4. *Discussion points* – several points or areas for discussion will be highlighted by the chairperson prior to the start of the meeting. Copies of the agenda will be circulated before the meeting to allow the client, architect, contractors, etc. to prepare for the areas for discussion and prevent the meeting overrunning its intended completion time.
5. *Actions* – discussion points at meetings usually raise issues to be resolved and therefore actions for individuals. These action points are usually agreed between the parties involved at the meeting and a deadline set for the points to be met; the actions are then reviewed at the next site meeting.
6. *Date for the next meeting* – dates are usually agreed at the end of a site meeting to ensure everybody is available.
7. *Any other business (AOB)* – usually members of the meeting are given the opportunity to raise and discuss any issues that are not on the agenda.

ACTIVITIES

Activity 12 – Communicating workplace requirements efficiently

Read through the following questions and answer them as fully as you can to help you develop your underpinning knowledge of this subject area.

1. List five members of the 'building team'.
2. What is the role of the 'clerk of works'?
3. What is an 'improvement notice'?
4. List two advantages of communicating via a fax machine.
5. Explain the purpose of a site diary.

MULTIPLE-CHOICE QUESTIONS

1 Which one of the following building drawing symbols shows brickwork?

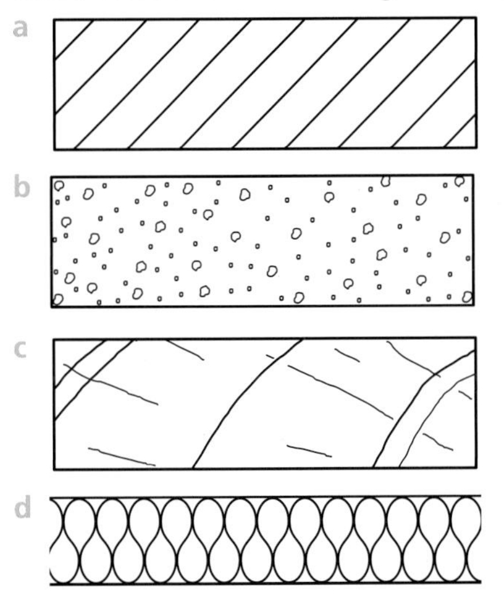

2 Instructions in the form of drawings and text included in the packaging of a night latch are referred to as
- a an invoice
- b a code of practice
- c an ironmongery schedule
- d manufacturer's technical literature

3 The **most** common form of communication used on site by carpenters is
- a text message
- b spoken word
- c written memo
- d hand gestures

4 The price given to the client by the builder for the work to be carried out is called a
- a quote
- b tender
- c estimate
- d guesstimate

5 A triangular sign with a black border and yellow background describes which one of the following sign types?
- a Warning
- b Mandatory
- c Prohibition
- d Information

6 Underestimating the quantity of materials required will
- a reduce work for the buyer
- b require less storage space on site
- c decrease the time required to complete the job
- d increase the time required to complete the job

7 Which **one** of the following parts of the Building Regulations covers Stairways, Ramps and Guards?
- a Part A
- b Part F
- c Part H
- d Part K

8 The contract document that contains a written description of the workmanship and materials is known as the
- a drawing
- b order sheet
- c specification
- d architect's instructions

9 Which **one** of the following regulations controls the layout and materials used in an office block to ensure strength and stability?
- a Building
- b Planning
- c Manual handling
- d Construction, design and management

10 A record of site deliveries would be made in the
- a schedule
- b site diary
- c daywork sheet
- d bill of quantities

BUILDING METHODS AND CONSTRUCTION TECHNOLOGY 2

CHAPTER 3

LEARNING OUTCOMES

By the end of this chapter you should have developed a knowledge and understanding of:

- the principles behind walls, floors and roofs;
- the principles behind internal work;
- materials storage and delivery of building materials.

INTRODUCTION

The aim of this chapter is for students to be able to recognise traditional and modern construction methods, and the principles behind them. It also looks at the some of the most common materials used in the industry, their methods of delivery and storage facilities. In addition, this chapter identifies and explains current Building Regulations applicable to this area of study, following all the relevant health and safety law and good working practices.

THE PRINCIPLES BEHIND WALLS, FLOORS AND ROOFS

The earliest stages of planning a building or structure usually begin with an idea and a brief; this is known as the 'conceptual design' stage. Clients will normally approach an architect to discuss their initial thoughts before developing the vision through to the 'preliminary design'. Architects will work closely with structural or civil engineers to ensure that the proposed designs will stand up to the forces and loads imposed upon them before the 'final design'. At this stage, the working drawings and structural calculations are normally submitted to the local authority for approval.

Buildings and structures are shaped by their design and construction methods because of the environment in which they exist. For example, a plot may sit in a historic village with listed buildings. The possibility of a high-rise concrete office block being granted planning permission would be slim. Good design ideas should to be sympathetic to the surrounding area and have as little impact on the environment as possible, while still meeting the needs of the client.

Developments in materials and construction methods have led to various types of structure being developed across the country, each one suited to demands of the location, local planning laws and Building Regulations.

TYPES OF BUILDING STRUCTURE

Building structures are usually designed with a combination of different building materials in order to perform the following functions:

- allow natural light and ventilation;
- allow people to work, live and play in them safely without risk;
- design (synthetic);
- provide security for the occupants and the materials contained within them;
- provide shelter;
- provide warmth.

Both residential (e.g. private dwelling) and non-residential (e.g. offices, shops) buildings are recognisable by the number of floors (known as 'storeys') over which they are constructed. Building structures can be divided into three categories:

1. high-rise (over seven storeys);
2. mid-rise (four to seven storeys);
3. low-rise (one to three storeys).

Note: low-rise buildings are further categorised into detached, semi-detached and terraced dwellings.

FIGURE 3.1 Types of building structure

SUBSTRUCTURE AND SUPERSTRUCTURE

The weight of any building or structure must be fully supported at ground level with adequately designed and calculated foundations. Elements of a structure below damp-proof course (DPC), including the ground floor and foundations, are known as the 'substructure'. All the internal and external elements of a building above the substructure are referred to as the 'superstructure'. The components of the superstructure will distribute the weight (loads) of the building safely through the roof, walls and floors to the substructure (see Figure 3.3).

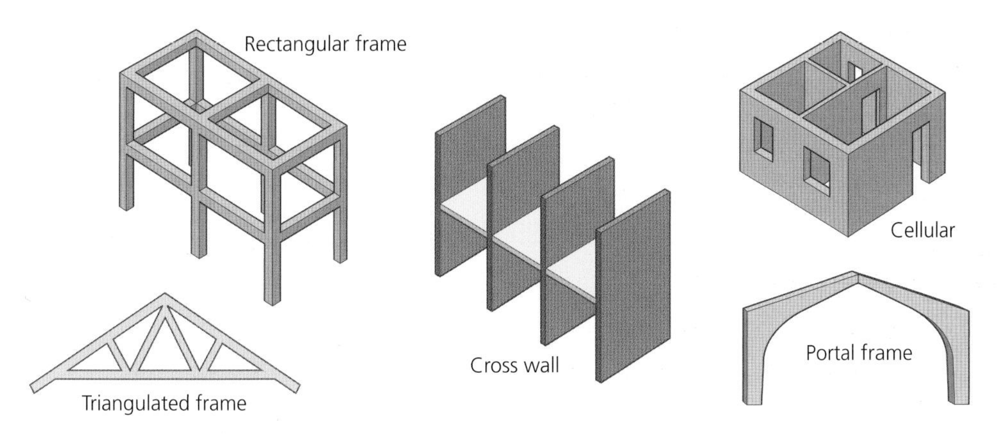

FIGURE 3.2 Elements of the superstructure

SOLID WALLS (WITHOUT CAVITY)

These are a fast, efficient method of building external walls. They have high 'U' values (thermal insulation), with the ability to retain heat in the winter months and keep the building cool during warmer periods. There are several methods commonly used to finish the surfaces on solid walls; these include:

- brick 'slip' systems;
- ceramic wall tiles;
- render;
- tile hanging;
- timber cladding.

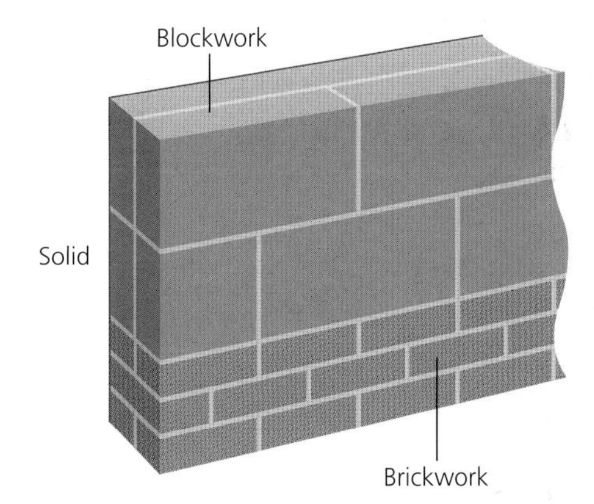

FIGURE 3.4 Solid wall

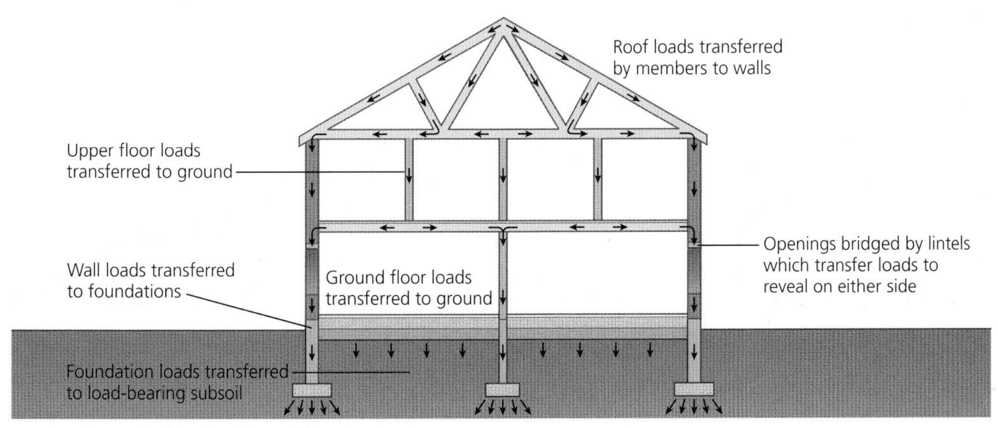

FIGURE 3.3 Safe distribution of loads via the superstructure

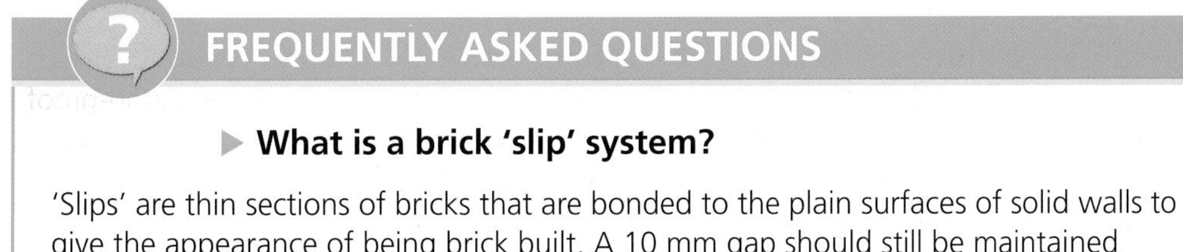

FREQUENTLY ASKED QUESTIONS

What is a brick 'slip' system?

'Slips' are thin sections of bricks that are bonded to the plain surfaces of solid walls to give the appearance of being brick built. A 10 mm gap should still be maintained between the slips and filled with mortar with an application gun.

CAVITY WALLS

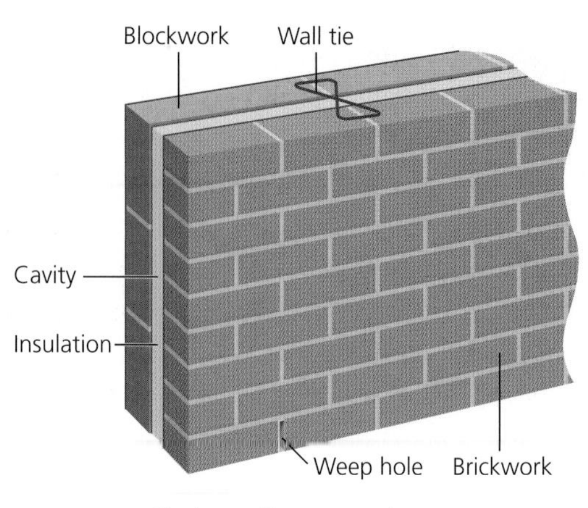

FIGURE 3.5 Cavity wall construction

Walls consist of an inner and outer wall divided by a void, known as the 'cavity'. The walls are usually constructed with bricks and blocks, or two skins of blocks with render applied, or another method of wall covering. The masonry bricks and blocks used are usually porous, which will allow the water to penetrate through the walls only to the point of the cavity. The water or moisture then runs down the inside of the cavity and disperses at ground level through 'weep holes'. Cavity walls provide very good sound and thermal insulation when lined with cavity wall insulation. Independent single-skin walls may become unstable when they are built over one storey high. To improve the stability of double-skin walls they are connected to each other via non-corrosive 'cavity wall ties'. These are simply built into the brickwork courses between the two walls as the building progresses. Most wall ties are fitted with a plastic collar that allows the cavity wall insulation to be held tight against the inner wall, therefore preventing 'bridging'.

FREQUENTLY ASKED QUESTIONS

What does the term 'bridging' mean?

'Bridging' is a term given to moisture or water penetrating external walls across the void in a cavity to the inner skin. Bridging across a cavity wall will cause damp patches to appear on the inner walls.

TIMBER FRAME

Houses were traditionally built with heavy structural sections to form a skeleton frame, and the frames were then in-filled with bricks to complete the 'shell'. Modern timber frame buildings are normally constructed with timber internal frames clad with plywood to add strength, and a single skin of external brick or blockwork supported with 'wall ties'. Alternatively, the depth of the timber frames can be increased to the full thickness of the wall, and the building can be constructed without the need for brickwork above damp-proof course level. In all cases a suitable moisture and vapour barrier must be included to prevent the ingress of water and the onset of rot in the timber frames, as well as thermal and sound insulation to ensure current Building Regulations are met.

FREQUENTLY ASKED QUESTIONS

What is a 'moisture barrier'?

Modern moisture barriers are paper-thin building materials, usually attached to the face of timber frame walls, or underneath the roof covering between the wall plate and ridge. They are used to ensure that the building remains dry by preventing water penetrating the exterior walls and roof. Moisture barriers also allow the building to 'breathe', by allowing stale moist air within the building to escape.

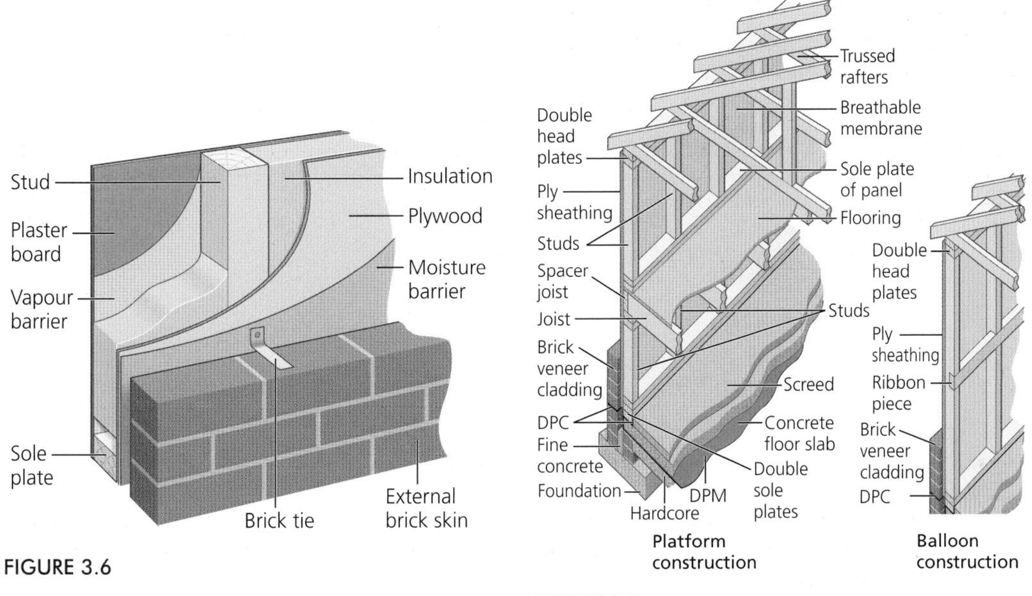

FIGURE 3.6

FIGURE 3.7

MODERN INSULATION MATERIALS

The early 1970s saw the development of a new idea of constructing buildings using 'structural insulated panels', or SIPs for short. The panels are a sandwich of two layers of oriented strand board (OSB) either side of a thick polyurethane foam core. The bond between all the materials results in extremely rigid panels, capable of supporting structural loads. SIPs are commonly used to construct whole buildings from the external walls and floors, up to the roof structure. Large SIPs require heavy lifting equipment to manoeuvre them into position on the building before they interlock with the other panels. There are several advantages of this method of construction; these include:

- speed of erection;
- no need for timber studs;
- no need for further insulation;
- no need for a vapour or air barrier.

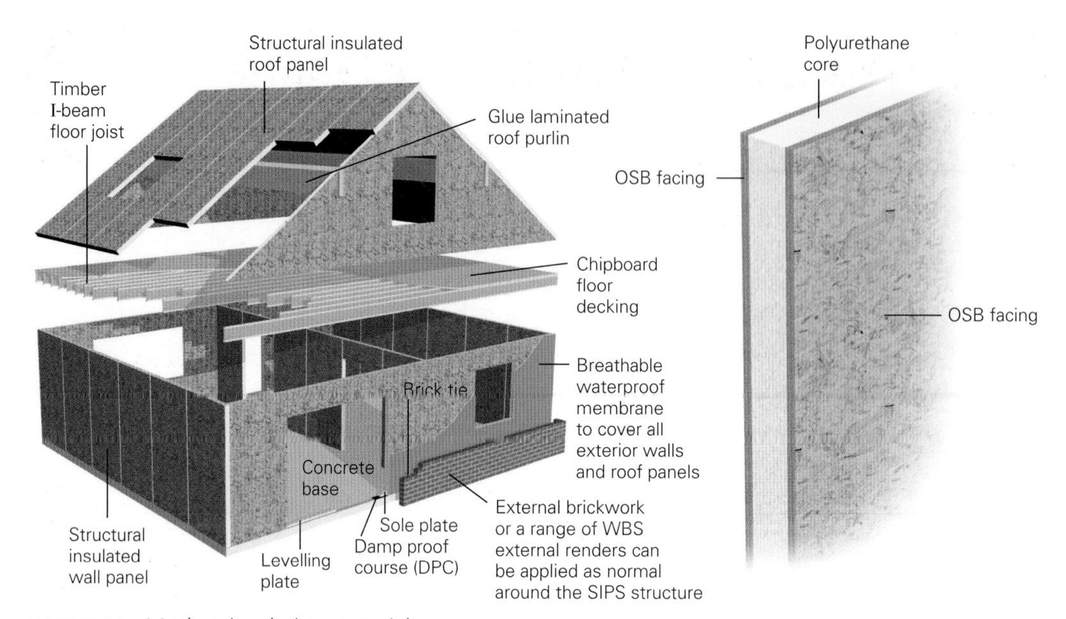

FIGURE 3.8 Modern insulation materials

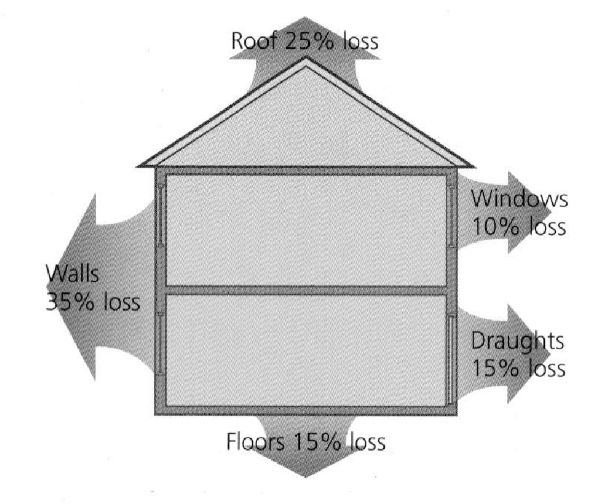

FIGURE 3.9 Energy loss

BUILDING METHODS

Structural and civil engineers are responsible for ensuring that the buildings and structures that they design in partnership with the architect are capable of withstanding the loads and stresses imposed on them.

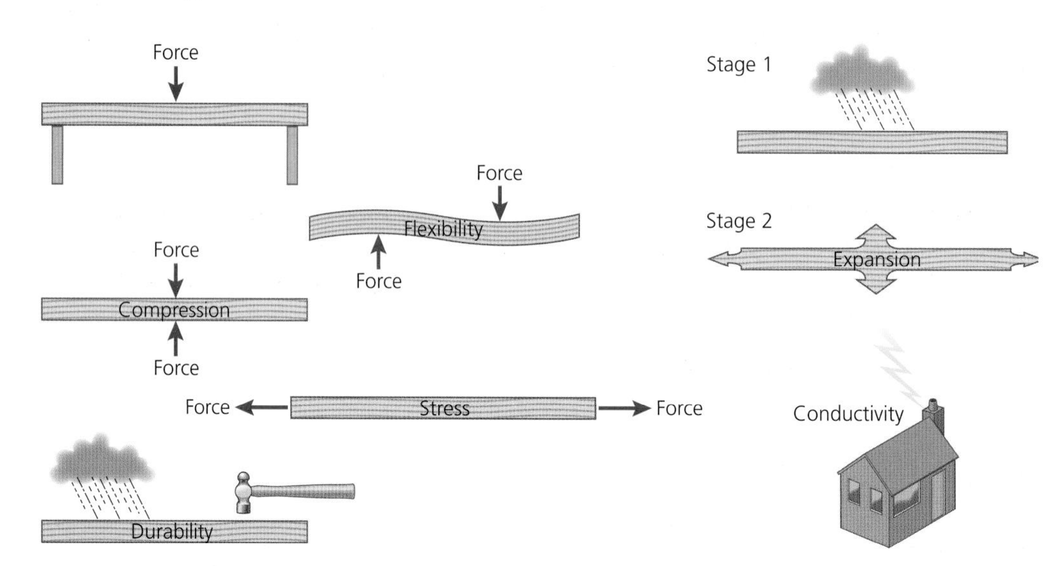

FIGURE 3.10 Loads and stress

SETTING OUT OF FOUNDATIONS AND WALLS

IDENTIFYING DIFFERENT TYPES OF CONCRETE FOUNDATION

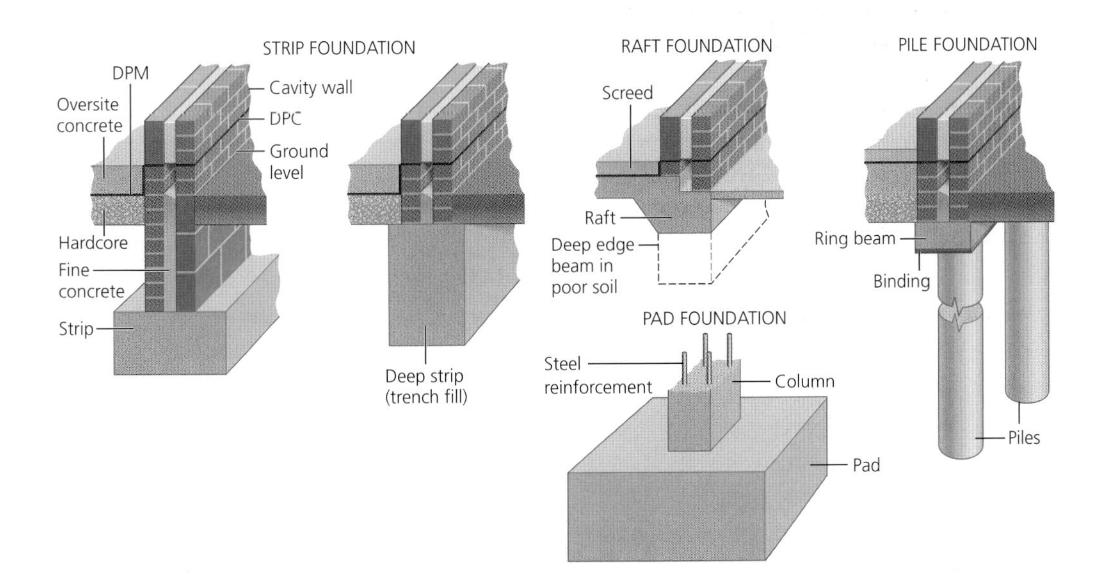

FIGURE 3.11 Different types of concrete foundation

There are several different types of concrete foundation commonly used in the construction industry. Each one is specifically engineered to support the dead load of the building, the type of soil in the ground and any other forces imposed upon the building. Strip foundations are designed to suit firm, stable ground conditions and are probably the most commonly used in the United Kingdom. Sites with soft ground or possibly on a hillside may require deep pile foundations to maintain stability.

SITE INVESTIGATION

A thorough investigation of a site will have to be carried out before any land can be built on, and the most suitable type of foundation can be determined. A surveyor would normally research the following areas before submitting a detailed report with recommendations to a structural engineer:

- previous use (brownfield site);
- contaminated land;
- mines or wells;
- type of soil;
- water levels in the ground;
- radon gas

FREQUENTLY ASKED QUESTIONS

What is 'radon' gas?

'Radon' is a naturally occurring gas that develops in some ground conditions, particularly in the Cornish area of the UK. Exposure to high levels of the odourless gas can be a health hazard that can potentially lead to lung cancer.

FLOOR CONSTRUCTION

The method of construction for a floor in a building is normally determined by its position (lower or upper), the type of soil, site conditions (flat/elevated) and the displacement of the load/weight. Weak or unsuitable foundations and floors may lead to movement in the ground and potentially cause subsidence or cracking.

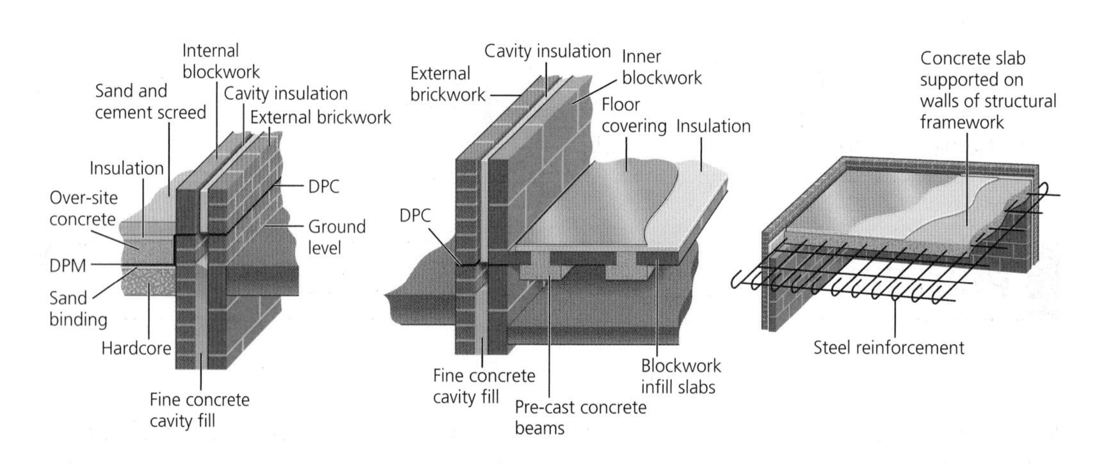

FIGURE 3.12 Methods of floor construction

HOLLOW TIMBER FLOORS

This method of floor construction is covered in detail in Chapter 6, Erect Structural Carcassing.

SCREED

Solid concrete, and beam and block floors are usually slightly uneven and unsuitable to accept a floor covering such as carpet. A screed is laid over the floor surface area and levelled manually with a long float, straight edge and spirit level. It consists of a semi-dry concrete mix of sharp sand and cement, usually containing an additive to slow down the drying time, therefore increasing the working time with the product.

DAMP-PROOF COURSE (DPC)

A DPC is normally built into the exterior brick and blockwork walls at least 150 mm above the exterior ground level. Its purpose is to provide a barrier to prevent moisture creeping up from the ground through the walls.

DAMP-PROOF MEMBRANE (DPM)

A DPM is normally built into the entire ground floor area of solid concrete floors. It is used to prevent moisture, damp and weed growth through the floors and lower portions of the internal walls.

ACTIVITIES

Activity 13 – The principles behind walls, floors and roofs

Read through the following questions and answer them as fully as you can to help you develop your underpinning knowledge of this subject area.

1. Sketch a section of a cavity wall and timber frame wall.
2. What do the initials DPC stand for?
3. Explain the purpose of a DPM.
4. Sketch a joist to demonstrate compression and tension.
5. Name one modern insulation material.

PRINCIPLES BEHIND INTERNAL WORK

PAINT COVERINGS

Internal walls are usually either lined with wallpaper or painted with water-based emulsion. Water-based paints are easily applied to walls and ceilings manually with a wide brush or deep pile roller. Working at height can be avoided in many cases when decorating high walls and ceilings if an extension pole is connected to the end of a roller. The disadvantage of applying paint with a roller or brush is the uneven results it produces. Many contractors now prefer to paint large surface areas with a spray system to achieve an even, smooth and flawless finish. All windows, mouldings and doors, etc. should be covered with masking tape and paper to prevent over-spraying and soiling. The time needed to prepare rooms for spraying is normally outweighed by the speed at which the non-toxic paint can be applied. Durable paints such as oil-based paints are usually applied with a brush and should not be applied with a spray system on site because of the risks the airborne paint can have to health.

BUILDING MATERIALS

AGGREGATES

Aggregates is a term used to describe different-sized crushed stone and minerals (e.g. fine sand, coarse sand, chippings). These materials are commonly used in the building industry to construct foundations, floors, beams and walls.

PLASTER

Information regarding the different types and uses of plaster are detailed in Chapter 7, Maintenance.

PLASTERBOARD

Information regarding the different types and uses of plasterboard are detailed in Chapter 4, First Fixing.

CONCRETE

Concrete is a mixture of fine and coarse aggregate, water and cement, mixed to specified ratios to make an extremely strong material once cured. Pre-cast structural beams and lintels are formed with high-tensile steel reinforcing bars running through just above their lower face to increase their tensile strength and prevent cracking under load. Solid floors are also strengthened using similar methods, only this time the steel bars are bound together to make a web over the entire area before it is submerged under concrete.

METALS

Metals are commonly used in framed substructures of industrial buildings, offices and warehouses. The preformed metal sections are bolted and welded together in situ for the speedy erection of a building. This then allows the inner walls to be built in between the rigid 'skeleton' frame in preparation for the external covering. Steel joists are also commonly used in the industry to span long distances between walls to provide adequate support for shorter timber joists. They may also be used as lintels over doorways or window openings to support the imposed loads from the brick or blockwork above.

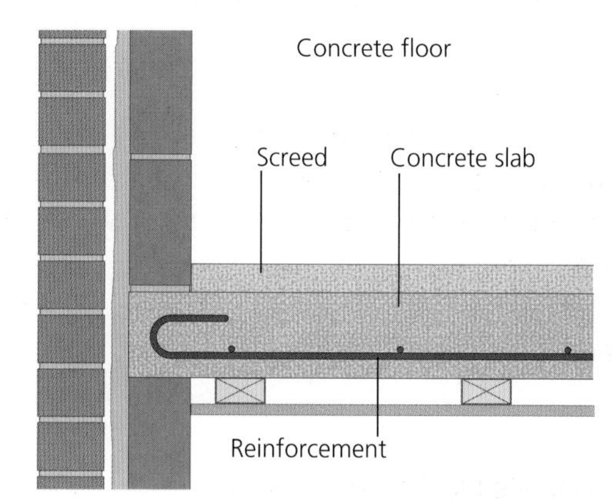

FIGURE 3.13 Concrete floor

SOFTWOOD AND HARDWOOD

These resources have been used to construct buildings for thousands of years because wood is a renewable building material and readily available. Timber-framed buildings have been commonly used to construct modern homes in the USA and Europe for many years. In recent years the building industry has grown in confidence with the resource and many companies and clients are experiencing the benefits. These include:

- increased productivity and speed of erection (example: a standard two-storey detached property frame can be erected and made watertight within five days);
- less disruption to the client;
- reduced building costs;
- less environmental impact.

GLASS

Information regarding different types of safety glass, single and double glazing, and installing glass are detailed in Chapter 4, First Fixing.

> **TRADE SECRETS**
>
> *Remember* – the difference between softwood and hardwood is *not* the strength of the timber. Timbers are categorised by their cellular structure, characteristics and species:
>
> - softwood trees have narrow needle leaves (coniferous);
> - hardwood trees have broad leaves that usually drop in the autumn and winter months (deciduous).

BRICKS AND BLOCKS

THERMAL BLOCKS

Thermal blocks are commonly used to construct the internal walls of offices, factories and houses. They are considerably lighter than conventional building blocks and easily cut with a handsaw, providing an efficient method of constructing partition walls. The exceptional thermal properties of the blocks mean that less insulation is required in the building. Although the blocks are lightweight, they can still be used to provide structural strength to a building. (*Note*: the exact load-bearing capacity for a wall constructed with thermal blocks should be obtained from a structural engineer.)

COMMON BRICKS

Common bricks are available in a vast range of different materials, colours and finishes to match existing brickwork or construct new. They are generally used only to build walls and columns if they are exposed because of the increased labour and material costs compared with other building resources. Careful consideration should be given to the type of bricks used in a given situation because some bricks are less dense and more porous than others, and will also be able to support heavier imposed loads.

CONCRETE BLOCKS

Concrete blocks are relatively inexpensive compared with facing bricks and thermal blocks, hence the reason they are commonly used in the construction industry to build the internal skins of cavity walls, etc. Concrete blocks are available in a range of different densities to suit the weight imposed upon them. A standard concrete block measures 440 × 210 × 100 mm excluding the mortar joints.

ENGINEERING BRICKS

Engineering bricks are normally used to construct load-bearing columns and brickwork below damp-proof course (DPC) level because their superior strength and density make them resistant to frost attack. All engineering bricks should comply with British Standards before being used structurally. Both engineering and common bricks are manufactured in standard dimensions of 215 × 102 × 65 mm, although more expensive handmade and recycled bricks may vary slightly.

ACTIVITIES

Activity 14 – The principles behind internal work

Read through the following questions and answer them as fully as you can to help you develop your underpinning knowledge of this subject area.

1. List three methods of applying paint finishes.
2. Describe the difference between concrete and thermal blocks.
3. Describe the difference between facing and engineering bricks.

MATERIAL STORAGE

ORDERING PLANT, EQUIPMENT AND MATERIALS

As materials, plant and equipment are required at different stages throughout a building project, the site manager or company buyer will refer to a 'programme of work' (programmes of work are covered in more detail in Chapter 2, Information, Quantities and Communicating with Others 2). Programmes of work are documents containing essential information about building projects, and include the following details:

- start and expected completion dates;
- periods of time each trade is expected to be working on site;
- actual time spent on each aspect of the build;
- significant phases of the work – for example, ground work, foundations, brickwork to damp-proof course (DPC).

While the programme of work is being drawn up, the 'lead' time for each item should be taken into account. In some cases suppliers may not have all the items or equipment in stock, so a period of delay may be incurred between the requisition (order) and the actual delivery to site. A site manager with good organisational skills will consider this factor, and make a note of the actual lead time required before ordering the items in advance, to prevent delays on site. Other factors to consider when organising a building project may include the following:

- availability of equipment and materials;
- availability of plant, due to breakdown, servicing or repairs;
- weather conditions/daylight;
- suitability of equipment or plant for the site conditions;
- availability of suitably trained personnel to operate the plant or equipment;
- access to the site and space to operate the plant or equipment;
- availability of a power source to operate the equipment;
- adequate labour and resources to unload the delivery;
- suitable assessments of the potential risk caused by unloading materials and equipment, and adequate controls in place to eliminate or reduce the danger and to ensure health and safety;
- suitable storage areas.

PHASED DELIVERIES

As mentioned previously, phasing the order of deliveries to site will prevent delays due to operatives waiting for materials, equipment and plant. These delays may have a knock-on effect that could result in the project overrunning the expected completion date. On the other hand, materials ordered too early in a construction project may result in there being insufficient space to store them and a potential obstruction or hazard. Materials inadequately stored may have to be moved several times before they are used, which normally results in them getting damaged, lost or stolen.

Construction materials such as paints, adhesives, plaster and cement. have a limited shelf life before they are opened, and an even shorter usable time after opening. Following these expiry dates the materials will start to deteriorate and 'go off' (set, cure or harden to a condition where they are no longer usable for the purpose for which they were intended). Such items should always be stored in date order, with the oldest at the front and then replaced with new stock at the back. This method of stock rotation is referred to as 'first in first out' (FIFO), and is used to prevent materials being stored at the rear of the stores or on shelves until they exceed their use-by dates. Large construction companies may employ a storeperson to control the stock levels and account for materials used. The storeperson will either use a 'tally book' to record the delivery of materials or a computer system on bigger sites. Materials requested by employees from a company's central store should be recorded on a 'requisition' (order form).

RDJ contractors
Deliveries Record

35 North Road
Bristol
BS12 7KT

Date 15 January 2009

Job Name Drysdale Avenue

Supplier	Delivery information	Delivery note no.	Office use only	
			Rate	Value
D. J. Building Supplies	Ready-mix concrete	3568		
D. J. Building Supplies	Plasterboard	3569		
			Total	

Site manager/foreman ____________

Delivery records to be sent to head office weekly with delivery notes.

FIGURE 3.14 Deliveries record

Good communication between the site manager, company buyer and the suppliers could avoid materials and equipment being delivered at lunchtimes or last thing at the end of the working day when only limited labour is available. Some developments may be located in built-up areas or city centres where there is an increased risk of endangering the general public. The 'pre-tender health and safety plan' should identify the busiest times of day so that deliveries can be scheduled around these periods.

TAKING DELIVERY OF GOODS

Upon arrival at the delivery address, the supplier's driver will be expect to 'drop off' the goods and receive a signature on a delivery note or ticket as evidence that the delivery has taken place. In most cases delivery drivers are employed directly by the suppliers or delivery companies to deliver multiple 'drops' (deliveries of items to several addresses). The drivers are usually keen to unload the deliveries without delay before progressing on to complete their duties, so it is important that the person taking the delivery is prepared. In preparation, the site manager should ensure that adequate lifting equipment and labour are ready to unload the delivery with the correct safety measures in place.

TRADE SECRETS

Remember – some plant and equipment may have to be hired from suppliers. It is just as important to notify the suppliers of a convenient collection date as it is to order it in the first place. Failure to keep suppliers informed could result in the items having to be stored on site longer than necessary and additional cost incurred.

Manually unloading heavy items or large quantities of goods from the delivery vehicle should always be a last resort after considering all other mechanical options. Before personnel move or lift any delivered items manually, the site manager should ensure that they are aware of the risks involved by drawing their attention to the relevant method statements. All employees should take adequate measures to reduce the risk of harm to them by wearing the correct personal protective equipment (PPE) to comply with the Manual Handling Operations Regulations 2002. Suitable items of PPE may include the following items:

- safety footwear;
- hard hat;
- high-visibility clothing;
- gloves or gauntlets.

Before carrying out manual handling operations it is strongly recommended that suitable barrier cream be worn to protect against some of the products being handled. Upon completion and before eating, hands should be thoroughly washed with soap to remove any residue, and replenishing cream rubbed into hands to replace any lost moisture. Further manual work would require the application of additional barrier cream to seal and protect the surface of the skin. Remember, any item of PPE, including barrier cream, moisturiser and soap, should be provided free of charge by employers to their employees.

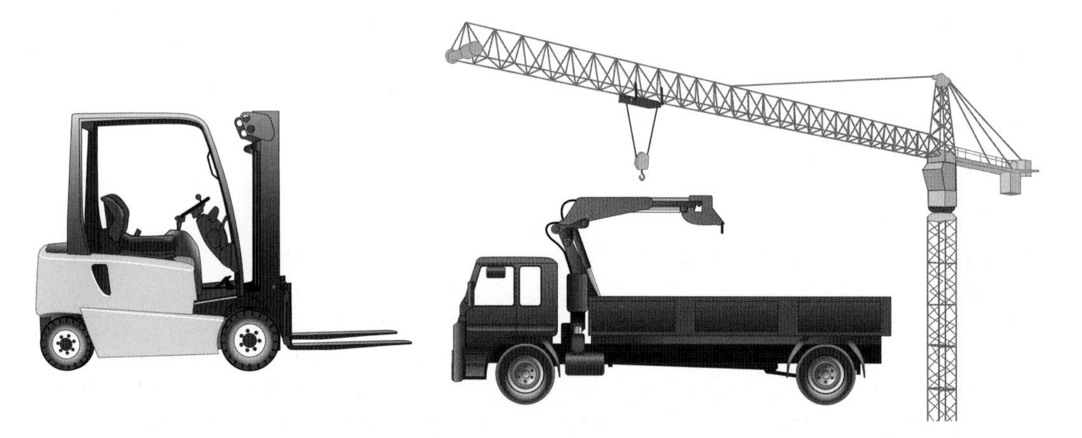

FIGURE 3.15 Unloading: mechanical options

Dorset Timber Supplies

Unit 4B Warick Industrial estate, North Road, Birmingham, B74 9TZ
Email sales@dorset.co.uk Tel. 0121 437 6541 Fax 0121 437 6542

RDJ Contractors
35 North Road
Bristol
BS12 7KT

DELIVERY NOTE
DN00000012/5

Order No J2455
Customers Order No 104
Account No 0002
Despatch Date 14 June 2009

INVOICE ADDRESS
RDJ Contractors
35 North Road
Bristol BS12 7KT

Line	Quantity	UOM	Part Number/Decription/Specification		Location	Net Mass (kg)
1	10	1	0002	Ready-mix concrete	A	10,000.00
2	8	1	0003	Plasterboard	A	8,000.00

Comments

Total Mass (kg) 18,000.00

FIGURE 3.16 Delivery note

Further details about delivery notes and checking deliveries are covered in Chapter 2, Information, Quantities and Communicating with Others 2.

PROTECTION OF MATERIALS

AGGREGATES

Large quantities of aggregates are normally delivered on the back of a truck and offloaded into divided areas known as 'bays'. This form of storage prevents the different types of aggregate mixing together to contaminate the products. Whenever possible, the base of the bays should be hardwearing, with a slight incline to allow water to escape. Alternatively, smaller quantities can be off loaded in bulk bags. This method avoids the need for bays to divide the different aggregates and the bags are simply cleared away after use.

BRICKS AND BLOCKS

Packs of bricks and blocks are usually delivered on pallets with plastic binding and wrapping to offer reasonable protection against the elements, including frost. The raised wooden pallets prevent them being stored directly on the floor, which may result in staining from salts contained in the soil, and moisture ingress. They also permit the forks on pallet and fork-lift trucks to slide underneath without obstruction, allowing future transportation from the storage areas closer to the building work. If individual bricks or blocks are delivered, they should be stacked in an alternating pattern no higher than the wrapped packs or that stated in the relevant method statement. Extra care should be taken when manually stacking and storing bricks and blocks to ensure that fingers do not get trapped and pinched between heavy blocks as they are being placed.

CEMENT, PLASTER, SAND, ETC.

Products affected by moisture, such as cement and dry plaster, are normally supplied in 25 kg moisture-resistant bags. These are the maximum weight bags that should be lifted manually, so only one bag should be carried at any one time. It is recommended that these types of product are stored off the ground in a dry, ventilated, lockable container. Generally, the bags should be stacked flat to prevent them from being perforated, and a maximum of five high, with a clear walkway between to gain easy access, in order to follow the 'first in first out' method.

DOORS

Expensive joinery products such as doors should always be stored in a dry lockable container until they are ready to be hung. Untreated soft and hardwood doors will twist, warp and absorb moisture very quickly if they are left open to the environment or stored on uneven surfaces. Some of this damage may not become apparent until after installation, when the doors may dry out, resulting in shrinkage, damage to the joints and warping. Mass-produced doors are usually covered with protective plastic with corner protectors, and should remain so until they are ready for installation. Whenever possible they should be stored flat, off the ground and supported on at least three 'bearers' with layers of cardboard in between each door to protect the faces. Purpose-made doors will usually be supplied with the 'horns' left on for removal on site just before their installation (further information regarding 'horns' can be found in Chapter 5, Second Fixing).

FLAMMABLES

Special precautions are required to store and handle pressurised containers such as 'gas' because of the risk of the substances leaking from the valves and potentially exploding. Incorrectly fitted valves and regulators, inadequate storage facilities and poor handling of pressurised cylinders are all hazards waiting to happen unless adequate controls are put in place. Every employer should ensure that assessments of the risks to health and safety are carried out before eliminating them as far as practically possible, or reducing them to an acceptable level.

Flammables should be purchased and stored on site only as and when they are needed, and should never be kept for long periods of time. When new containers are delivered to the site storage facilities, older cylinders should be rotated to the front. All pressurised cylinders should be stored vertically unless otherwise stated, and clearly labelled with the contents on their casing. They should also be stored:

- on a flat surface;
- under cover;
- with adequate ventilation;
- away from sources of heat and ignition;
- with restraints around the cylinders to prevent tipping;
- protected with 'valve caps' to prevent damage to the valves in the event of being dropped;
- in a well-signed area warning of hazards and precautions to be taken in that area.

GLASS

Clean and dry purpose-made racks should be used to store glass in a vertical position. Glass racks (also known as 'frails') should also have a slight lean on them to prevent the glazing tipping forward and to provide adequate support. Small foam sticky pads are normally attached to one side of the glazing panels before delivery to site. This prevents them sticking to each other when they are in storage, and reduces the risk of scratching occurring as a result of trapped dirt or grit.

INTERNAL TIMBER

Skirting, architraves, door linings, etc. are all examples of joinery quality timber used for second fixing. Internal timber is normally seasoned and moulded before installation to prevent the effects of shrinkage occurring. Whenever possible, they should be stored in the room or area where they are going to be fixed prior to cutting and fixing (second seasoning). They should remain in the building until further movement has occurred in the timber and the 'equilibrium' moisture content has been achieved. All internal timber must be stored flat, off the ground on bearers, with piling sticks in between each layer to allow the air to flow around the stack.

IRONMONGERY AND FIXINGS

Desirable and expensive items such as ironmongery and fixings are likely to be stolen if they are left in unsecured areas for any length of time, or without an adequate store control system in place. If these items are stored in damp or wet conditions they may also start to show signs of deterioration. To prevent this happening they should be stored in dry, lockable containers and off the floor on shelves. The storage shelves should be clearly labelled for easy identification at a later date when they are required for use.

TRADE SECRETS

Whenever possible, heavy items such as ironmongery and fixings should be stored on the lower shelves in a storage area. This avoids unnecessary heavy lifting from the ground to the upper shelves.

PAINT, ADHESIVES, STRIPPERS, ETC.

Hazardous materials used in the construction industry are covered by the Control of Substances Hazardous to Health (COSHH) Regulations. The regulations place legal duties on employers to control the exposure of employees to chemicals and other substances. Employers should assess the risks involved with each product and provide employees with training, information and the correct PPE in order to use the substances safely without risk to themselves and/or others. Manufacturers and suppliers are required by law to provide 'safety data sheets' with each product. The safety data sheets are used as references for information regarding the safe handling, storage and use of a particular substance hazardous to health.

Hazardous materials decanted (poured) from large containers to smaller ones should be adequately labelled with their content and safety requirements.

PLASTERBOARD

Sheet materials should be stored under cover in a dry environment, flat, off the ground and on bearers, evenly spaced to prevent sagging. Alternatively, plasterboards can be stored vertically, provided that the bottoms of the boards are elevated above the floor. The disadvantage of this method is the potential to cause damage to the delicate edges and corners. If plasterboard is incorrectly stored and allowed to become damp, the plaster will become soft and difficult to cut cleanly when it is being installed and secured.

TIMBER-BASED SHEET MATERIALS

Timber-based sheet materials should be stored using the same methods as with plasterboard, although decorative or veneered boards should have their faces protected with cardboard. Alternatively, the best faces of the boards should be placed together to prevent scratching or staining of the surfaces.

WINDOWS

Plastic, aluminium and timber windows are rarely stored on site for long periods. Generally, they are installed by specialised window fitters who install the frames at the same time they deliver them to site. This prevents the items taking up valuable storage space for longer periods than necessary, and the potential to get damaged while on site. Any window frames stored on site should be secured under cover to protect against the elements, and adequately supported to prevent against twisting.

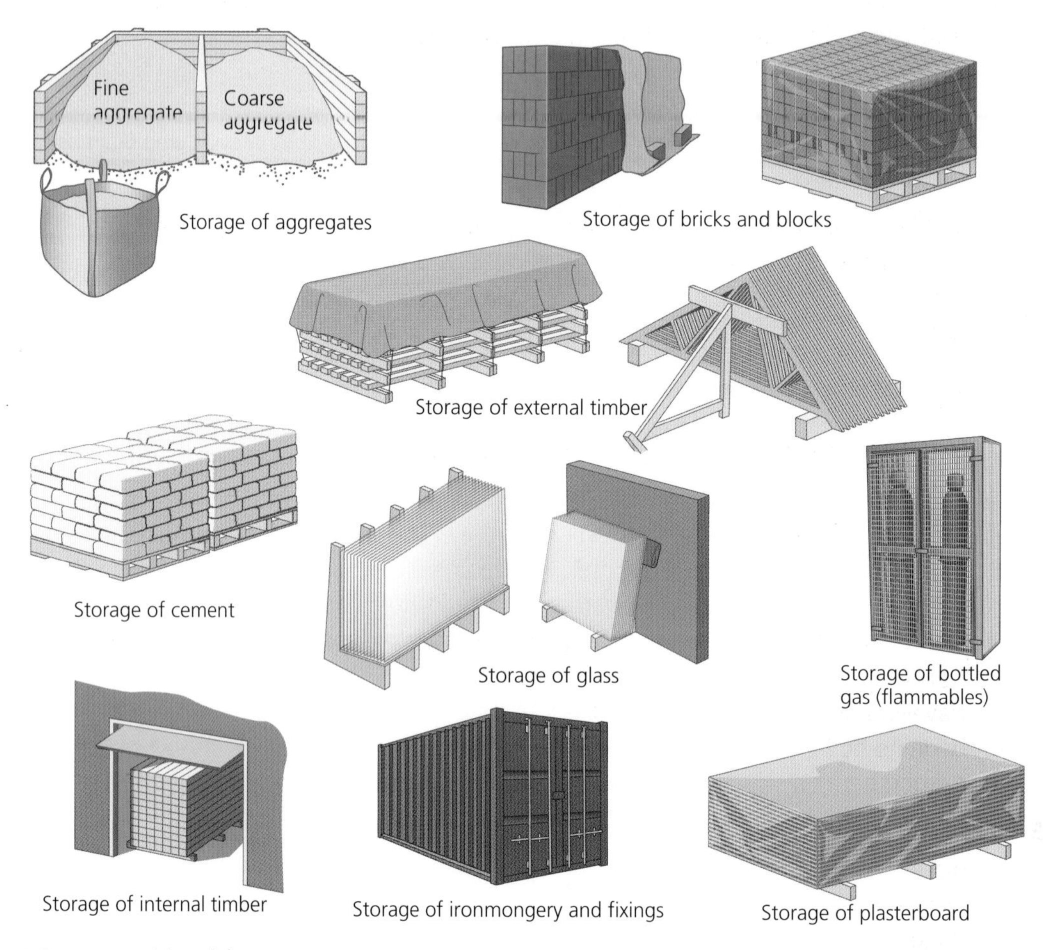

FIGURE 3.17 Materials storage

ACTIVITIES

Activity 15 – Materials storage and delivery of building materials

Read through the following questions and answer them as fully as you can to help you develop your underpinning knowledge of this subject area.

1. Explain the purpose of stock rotation.
2. Explain the effect on plaster if stored in damp conditions.
3. Why is it advisable to phase deliveries to site?
4. List the checks that should be carried out when a delivery is made to site.
5. How should internal joinery quality timber be stored on site?

MULTIPLE-CHOICE QUESTIONS

1 An inner and outer wall, separated by an insulated void, **best** describes which one of the following types of wall construction?
   a Mass
   b Cavity
   c Concrete
   d Timber framed

2 A reinforced concrete slab, spanning the full length and width of a building, best describes which one of the following foundation types?
   a Pad
   b Raft
   c Strip
   d Piled

3 The measurements for setting out foundations are taken from the
   a estimate
   b drawings
   c schedules
   d specification

4 The position of the damp-proof course within a wall prevents water
   a travelling up from the ground
   b travelling down from the roof
   c bridging between the lintel and sill
   d passing between the outside and inside skin

5 Common rafters are fixed at their upper end to the
   a hip
   b crown
   c ridge
   d purlin

6 The sheet material used to cover stud partitions is
   a panelboard
   b fibreboard
   c battenboard
   d plasterboard

7 Concrete is made from a mix of water,
   a sand, bricks and cement
   b steel, fine aggregate and cement
   c fine aggregate, course aggregate and cement
   d fine aggregate, course aggregate and plaster

8 Which of the following materials would **not** be fit for purpose if left out in the rain?

a Tiles
b Bricks
c Blocks
d Plaster

9 The **most** suitable type of paint for internal walls is

a vinyl
b gloss
c emulsion
d micro-porous

10 Which one of the following statements is true about flammable materials? They

a can't be stored outdoors
b have to be stored indoors
c should be stored in large tubs
d have special storage requirements

CHAPTER 4

FIRST FIXING

LEARNING OUTCOMES

By the end of this chapter you should have developed a knowledge and understanding of:

- fixing frames and linings;
- fitting and fixing floor coverings and flat roof decking;
- erecting timber stud partitions;
- assembling, erecting and fixing straight flights of stairs, including handrails.

INTRODUCTION

The aim of this chapter is for students to be able to recognise first fixing components and materials, and the methods used to install them. It also identifies and explains current Building Regulations applicable to this area of study, following all the relevant health and safety law and good working practices.

DOOR FRAME IDENTIFICATION

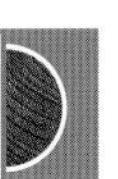

Door frames are often confused with door linings, but there are clear differences between the two:

- door frames are used in mainly an exterior position (the division between the inside and outside of a building);
- they have a thicker section than door linings (approximately 95 × 58 mm). This will provide strength to the frame and increased security to the building;
- they have a rebated profile or section rather than 'planted on' door stops;

FIGURE 4.1 A door frame in situ

- they usually have draught proofing or weather strips. Alternatively, the door frames will have capillary grooves to prevent water penetrating between the door edge and the inner edge of the door frame.

As mentioned previously, door frames are normally positioned in the openings between the brickwork on the exterior walls of a building, although this is not the only place they could be located. Conventional domestic (residential) buildings will be constructed with cavity walls on the exterior of the building (Figure 4.2). These are inner and outer walls with a space between the brickwork; this area is known as the cavity. The cavity prevents moisture penetration between the two walls and reduces heat loss from the building.

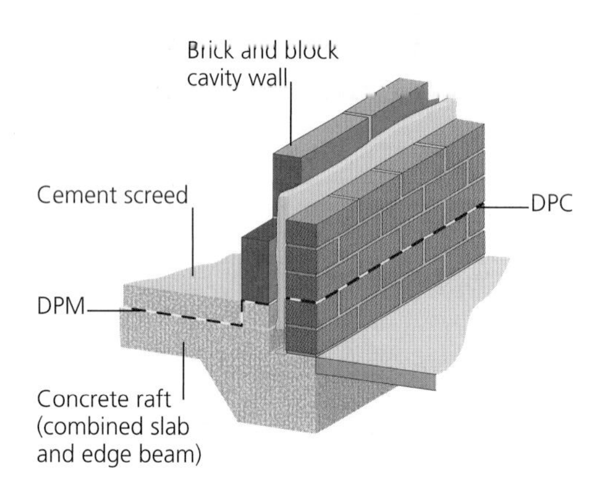

FIGURE 4.2 Cavity wall

FIGURE 4.3 Vertical DPC between a cavity wall and jamb

DOOR FRAME SELECTION

DOOR FRAMES/LINING WITH FANLIGHTS

A room lit with natural daylight is far better for its occupants' health than an artificial source (e.g. fluorescent strip lighting). Some buildings may not have the room or wall space to accommodate extra windows or roof lights. It may also be possible that some rooms may be situated in the centre of a building, and therefore have no external walls for windows to be inserted. There are several methods of producing natural light in these areas, as described below.

Reflective sun tubes

These are usually installed through an opening in the roof to accommodate a pipe, lined with reflective foil. The pipe has a translucent plastic dome over the exposed top that projects from the surface of the roof to prevent moisture and rain entering into the property. Natural sunlight reflects off the inner lining of the sun tube and down through its length until it reaches the end of the pipe. At this point the light is bounced around the room through a *convex* (curved) lens, which is screwed flush to the ceiling.

Borrowed lights built into the frame/lining

Borrowed lights are basically created by extending the length of the jambs, past the head of the door lining or frame, to form an addition to the frame. The area above the lining/frame is known as the 'fanlight'. The fanlight can either be directly glazed into the rebates or against the loose beading; alternatively, this area could house an opening in the frame by adding a sash.

DOOR FRAME POSITIONING

Door frames are most commonly used in the openings formed in external cavity walls. These walls are substantially thicker than internal walls due to their double skin of brickwork. The door frames normally are positioned set slightly back from the face of the brickwork. The exact position of the frame will normally be specified on the architect's drawings, although consideration should be given to door frames with thresholds.

A threshold is the section of timber fixed on the bottom of the two vertical sections of the frame, known as the 'jambs'. The threshold provides weather protection to the building by deflecting rainwater that runs down the face of the door away from the building. Water that travels down the slope or 'weathering' shaped on the front of the threshold may run underneath. To prevent water entering the building under the threshold, a drip groove must be formed. The 'drip' on the bottom of a threshold must be situated over the brickwork to work effectively, when it is being positioned during fixing.

DOOR FRAME PROFILES

The exact shape or 'profile' of door frames will vary between manufacturers, but there are several distinct features that should remain the same in order to function properly. Each door frame will have a rebated section removed to allow a door, or pair of doors, to sit within the frame. It would *not* be acceptable to 'plant on' (loosely fix) door stops to a squared section of timber to form the rebate. It would be difficult to prevent water leaking through the frame using this method, because the joint between the door stops and the frame would cause capillary action. This type of construction may also lead to a breach of security because of the joint between the frame and planted-on door stops (see additional information on capillary action, page 105).

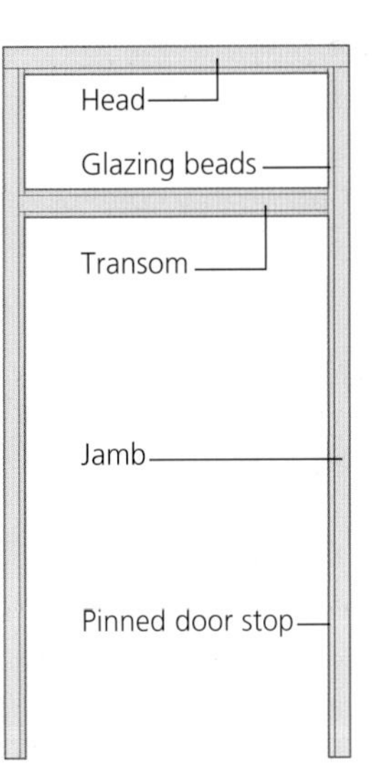

FIGURE 4.4 Parts of a door frame

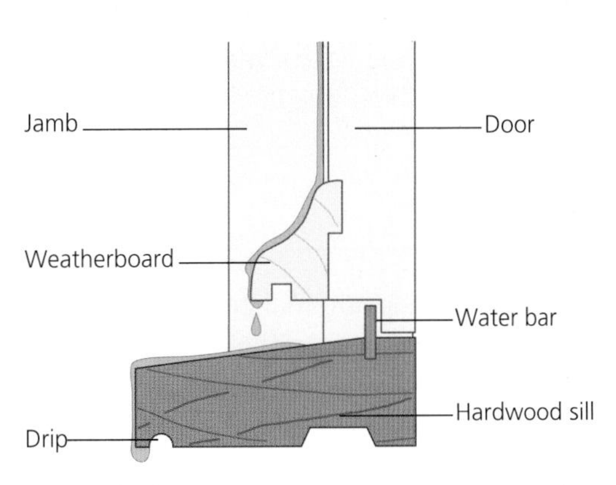

FIGURE 4.5 Threshold

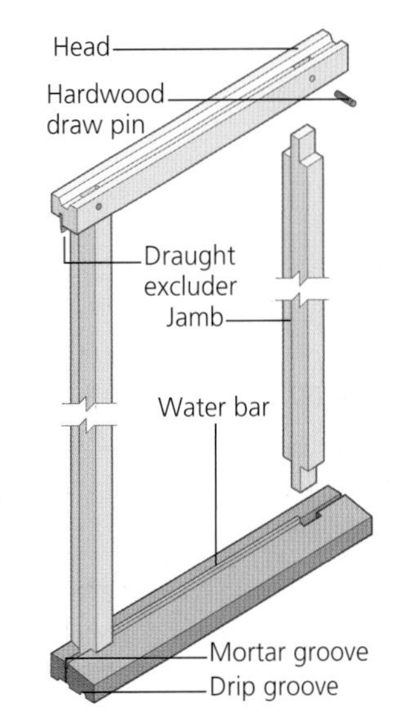

FIGURE 4.6

Door frames may also be used internally in situations that may require additional security or where a heavier door requires supporting. A good example for the use of a door frame in an internal position is a fire rated door, as these are normally substantially heavier than standard lightweight or 'hollow core' doors. In this case the door must withstand the extra load on the hinged side and the weight of the door closing up against the rebates on the locking side.

FREQUENTLY ASKED QUESTIONS

What are hollow core doors?

'Hollow core' doors are, as the name suggests, doors with a relatively hollow core. Hollow core doors are constructed with two outer faces made of either plywood or hardboard. The 'core' or centre of the door must have some form of support to add stability to the door and prevent it from collapsing. The cost-effective way of filling the core is to use a web of cardboard on its edge. The web is sandwiched between the two faces of the door and held in position with adhesive. The hollow core door will also contain a section of timber known as a 'lock block'. The lock block is positioned in the middle, on one side of the door, to allow the door lock or latch to be mortised into. The lock block is identified with a symbol on either the top or edge of the door.

FIRE DOOR FRAMES

Frames used in an internal position to hang fire doors will have a slightly different profile. They will not require grooves around the inner face of the rebates for draught excluders or weather strips due to their positioning within the building. Some fire rated door frames may need to have smoke or fire seals incorporated into the frame; this is normally achieved by having a machined slot around the frame for the seals to fit.

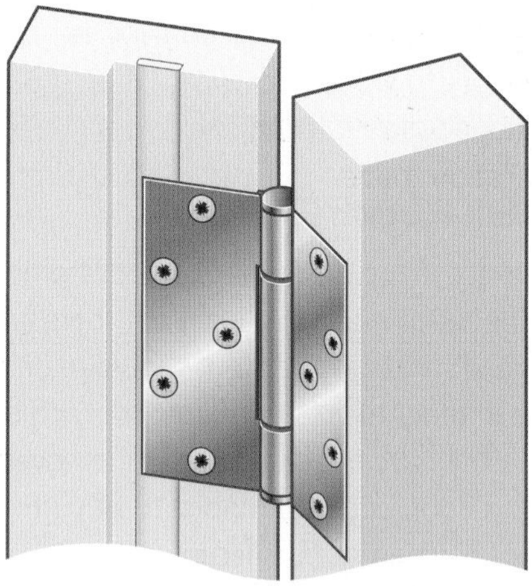

FIGURE 4.7 A door frame suitable for fire protection

DOOR FRAME CONSTRUCTION

There are three main components needed to complete a door frame.

It is common practice to connect the jambs to the head and sill with mortise and tenon joints. The head and threshold should run past the jambs (these are known as the 'horns'), allowing them to be jointed together while the weathering still provides weather protection across the entire front of the frame.

There are several methods of securing the mortise and tenon joints together on door frames. Traditionally, the mortise and tenon would include a haunch and wedges to secure the joint. Modern practice would normally include a frame, either pre-assembled in a joinery workshop or factory with through mortise and tenon joints, or all the components supplied in 'flat pack' form. These joints are secured with adhesive and either nails or screws. An alternative method of pulling the joints together on a door frame is to use draw dowels, which will eliminate the need for long clamps.

FREQUENTLY ASKED QUESTIONS

What does the term 'flat pack' mean?

To reduce the transport costs of large joinery items, they are sometimes delivered to site unassembled. Items of joinery that are delivered in this condition are known as 'flat pack'.

WEATHER PROTECTION FOR DOOR FRAMES

Traditionally, exterior door frames would have the head, jambs and sill shaped to prevent rainwater passing from the outside of a building to the inside, without the use of weather seals. Although the profiles on these types of door frame are functional in restricting some of the elements through the door frames, they do not prevent draughts and heat loss. Modern exterior door frames are complete with draught excluders made from PVC and foam, or rubber around the inner edges of the rebated frames. The exact position and shape of draught excluders will vary between manufacturers. Figure 4.8 shows some alternative draught excluder profiles commonly available and the position in which they should be used.

Exterior door frames have metal thresholds screwed along the top edges of their sills. Some metal thresholds are supplied with sections that must be secured to either the face of the bottom of the door or along the bottom edge. These combination thresholds work in conjunction with each other, and should always be fitted following the manufacturer's installation guide.

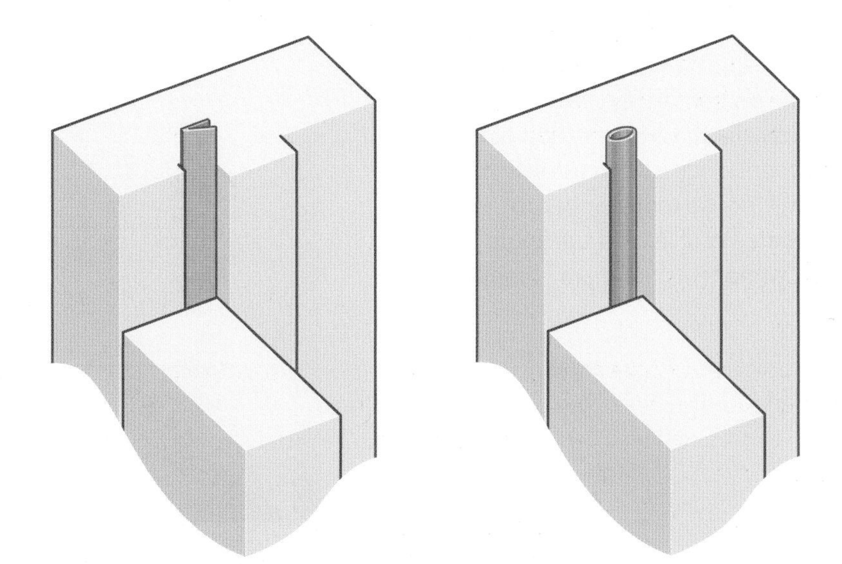

FIGURE 4.8 Some draught excluder profiles

DOOR LINING IDENTIFICATION

Door linings differ from door frames because they can only be used internally and are smaller in section (26–38 mm thick). The width of door linings is normally determined by the thickness of the internal walls to which they are going to be fixed. A range of standard-width door linings is available from building suppliers. The most common widths are:

- 115 mm;
- 138 mm.

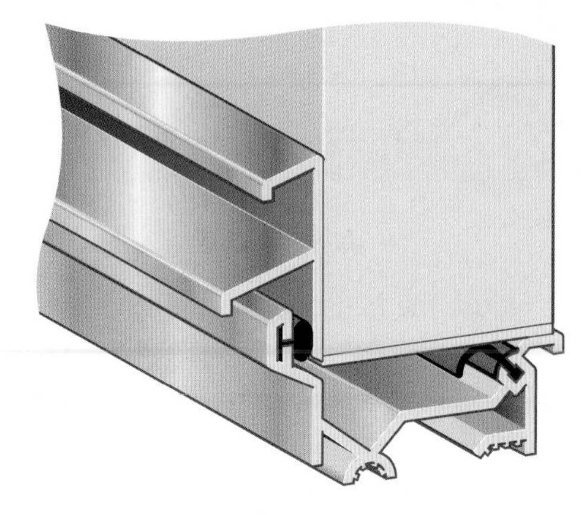

FIGURE 4.9 Metal threshold

'Off the shelf' door linings are produced to accommodate the most commonly used door widths. The head of a standard door lining is manufactured with housing joints on either side of the timber; this reduces the need for suppliers to stock two different-width linings and reduces costs. Traditionally, the jointing between the jamb and head lining would use a bare-faced tongue and trench (housing), and be secured with either screws or wire nails.

DOOR LINING PROFILES

The most commonly used profile for door linings is squared section material with loose door stops. The loose door stops are nailed to the door lining to form a rebate for the door to close against. The door stop is normally attached to the door lining using 40 mm oval nails, after the door lining has been installed and the door hung and securely fixed.

Door linings that have loose door stops have the added advantage of allowing the stops to fit to the exact thickness and shape of the door. This will ensure a perfect closing action and the door finishing flush within the face of the lining.

Door linings may also have solid rebates machined in the sections of the lining prior to assembly and installation. This method of constructing door linings is less cost-effective, as the lining will require more timber to achieve the rebated profile, and the joints between the components are more complex.

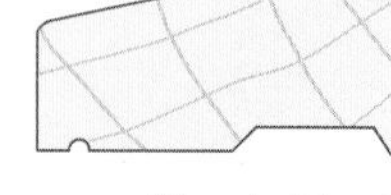

FIGURE 4.10

FIGURE 4.11 Door lining

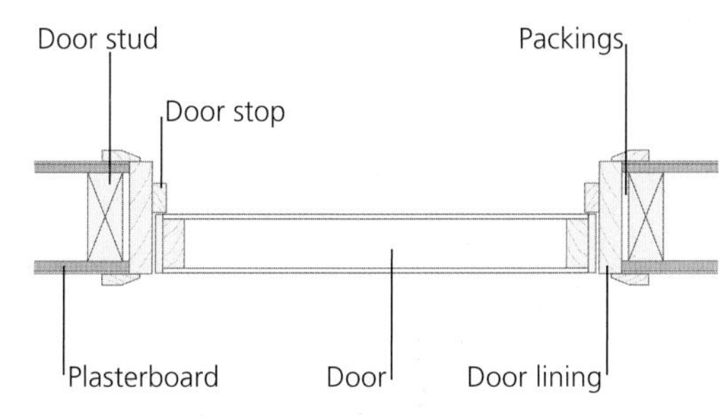

FIGURE 4.12 A standard lining

The advantages of using rebated frames are that they are stronger and provide a neater finish on the face of the lining because there are no unsightly joints or redundant rebates (rebates that are not used).

DOOR LINING CONSTRUCTION

Door linings with loose door stops have very simple methods of construction; they can be manufactured with either through housing joints or tongued housing joints. Through housing joints are simpler to construct and are mostly used in the mass production of 'off the shelf' joinery. Through housing joints in the head of a door lining will locate and secure the jambs as the frame is being assembled.

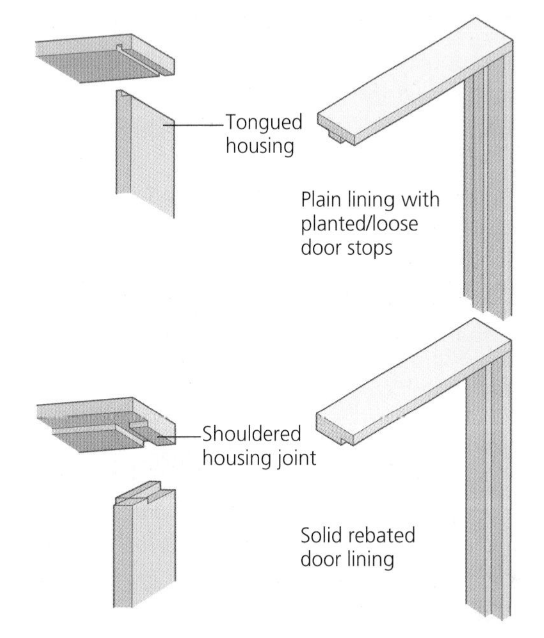

FIGURE 4.13 Door lining construction

FIGURE 4.14 A tongued housing joint between the head and jamb of a door lining

TRADE SECRETS

When installing door linings, consider how the fixings can be concealed. This method of fixing will not only reduce the amount of time needed to cover or fill fixing points, but will also provide a professional finish. Fixing points can sometimes be hidden behind fire or smoke seals, door stops or even the door hinges.

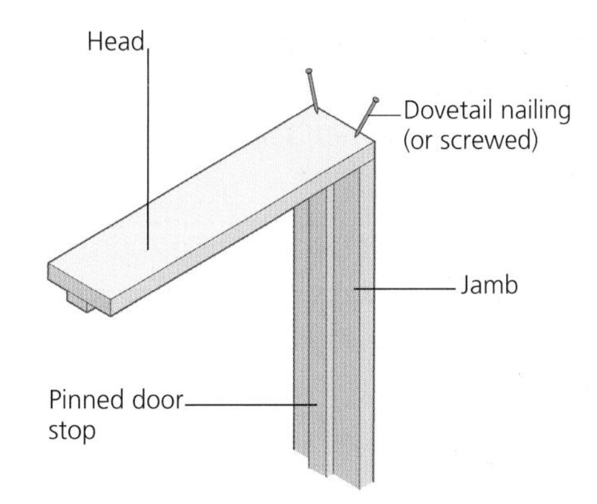

FIGURE 4.15 Door lining fixing

This type of joint may weaken after the horns are removed before installation due to the lack of support on the outer edge of the housing joint. A more secure method of jointing the head of a door lining to the jambs is the tongued housing joint. This method still offers support to the housing joint after the horns have been removed to allow for fixing.

DOOR FRAME/LINING MATERIALS

Door frames and linings can be produced in a vast amount of different materials. Most suppliers will carry a limited range of softwood frames and linings; non-standard frames built from hardwood are usually supplied to order. Softwood door frames usually have an optional low-cost hardwood sill; this will provide durability and protection from the traffic that passes over it.

Man-made timber-based sheet materials are widely used in all areas of construction because they:

- are relatively inexpensive;
- are easy to cut and shape;
- have no natural defects;
- do not shrink, expand, twist or warp like solid timbers;
- have smooth surfaces ready to apply finishes such as paint or stain.

Door linings are also manufactured from 30 mm thick medium-density fibreboard (MDF). They are usually supplied with several coats of primer/undercoat to seal and protect the MDF from moisture.

MDF mouldings and linings are also available with 'foil' or veneers applied to the exposed faces to simulate solid timber. 'Veneers' are thin layers of real wood; they are usually about 1 mm thick and can be sanded and painted just like solid timber. 'Foil' is basically a printed image of wood grains or solid colours that is applied to paper; the foil is extremely flexible and can be wrapped around any shape. Foil-wrapped mouldings are pre-finished so they do not require any sanding or painting; they are also very durable. They are not, however, suitable for areas where the bottom ends (feet) are likely to become wet as the MDF will swell, for example commercial premises where floors are washed daily, as the feet will soak up the water.

PREPARING DOOR LININGS FOR INSTALLATION

Door linings for low-cost work are normally assembled on site, especially if they are purchased 'off the shelf'. It is vitally important that the sequence of preparing a lining for installation is followed; this will benefit the carpenter while fitting the lining because the lining will:

- be sanded and prepared for painting;
- have the two jambs parallel to each other;
- be square;
- have strong joints between the head and the jambs.

Prepare the area

Before any part of a lining is assembled, an area must be prepared to provide adequate room to move, assemble and store the completed frames. A well-prepared assembly area will save time and money, prevent damage to the parts, and reduce the risk to health and safety.

Preparing the lining

Door linings and frames should have their inner faces and rebates sanded and the sharp edges (arrises) removed along their longest edges, prior to gluing and assembly. This will prevent any difficulty when trying to sand the jointed corners after the linings or frames have been secured together.

PREVENTING DAMAGE TO THE LINING

Door linings should *never* be assembled directly on the floor because this may damage the surfaces of the timber and possibly leave the frame twisted. Ideally, door linings should be assembled on a bench, although this is not normally possible on site. Alternatively, linings can be laid across bearers that have been positioned on a flat, level surface (see 'Boxed frame assembly' in Chapter 7, page 286, for more detail on bearers).

SQUARING A DOOR LINING

When preparing a door lining for installation, it is important that the jambs are both parallel to each other and square to the head of the lining.

The best method of aligning and securing these components in position before installation is listed in the following steps.

- Step 1 – Check the length of the jambs.
- Step 2 – Secure the housing joints between the head and the jambs with strong wood adhesive and either screws or round head nails.
- Step 3 – Fix a stretcher on either side of the door lining, about 150 mm up from the 'feet' of the jambs. This will prevent the jambs of the lining from twisting and keep them parallel to each other.

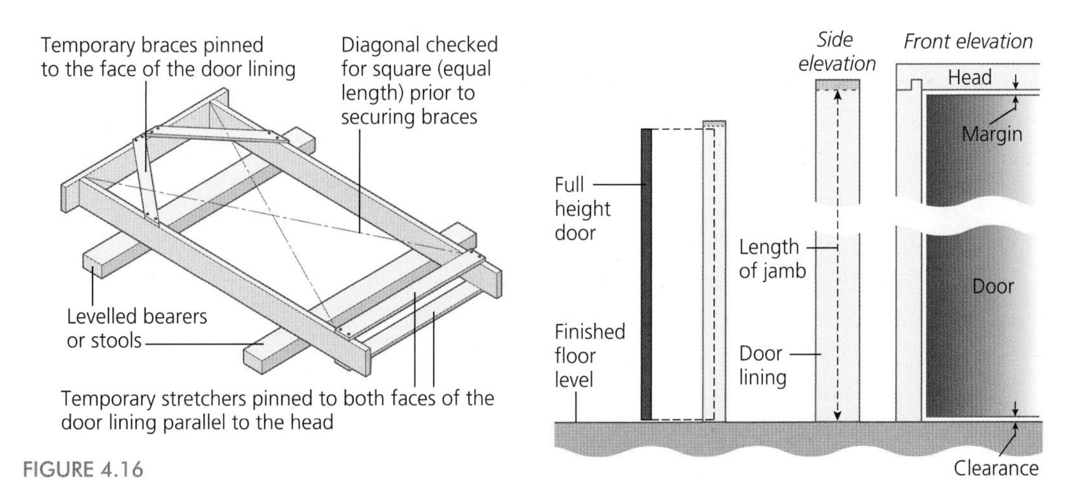

FIGURE 4.16

FIGURE 4.17

TRADE SECRETS

The length of the stretchers is determined by the overall width of the door lining. The most accurate way of achieving this is to transfer the width onto the stretcher from the head of the lining before fixing.
Use round head nails to temporarily fix the stretchers to the lining. The heads of the nails should be left above the surface of the stretchers to allow for easy removal after the lining has been installed.

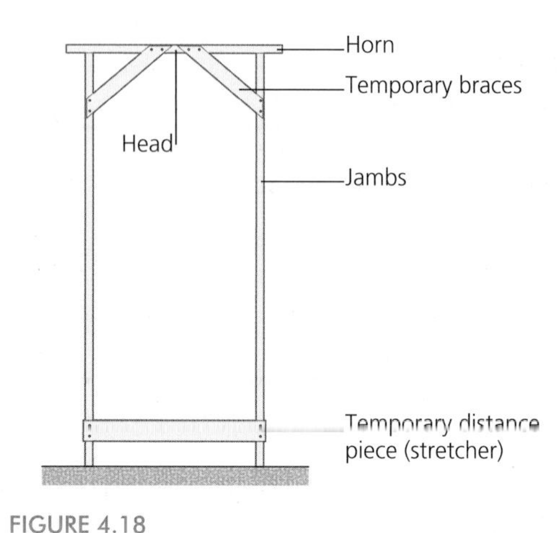

FIGURE 4.18

FREQUENTLY ASKED QUESTIONS

What are the 'feet' of a door lining?

The term 'feet' refers to the bottom of the jambs on a door lining. The term may also refer to door frames, architraves and other similar arrangements.

- Step 4 – The most accurate method used to check that a frame is square is to measure diagonally across the frame between the stretchers and the head, and repeat the process across the other diagonal. This method is known as squaring from 'corner to corner'. The measurements between the diagonal corners should be equal to each other. If they are not, then pressure will have to be applied between the longest opposite corners to adjust the frame until the diagonal measurements are equal. A brace should be prepared with 45° corners cut on both ends. The brace should have a nail started at each end ready to be driven into the face of the lining. When assembled, the brace can be positioned and the nail driven into the jamb. When the frame has had the diagonals checked and adjusted to square, the other end of the brace can have its securing nail driven in. This will hold the frame square during storage and transportation. The brace should not be removed until the lining has been fixed.

FIGURE 4.19 Using a squaring rod on a door lining

FIGURE 4.20 Adjusting a door lining to square and attaching the brace

When a door lining is being checked to see that it is square, you may use a tape measure to compare the diagonal sizes. This method usually requires two people: one to hold the tape measure tight in one corner and the other to read the measurements from the tape. Using a tape measure to check that a frame or lining is square can sometimes be difficult and inaccurate. This is because the end of the tape measure cannot fit completely into the square corners of the lining and it could also sag and twist.

- Step 5 – Secure the door lining in its square position with diagonal bracing between the head and one of the jambs. The bracing should be nailed in place at an angle of approximately 45°; if the bracing is fixed at a shallower angle than this it becomes less effective at retaining the frame in its square position.

Note: bracing and stretchers should be secured in place with two nails at each fixing point. This will help to prevent the frame or lining from twisting or racking on a single fixing point.

FIGURE 4.21 The hook end of a tape measure

TRADE SECRETS

The 'hook end' of a tape measure is securely fixed to the body of the tape with metal rivets through elongated slots. The elongated slots allow the hook end of the tape to slide backwards and forwards the thickness of the bent metal. This movement in the hook end of the tape measure allows for accurate measurements to be taken. Continual use of a retractable tape measure will wear the elongated holes and the tape will no longer give accurate measurements. Accurate measurements can still be taken from a worn tape by holding the end on the 100 mm measurement and extending the tape to the required length. It is essential that the 100 mm measurement has been accounted for when reading the extended tape.

INSTALLING DOOR FRAMES AND LININGS

Door frames and linings can be installed using the same methods as for installing window frames (see more detail in 'Installing windows', below).

Traditionally, door linings would have been fixed into wooden pads/pallets (four up each side) placed in between the brickwork bed (horizontal) joints by the bricklayer as the opening for the lining was created.

Replacing door linings that have been previously fixed into the wood pads can sometimes be difficult, because the original fixings normally pull the pads from between the brickwork joints as the lining is being

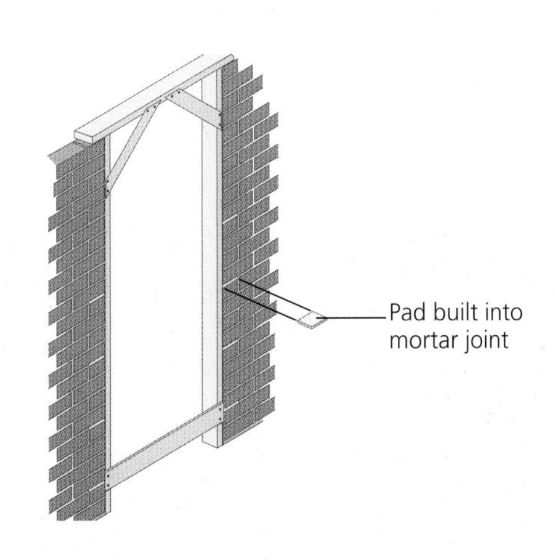

FIGURE 4.22

removed. This usually means that there is a void between the brickwork at the points you would normally secure the new lining.

This problem can be overcome with the use of 'propeller (twisted) wedges'.

Modern methods of installing door linings and frames have been developed to replace traditional fixings. This is partly due to the development of new construction materials used in the industry, as well as reducing labour costs and increasing productivity.

SEVEN STEPS TO INSTALLING A DOOR LINING

1. Check the sizes and lining reference (as appropriate) to ensure it is the correct lining for the opening.
2. Remove the horns on the head of the door lining and cut jambs to finished floor level (FFL), ensuring there is sufficient clearance between the underside of the head and the floor (this would be calculated by adding the height of the door, the top and bottom clearances, plus the thickness of the floor covering; typically this would be 1981 + 3 + 3 + 18 = 2005 mm).
3. Position the lining in the opening; use pairs of folding wedges between the top of the door opening and the top of the lining to secure the lining down tight to the floor. Check that the head is level. Wedging down will allow for adjustment to the faces of the door lining to ensure plumb without them moving. The lining should have an equal overhang on both faces of the wall to accommodate the plaster finish.
4. Put the first fixing approximately 100 mm down from the top corner of one of the jambs, and check the face and edge of the jamb are 'plumb' with a long spirit level or plumb rule. Use packers to adjust the jamb accordingly, and fix through the jamb and packers to avoid them slipping out at a later stage. The second fixing should be about 150 mm up from the floor level on the same jamb. Two intermediate fixings should then be made, ensuring the inside face of the lining is straight. Check the

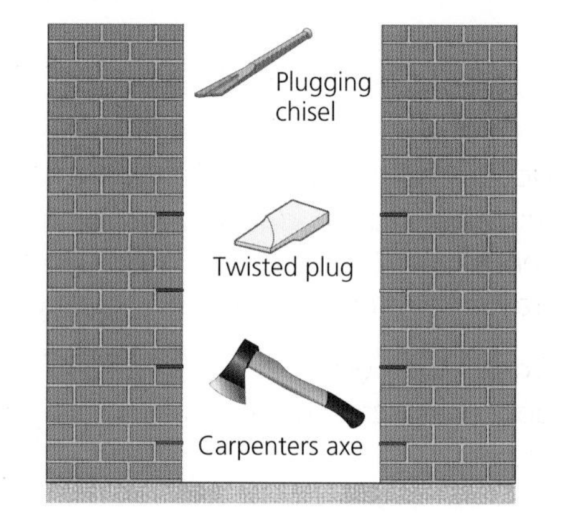

FIGURE 4.23

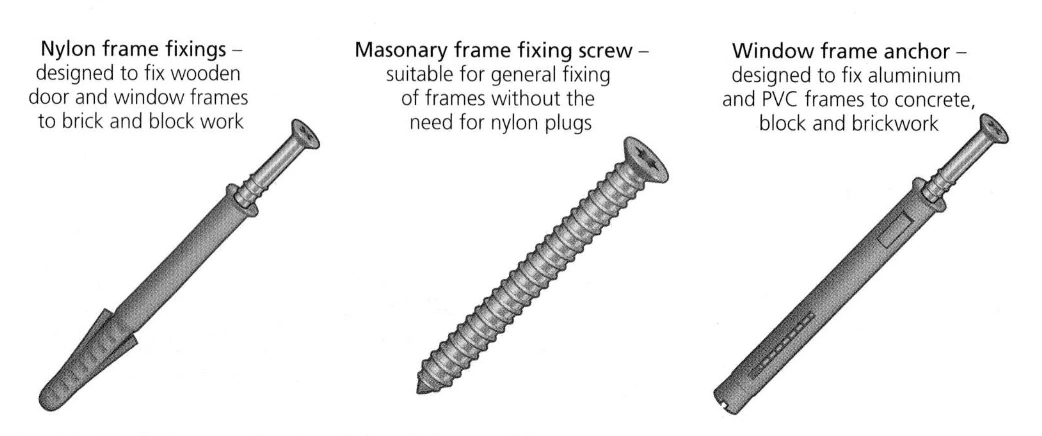

FIGURE 4.24 Fixings used to install door linings and frames

other jamb has an equal margin either side of the wall, and check it is plumb on the face and edge.

5. Check the lining is not twisted. This is achieved by 'sighting through' the frame at the top and following the edges down to the feet. The outer edge of one jamb should be parallel to the inner edge of the other; if not, then this jamb must be adjusted to correct.
6. Install all the fixings, making sure they will not interfere with the installation of the hinges, striking plate, etc. Check all the faces and edges are straight, plumb and level before removing the bracing and stretchers.
7. The door stops can be loosely nailed into position in preparation for the installation of the door(s) (it is good practice to always install the door stop on the head first; the other stops, on each of the jambs, should be butt jointed to this one).

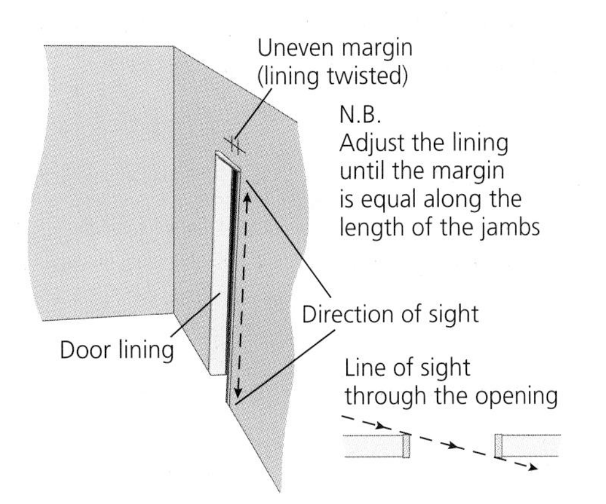

FIGURE 4.25 'Sighting through' a door lining

FREQUENTLY ASKED QUESTIONS

How many fixings should you put in the jambs of a door lining?

Door linings and frames should be securely fixed into their openings to prevent movement as the door(s) open and close. Door frames usually have four fixings evenly spaced along each jamb. The section of a door lining is usually wider than the door frame, so it requires two fixings side by side at four locations on each jamb.

Door frames without a sill attached can be secured at their feet by inserting galvanised dowels into the end grain and leaving them proud. The frames can then be secured by raising the floor to the required height with concrete 'screed'.

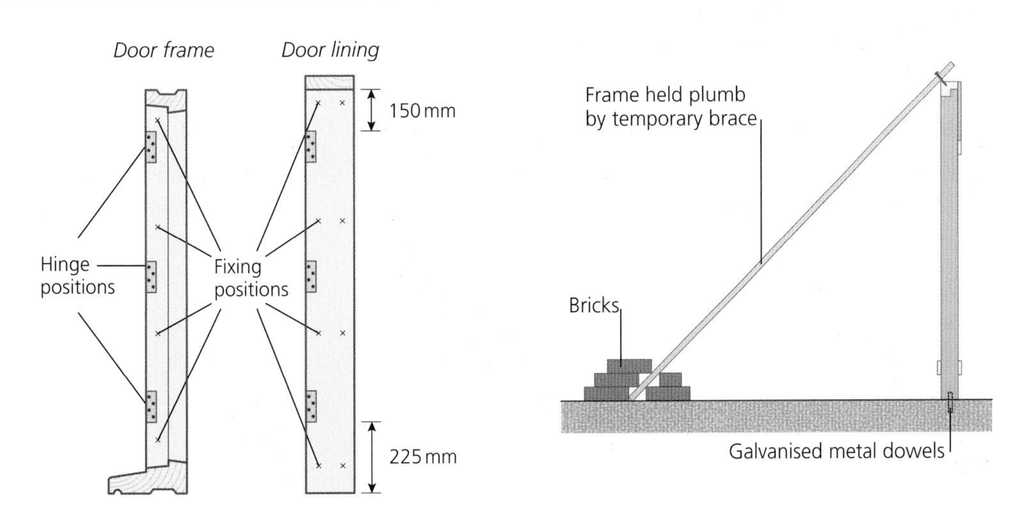

FIGURE 4.26 Fixing positions

FIGURE 4.27

PROTECTING COMPLETED DOOR FRAMES AND LININGS

Door linings and frames are usually some of the first items of joinery to be installed into a building under construction. The exact sequence of installing door linings and frames will be determined by the site manager, who would have produced a 'programme of work'. The 'programme of work' will be referred to when ordering the building materials, plant, equipment and labour.

Phasing the delivery of vulnerable items, such as door linings and frames, will reduce the risk of damage when the items are on site. Once they are installed permanently, the corners should have edge protection applied.

FUNCTIONS OF WINDOWS

Generally, the main functions of windows are to provide light, ventilation and a means of escape in the event of an emergency. Creative design by the architect at the planning stage can allow the shape, outlook and finish of the windows to enhance the appearance of a dwelling while maintaining their main functions. The size of each window built within a room will be stipulated on the architect's drawings and the accompanying window schedule. Building Regulations contain specific details for the area of light and ventilation needed for a room; this is normally a proportional size of the floor area and, in most cases, one-twentieth is required to adequately ventilate an area.

Although Building Regulations specify a minimum amount of light and openings needed in a building, planning permission may restrict the maximum amount of window space used. The use of large window openings within buildings has become very popular with designers and architects; however, these large expanses of glazing can allow the loss of heat unless careful consideration for the use of the following is employed in their construction:

- stable materials;
- double or triple glazing;
- draft excluders.

Windows can be constructed from a number of different materials, including softwood and hardwood, uPVC (polyvinyl chloride) and aluminium or steel. Metal and uPVC windows are relatively maintenance free and cheap to manufacture due to their simple construction methods; for these reasons they have become very popular in recent years. The use of materials and style of replacement windows will be strictly controlled by the local authority. The restrictions imposed upon developers and builders to control the use of plastic windows etc. will preserve the aesthetic appearance of listed buildings and be 'in keeping' with local historic areas.

The use of any wood-based material in an exterior position will require regular maintenance to preserve its integrity. There are a number of optional methods used to protect exposed woodwork from the elements; generally the use of preservatives – brushed, dipped or pressure treated – will offer the best protection, together with a good-quality paint or stain.

Figures 4.28 and 4.29 demonstrate the appearance of uPVC and timber windows from a front perspective. The uPVC windows provide a substantially larger frame than the timber and metal alternatives. Although metal frames provide a window that is very durable and small in section, it remains functional and less decorative in its appearance. Timber windows will provide a compromise between metal and uPVC window frames, and can be designed to suit virtually any shape or design.

FIGURE 4.28 A uPVC window

FIGURE 4.29 A timber window frame

FREQUENTLY ASKED QUESTIONS

How long will timber windows last?

The life expectancy for windows that have been treated with preservative, prior to having a sequence of primer, undercoat and top coat applied for a painted finish, or a good-quality base and top coat of stain, is:

- 50 years for softwood;
- 75 years for hardwood.

Note: the exact life span of a window frame will depend on the type of timber used and regular maintenance.

WINDOW IDENTIFICATION

All windows are identified by the material from which they are constructed and the positioning of the casement within the frame; for example, traditional casements (casements = the opening part of a window) sit flush within the window frame. Windows are further classified by the way that the casements or sashes operate within the window frame.

FIXED SASH, COMMONLY KNOWN AS A 'DEADLIGHT'

Fixed sashes are constructed in the same way as if they were to open but, as the name suggests, they are screwed securely into position (see 'Traditional' casements, page 104). Alternatively, a window can be directly glazed ('direct glazing') into the rebates of the frame, therefore offering a cheaper option.

SIDE-HUNG CASEMENT

Side-hung sashes can be hinged on the stile of the casement with the use of butt hinges for traditional windows or cranked hinges for storm-proof windows. Side-hung sashes can also pivot with the use of friction hinges fixed to the top and bottom rails on the casement (see 'Window ironmongery', below).

FANLIGHT

Top-hung sashes (fanlights) are hinged using the same methods as the side-hung sashes, only this time they swing from the top portion of the casement to open at the bottom.

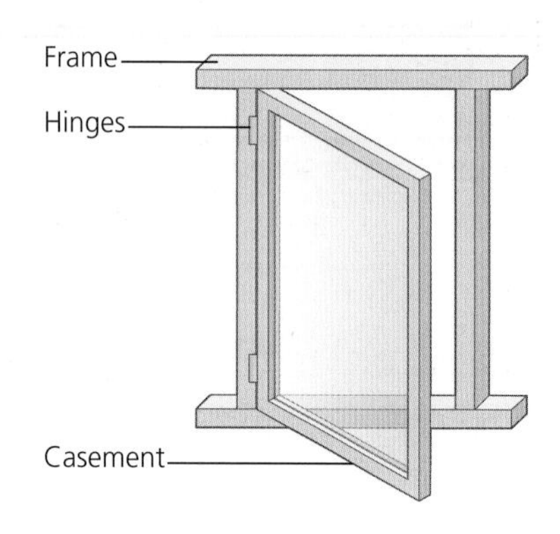

FIGURE 4.30 Side-hung sash

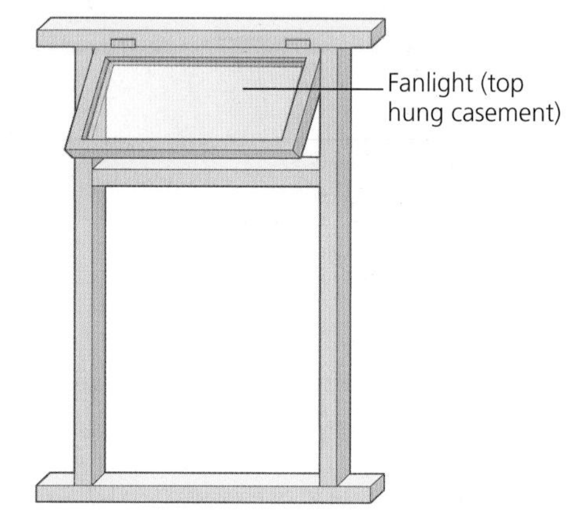

FIGURE 4.31 Top-hung sash (fanlight)

SLIDING SASH/BOX FRAME

These operate by sliding up and down in a groove formed by the construction of the jamb. The sashes are suspended by sash cords and counterbalanced by weights that run in the box, formed as shown in Figure 4.32. An alternative method is to use 'spring balances' (Chapter 7, 'Maintenance', should be referred to for further information).

Sash windows can also be designed to slide horizontally along the window frame, although these are not so common. This style of window frame is known as 'Yorkshire lights'.

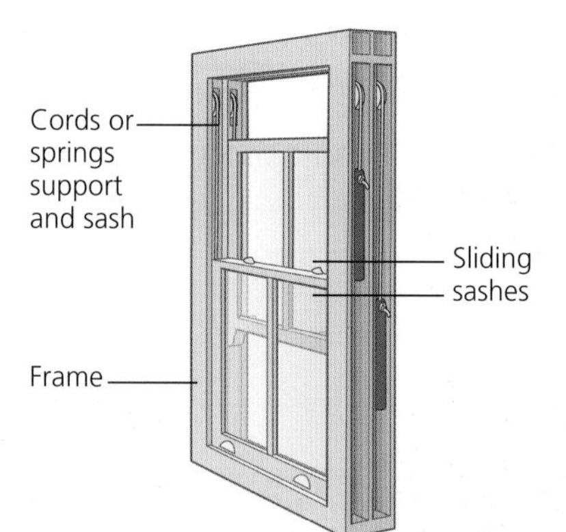

FIGURE 4.32 Vertical sliding sash (box frame)

PIVOTED SASH

This style of casement is pivoted in the centre of the sash, using either a 'window pivot' for traditional windows or 'friction pivots' on storm-proof windows (see 'Window ironmongery', below).

TILT AND TURN

This style of opening can be used on both windows and the top sections of 'stable doors' (a stable door is a single door split across the middle to form a top and a bottom opening section). This method of opening will allow the frame to hinge in two directions by bolting one side of the mechanism in the closed position while the other operation takes place.

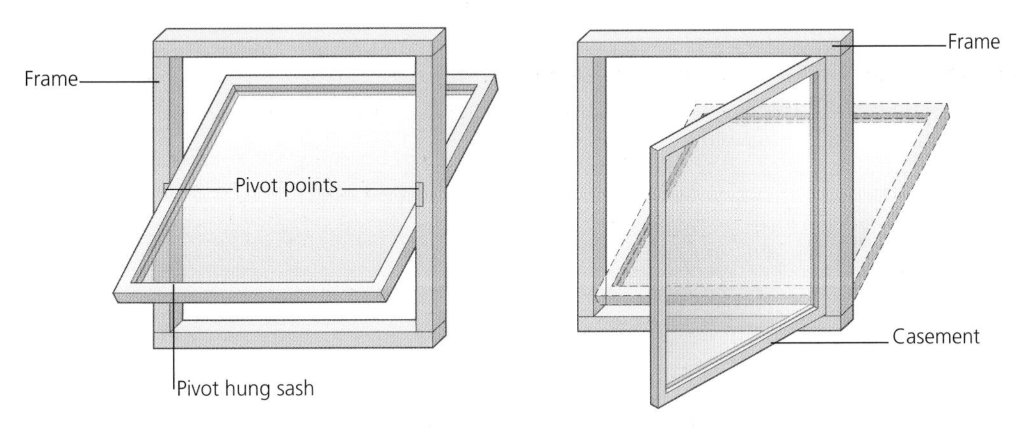

FIGURE 4.33 Pivot-hung sash

FIGURE 4.34 Tilt and turn

BAY WINDOWS

These can be constructed from either box frames or casement-type windows, and consist of a combination of several window frames fixed together. The number of frames used to construct a bay and the positioning of each frame will vary from job to job; these can be categorised as follows:

- square;
- splayed/cant;
- segmental.

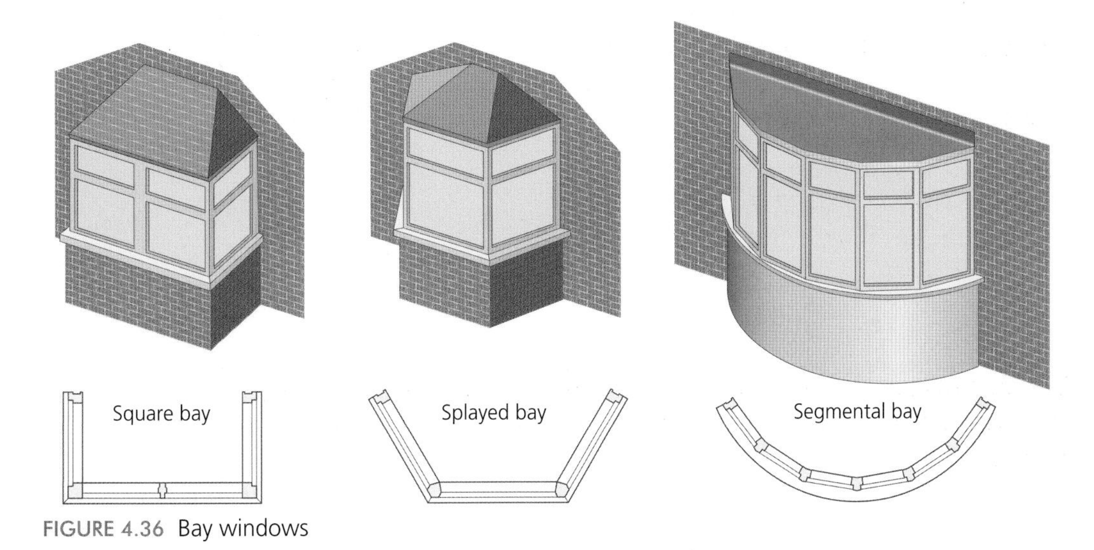

FIGURE 4.36 Bay windows

ROOF WINDOWS (SKYLIGHT AND DORMER)

Roof windows are most commonly used when converting loft spaces into living accommodation; they will allow natural light into the roof area without altering the shape of the roof. Roof windows will require 'flashing' around their perimeter to provide a weathertight seal; these are normally purchased in a kit with installation instructions.

FIGURE 4.36 Roof windows

FREQUENTLY ASKED QUESTIONS

What is 'flashing'?

Flashing is a term used for the components used to seal the joint between exterior surfaces. Suitable materials for flashing must be durable, flexible and weather resistant. The following materials are commonly used as flashing on roofs:

- lead;
- copper;
- zinc;
- galvanised sheeting.

WINDOW SELECTION

STORM-PROOF WINDOWS (TRADITIONAL AND HIGH PERFORMANCE)

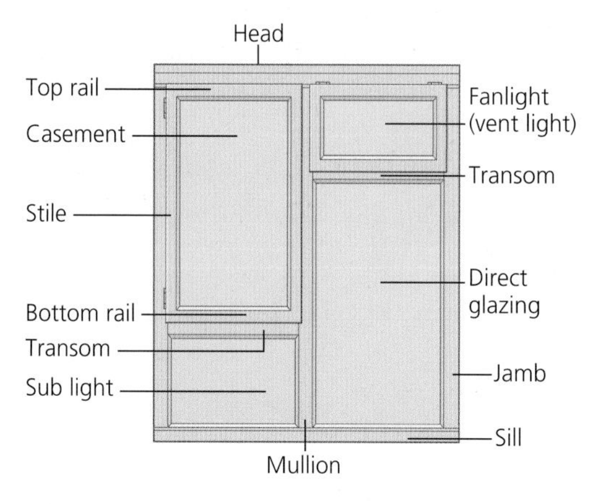

FIGURE 4.37

High performance stormproof window

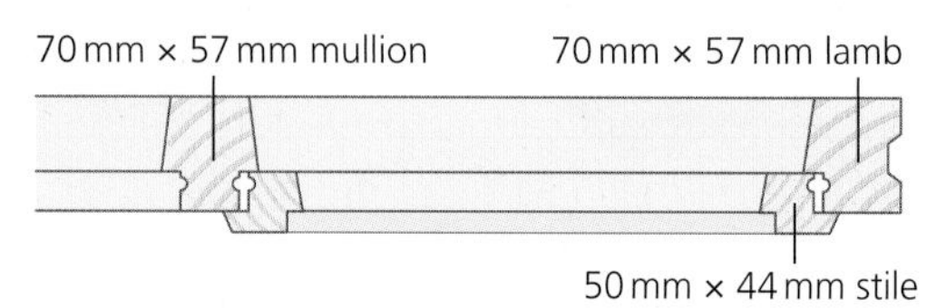

Traditional stormproof window

FIGURE 4.38

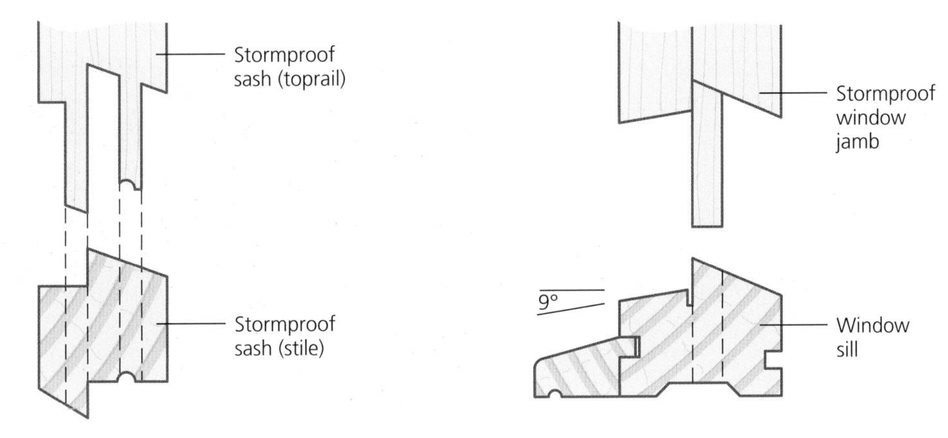

FIGURE 4.39

FIGURE 4.40

As the name suggests, storm-proof windows are constructed to give superior weather protection compared to traditionally constructed window frames. Storm-proof windows have casements that are rebated over the main frame of the window. The rebates will protect the window opening from direct rain passing through the joint between the sash and the frame. The use of draught excluders between the opening sashes and the window frame will prevent warm air escaping the building in the winter and cold air entering the property.

For a property to become energy efficient, consideration must be made for the openings within the building, in particular door and window openings. The loss of warm air through window openings can be almost eliminated with a well-designed window frame and the use of stable and durable timbers, combined with double or triple glazing.

In practice, a property that is completely sealed from any air flow through the dwelling has proven disadvantages, including stale air. Exposure to stale air for long periods of time may affect the occupants and their respiratory organs, due to a build-up of moisture causing condensation; this in turn causes mould and bacteria growth. To avoid any potential long-term health problems to the occupants, small openings are formed either through the head of the window frame or through the top rails in the sashes; this allows a controlled amount of air to pass through the window.

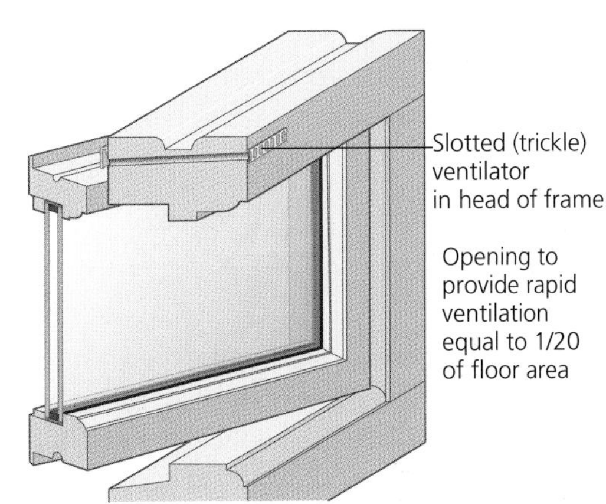

FIGURE 4.41 Ventilation

The small opening in the frame is normally covered with a plastic trickle vent. This will prevent water passing through the window and control the amount of air flow by opening or closing either all or a portion of the vent from the inside of the window frame. Trickle vents are available in a wide range of solid colours to match the window frames.

TRADITIONAL CASEMENTS

Although the component names remain largely the same as for storm-proof windows, the sections vary. In simple terms, the difference between traditional and storm-proof windows is the detailing of the positioning of the casements within the window frame. Traditionally, the full thickness of a casement would be positioned flush to the main window frame; this is achieved by rebating the thickness of the casement within the frame section.

Figure 4.42 highlights the use of weathering on the transoms and sill sections. Any water that penetrates between the sash and the window frame would advance only to the 'capillary' groove.

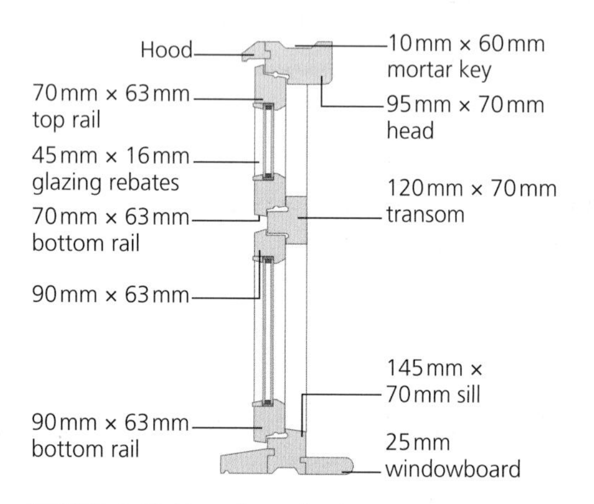

FIGURE 4.42 Use of weathering

FIGURE 4.43 Traditional window frame

FREQUENTLY ASKED QUESTIONS

▶ **What do the terms 'weathering' and 'capillary groove' mean?**

Weathering is the slope that is moulded onto the timber during the machining stage of manufacturing. It allows water to run off the window section and prevents the onset of rot.

Capillary grooves are the mouldings machined around the edge of the casements and around the rebate on the frame of the window.

Traditional windows would normally contain casements in all of the rebated sections of the main window frame, regardless of whether the casements were opening or not. Although a costly alternative, a traditional window frame with this arrangement of the casements offers a uniform, balanced and aesthetically pleasing appearance. It is common practice for modern styled windows to only contain casements within the opening sections of the window due to the cost implications of the alternative. This method of glazing a window is known as 'fixed' or 'deadlights'.

CAPILLARY ACTION

The demonstration in Figure 4.44 shows how water can be drawn upwards between two sections, in this case glass panels. The use of anti-capillary grooves between a sash and window frame will stop this happening and thus prevent water entering through the window. Water entering this area of the window frame runs down the outer edge of the sash and is then directed out from the window by the weathering on either the transom or sill, depending on the positioning of the casement. The drip moulds formed on the underside of the sill, transom and hood will prevent water travelling over the weathering and back underneath the sections of the window frame.

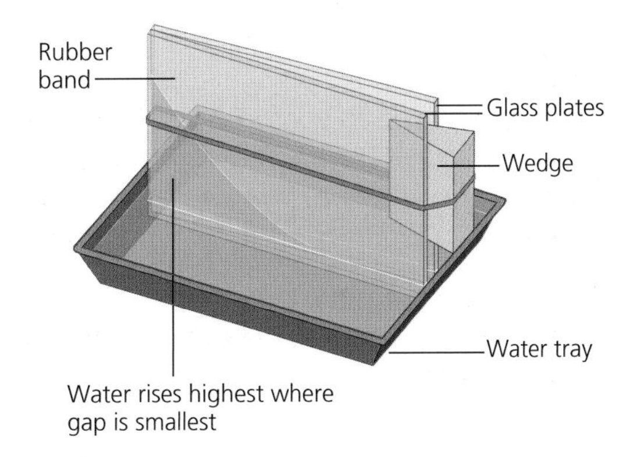

FIGURE 4.44 Capillary action

BAY WINDOWS

As mentioned previously, bay windows are simply made up from a series of window frames fixed together to form one assembly. It is essential that the joints between each window are weathertight; Figures 4.45–47 suggest some methods of connecting the window frames to corner blocks and infill pieces.

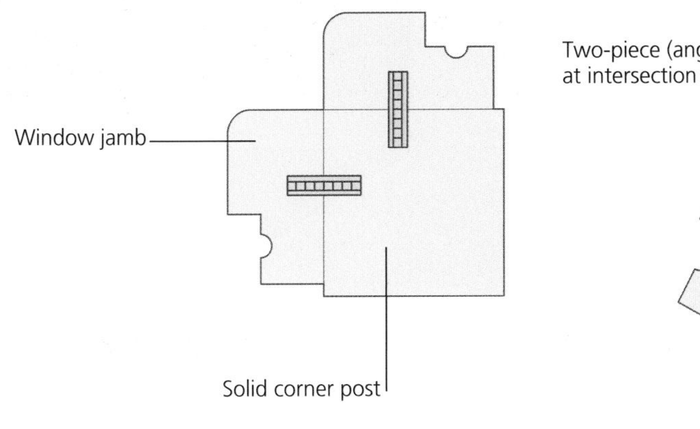

FIGURE 4.45

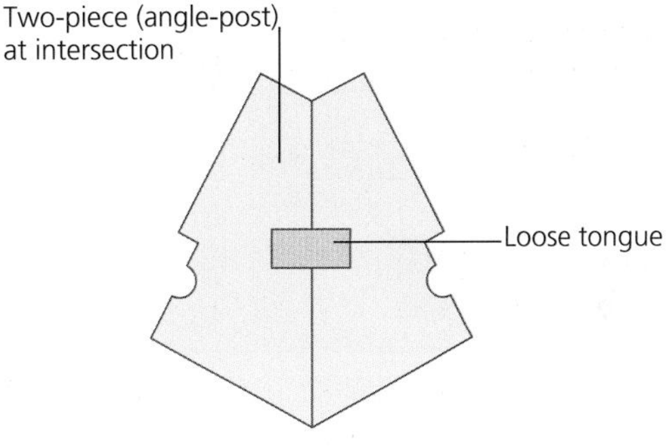

FIGURE 4.46

CONNECTING THE SILL

When installed, the bottom section of a window frame will receive the majority of the weather, so it is important that the sill is durable and strongly constructed. This will prevent any unsightly gaps appearing and moisture penetrating through the joints. The best method of connecting the sill at the joints is to use a handrail bolt and wood dowels.

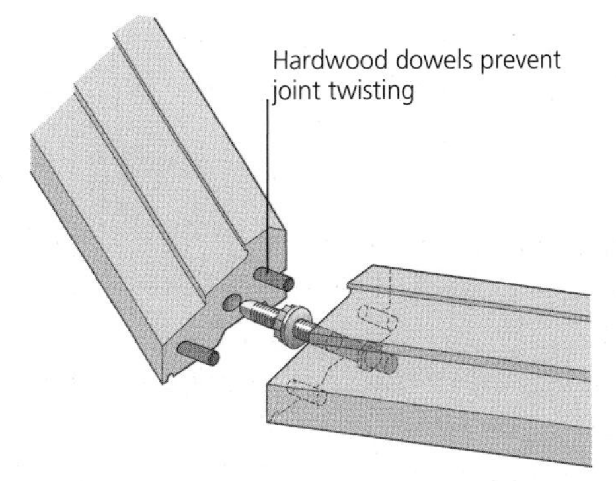

FIGURE 4.47

SHOP FRONTS

A shop front is made up from a combination of a window frame and a door frame. Its primary purpose is to house large sections of glass, contain an individual or pair of doors, and to attract customers. The majority of replacement commercial shop fronts are produced from aluminium box section frames because these are maintenance free. The disadvantages of using metal frames are the limited range of profiles available and the fact that they are usually limited to a choice of solid colours rather than decorative wood grains.

Wooden shop fronts should be used in conservation areas to preserve areas of historical interest. Timber moulded sections offer a versatile alternative and natural beauty that other materials cannot.

Designing timber-based shop fronts requires careful consideration to ensure that the combination of a window frame and a door frame section work together. The transition between the two frames can sometimes create difficulties with the methods of jointing. There are two ways that the two frames can be connected, as described below.

FIGURE 4.48 A shop front

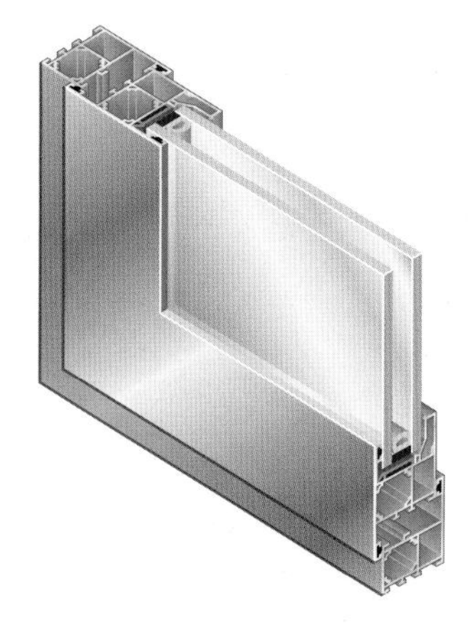

FIGURE 4.49 Box section frame

Individual frames jointed together

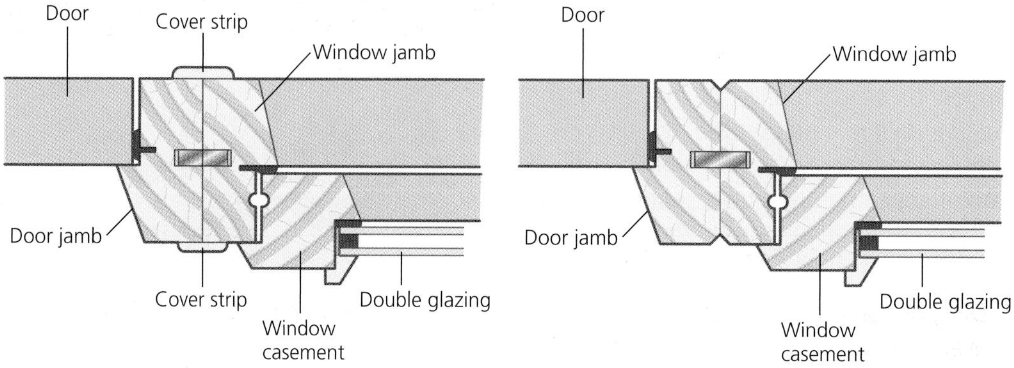

FIGURE 4.50

One Assembly

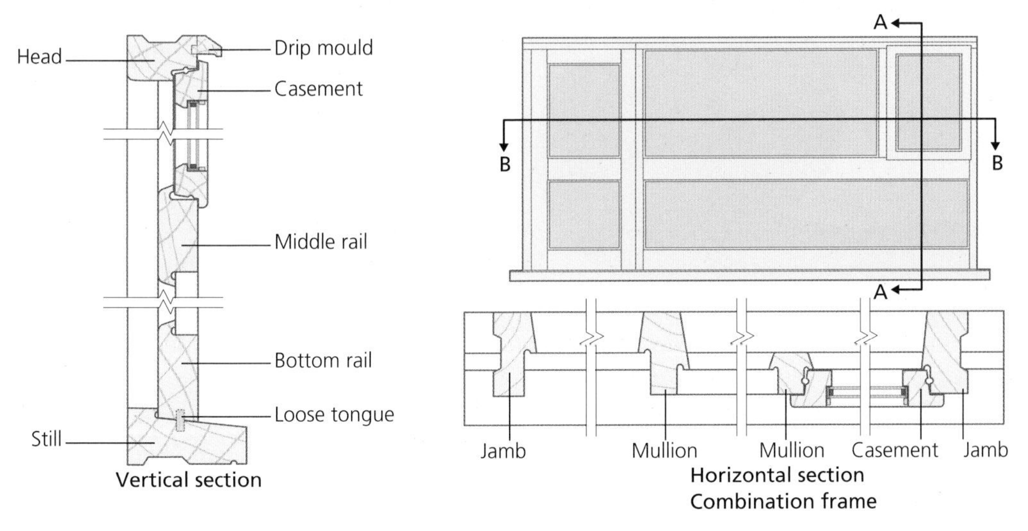

FIGURE 4.51 One assembly

INSTALLING WINDOWS

There are two methods that can be used to install windows in position:

1. built-in during the construction of the brickwork;
2. fixed in after the openings are formed.

BUILDING IN

The building in of frames is a process of positioning the window in a plumb and level position when the bricklayers have reached the height of the window sill. The sill should be placed on a bed of mortar with damp-proof course (DPC) in between to prevent rot.

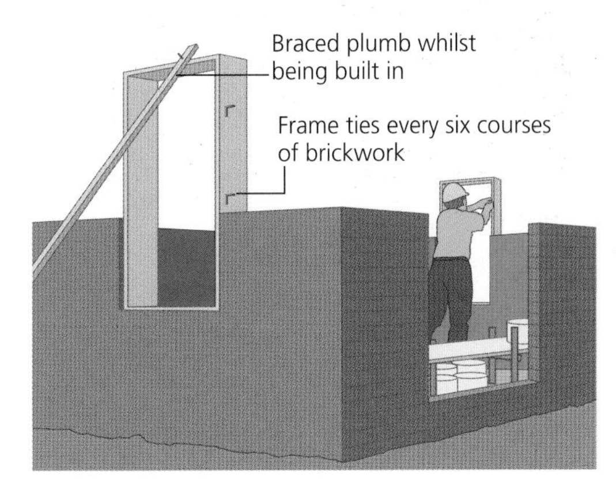

FIGURE 4.52 Building in of frames

FREQUENTLY ASKED QUESTIONS

What is DPC?

DPC, or damp-proof course, is a flexible roll of solid polyethylene (Figure 4.53). It is available in a range of widths from 100 to 900 mm. It is normally used by bricklayers to build in to the course of brickwork 150 mm up from ground level. Its purpose is to prevent moisture travelling up the brickwork.

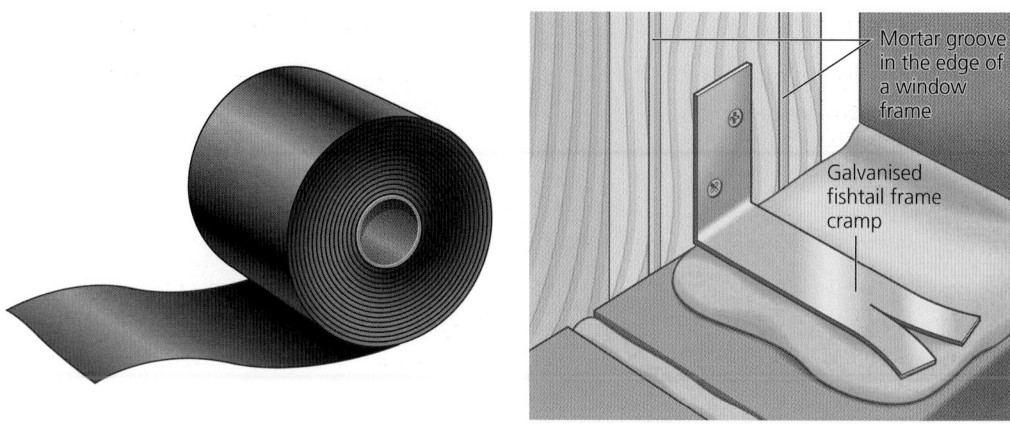

FIGURE 4.53 DPC

FIGURE 4.54 Galvanised cramps

A temporary brace is then attached to the frame by the carpenter and secured at the other end so that the frame will remain in the correct position as the bricklaying progresses. The bricklayer will attach three to four galvanised fishtail frame cramps to the outside of each jamb on the window, in line with the mortar joints in the brickwork. As the bricklayer progresses with the build, the ends of the bricks are buttered with mortar. As these are placed against the jamb the mortar 'keys' into the groove on the outer edge of the window frame. When the mortar dries it will create a secure method of fixing the window along with the frame cramps. Traditionally, horns of the head of the frame would also be built in to the brickwork for additional support, although this is a method rarely used on new construction work these days.

The *disadvantages* of building in items of joinery such as window frames are:

- exposed to the elements (water damage, etc.);
- increased risk of damage to the frames through work activities;
- possibility that the frames may be moved out of level by other operatives or poor weather conditions.

The *advantages* of building in are:

- no visible fixings;
- the brickwork opening is exactly the same size as the window frame;
- no need for trimming or making good/filling large gaps around the frame;
- reduces the likelihood of mistakes caused by the inaccurate setting out of the window sizes.

DUMMY 'BUCK' OR 'PROFILE' FRAMES

These allow brickwork openings to be formed prior to the arrival on site of the windows, thus preventing damage to the window frames as the build progresses. Dummy profiles are usually made from 100 × 50 mm carcassing timber with a diagonal brace to keep the frame square. Timber dummy profiles are relatively inexpensive but can sometimes expand if left uncovered in poor weather conditions. Swollen frames can sometimes become difficult to remove from the window opening without damaging the brickwork, and they rarely survive more than one build. Aluminium dummy profiles are also available to purchase in the most common sizes, and unusual sizes can be purpose built to order.

The *disadvantage* of aluminium frames is the initial purchase costs. The *advantages* are:

- they can easily be adjusted for removal without damaging the brickwork opening;
- they can be easily stored;
- they will not deteriorate;
- they are lightweight and very durable.

FIXED IN

The majority of new window frames installed on building sites are fitted after the roof has been fitted, thus protecting the window from damage while the build progresses. As soon as the windows and doors are fitted into a new building, it can then become secure and the 'drying out' of the building will begin.

FREQUENTLY ASKED QUESTIONS

What needs to 'dry out'?

At the early stages of a building project, poor weather conditions can delay progress. When the roof, windows and doors are fitted the weather can no longer affect progress, so the whole building can begin to dry, including the brickwork mortar joints, rendering and plastered walls.

When creating window openings in a building, the bricklayer will be using the architect's drawings as a reference for the window positions and sizes.

Although the bricklayer will sometimes form the openings as accurately as possible without the use of dummy frames, they can sometimes be built ± 3 mm/m under or over the specified sizes.

These tolerances are perfectly normal for bricklayers, but this variation can cause difficulties for the window fitters, unless a suitable allowance has been incorporated when the frames are produced.

It is vitally important that the window openings are checked before the production of the window frames; this is usually carried out during a site survey by the window manufacturers. The surveyor should check:

- the smallest height dimension;
- the narrowest width dimension;
- if the opening is square;
- if the sill, lintel and brickwork are level.

Alternatively, an allowance can be made by the bricklayer on the brickwork opening size to allow for easy installation.

When fixing in window frames you must try to hide unsightly fixing points; this will provide a professional finish. There are several ways that this can be achieved, as described below.

- Fix through the rebates in the jambs on the window frame. This way any fixings will be concealed by the closing sash. If possible, pull out any draught excluder in the rebated frame to hide your fixings behind. Draught excluders are normally dry fixed into a groove in the frame and can be simply removed and replaced.

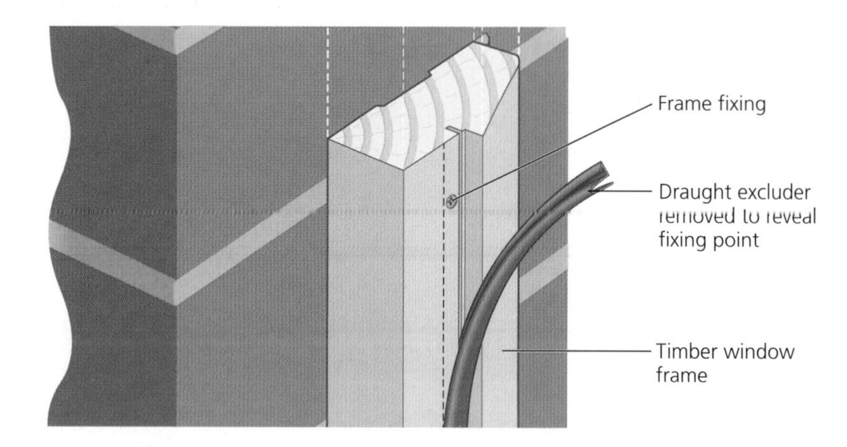

FIGURE 4.55 A window jamb fixed through the rebate into a wall

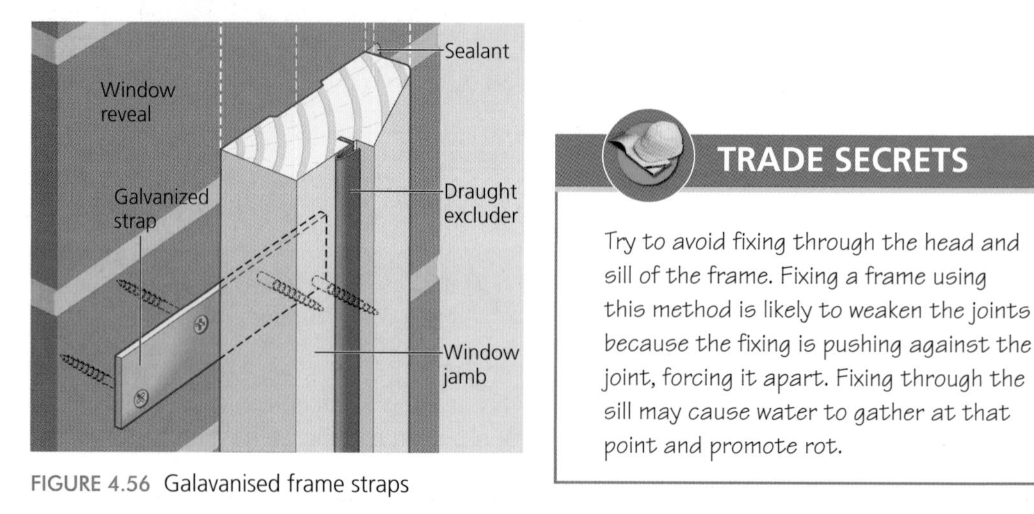

FIGURE 4.56 Galavanised frame straps

TRADE SECRETS

Try to avoid fixing through the head and sill of the frame. Fixing a frame using this method is likely to weaken the joints because the fixing is pushing against the joint, forcing it apart. Fixing through the sill may cause water to gather at that point and promote rot.

- Screw galvanised straps to the outer edge of the frame with approximately 400 mm in between. A fixing point can then be concealed by fixing back through the strap when the window frame is positioned. The galvanised straps will then be hidden by the plastered or dry-lined wall.
- It is not possible to fix boxed frame sliding sash windows using any of the above mentioned methods because of their hollow construction. The most effective method to fix this type of frame is to use folding wedges to secure the window from either side.

GLAZING

SINGLE GLAZING

Traditionally, windows would have contained one layer of glass (single glazed) within the casement of the window and would have been either 3–4 mm or 6 mm thick. It was common practice to use a 3–4 mm pane in windows (depending on the size of the glass) and, for increased protection against breakages, 6 mm panes in doors. While single-glazed windows still have to be used in some historic buildings to protect the heritage of the area, there are many disadvantages to using them, including:

- heat loss;
- condensation;
- sound insulation;
- security.

BUILDING REGULATIONS

Approved document N of the Building Regulations states that any glazing below 800 mm must be safety glass to protect against the added risk of accidental human breakage, and any doors containing glazing cannot contain any standard glazing below 1.5 m. Safety glass is produced in a number of different forms; each method offers protection if the glazing is broken. This may be achieved, for example, by fitting glass that breaks safely, small panes of ordinary glass or thicker ordinary glass, by protecting the glass with a permanent robust screen or using plastic glazing sheets.

LAMINATED GLASS

Laminated glass is simply two sections of ordinary glass bonded either side of a plastic inner layer. Upon strong impact, the plastic layer provides a barrier to which the broken segments remain attached, therefore reducing the likelihood of serious personal injury. Laminated glass is a cheaper alternative to toughened glass and is readily available from most glazing companies.

TOUGHENED/TEMPERED GLASS

Although toughened glass will give the same appearance as ordinary glass, it has been through an additional special heating process to give it a unique safety feature. The process of reheating the glass under special conditions to just under melting point, and then quickly cooling it down, will allow the glass to disintegrate into very small granular pieces on impact. Each of the granular pieces will have smooth edges, thus preventing serious injury. Toughened glass is up to *five times stronger* than ordinary glass and cannot be re-cut after being tempered.

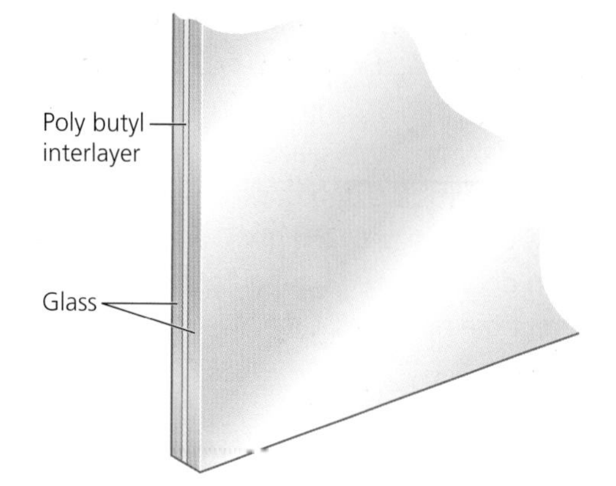

FIGURE 4.57 Section through laminated glass

FIGURE 4.58 Broken laminated glass

FIGURE 4.59 The British Standards Kitemark

FIGURE 4.60 Broken toughened glass

WIRED GLASS

Wired glass has a network of visible wires embedded into it; these wires provide the glass with its additional strength. Upon impact, the wired glass will break but still remain relatively complete; only when the glass comes under an extreme impact is its integrity compromised. Certain types of wired glass will provide fire protection as well as impact protection.

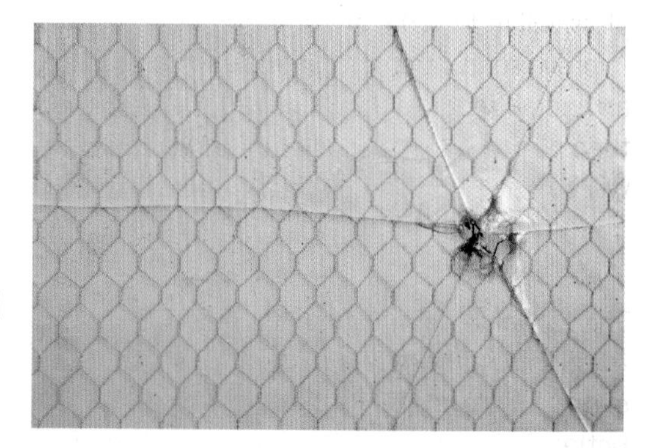

FIGURE 4.61 Broken wire glass

Further details of British Standards are contained within BS 6262: Part 4: 1994 Code of Practice for Glazing for Buildings.

- Doors – any glazing or part of that glazing in a door, which is between the finished floor level and a height of 1500 mm above the floor level, is in a 'critical location'.
- Side panels to doors – any glazing or part of that glazing, which is within 300 mm of either side of a door edge and which is between the finished floor level and a height of 1500 mm above the floor level, is in a 'critical location'.

- Windows, partitions, and walls – any glazing or part of that glazing, which is between the finished floor level and a height of 800 mm above the floor level, is in a 'critical location'.

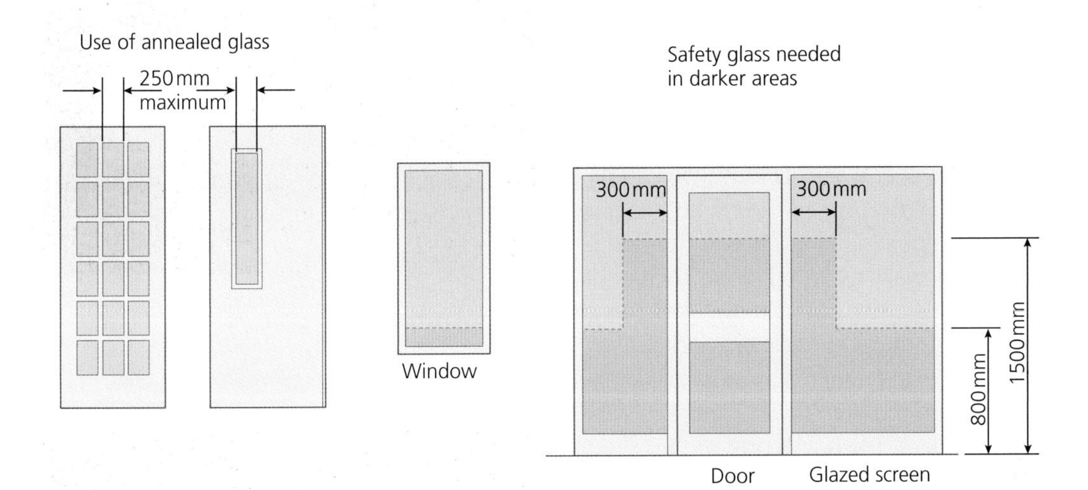

FIGURE 4.62

DOUBLE GLAZING

Double glazing is simply a glazing unit made up of two pieces of glass divided by an aluminium spacer, creating an air gap in between. The space between the two pieces of glass can be increased by using a thicker aluminium spacer around the perimeter of the glazing unit. The increased air space between the two pieces of glass will reduce the amount of heat loss and improve sound transfer through the window. If a window requires the glass to be obscured, then normally only one of the pieces of glass will be patterned to reduce costs.

'U' VALUE

The 'U' value is a measure of heat loss or gain through a material (e.g. windows, doors) and is expressed in 'units'. The lower the 'U' value, the better the resistance to heat transfer through that material or object, and therefore the more superior the insulating value.

TRADE SECRETS

- When installing obscured double glazed units, the patterned side of the unit should be on the inside of the window. This will prevent any dirt in the rainwater holding in the patterned surface and will also make them easier to clean.
- Use plastic packers to give an equal margin around the double-glazed unit when fixing it into the rebated window frame. This will prevent the aluminium spacer between the glass within the glazed unit being visible after installation.

WINDOW BOARDS

Window boards provide a level, decorative trim to complete a window frame. They are available in a range of materials, including:

- hardwood;
- softwood;

- uPVC;
- man-made timber-based sheet materials (e.g. medium-density fibreboard (MDF), plywood).

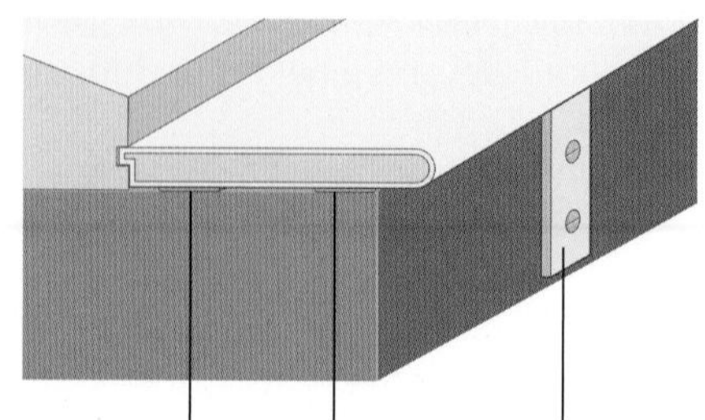

FIGURE 4.63 Securing window boards

Although all window boards will receive scratches and slight damage through regular use, the advantage that timber-based window boards have over plastic is that they can be sanded and refinished. Window boards have a bullnosed moulding along the front edge, which is normally returned along both ends, although this is not possible on plastic window boards. It is natural for timber window boards to have a minimal amount of movement in their width. To conceal this movement, the back edge of the window board will have a rebated edge forming the tongue, which slots into a groove in the back of the window sill.

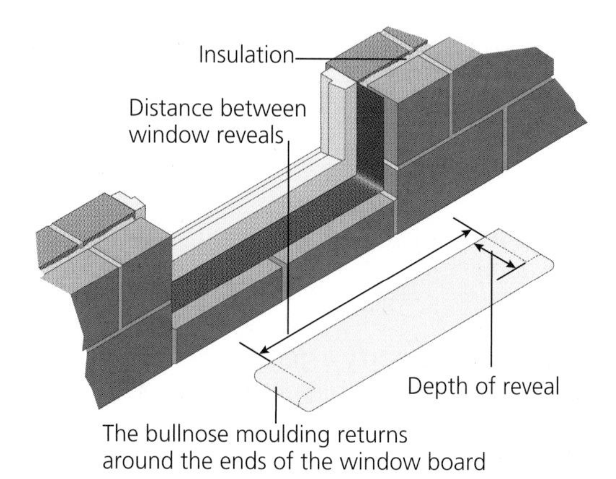

FIGURE 4.64

There are several ways to secure timber window boards into position; this may depend upon the finish on the window board. Screwing or nailing the window board into position through the face is common site practice but can be unsightly. Window boards can be secured by screwing a galvanised window board strap to the underside of the window board and then placing the window board into position. The window board strap will return from the underside of the window board down the face of the inner cavity wall; this is then fixed back to the wall using two or three screws.

The back edge of the window board should sit in a groove. This will prevent a gap from showing at the joint if the board shrinks.

ACTIVITIES

Activity 16 – Fixing frames and linings

Read through the following questions and answer them as fully as you can to help you develop your underpinning knowledge of this subject area.

1. What is the difference between a door lining and a door frame?
2. Explain the best method of squaring a door lining.
3. What three components may be used to construct a door frame?
4. Sketch a through housing and a tongued housing joint.
5. Explain the purpose of 'horns'.

WINDOW IRONMONGERY

Various types of window ironmongery are illustrated in Figure 4.65.

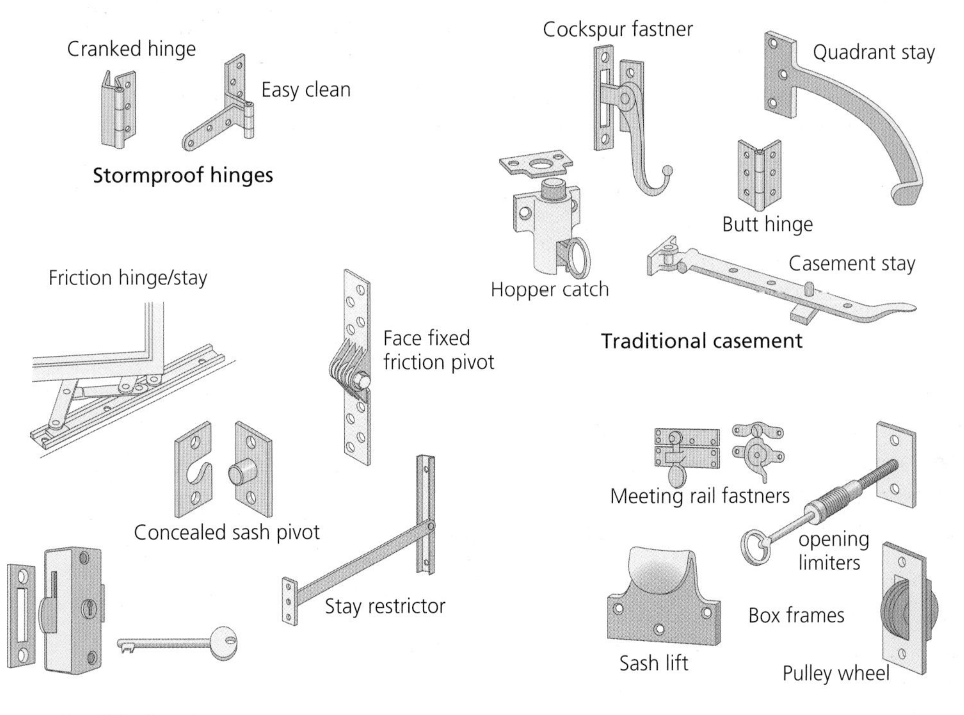

FIGURE 4.65 Window ironmongery

TYPES OF FLOORBOARDING

Traditionally, suspended hollow timber floors would have the floor joists covered in floorboards made of either hardwood or softwood. Developments in man-made timber-based sheet materials have led to materials that can be laid quicker and provide level surfaces that will maintain their shape. Solid timber floorboards that have not been installed correctly are likely to shrink and reveal gaps between the boards; they may also expand if the moisture content of the timber used is incorrect.

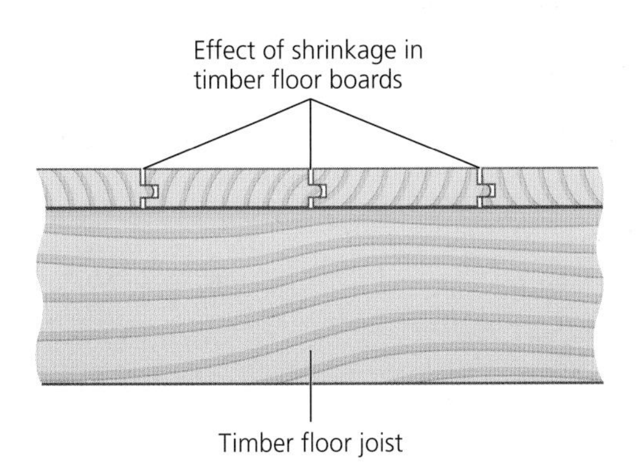

FIGURE 4.66 Solid timber floorboards

FIGURE 4.67 Man-made sheet material

Solid timber floors are sometimes left exposed to display the grain in the boards as a feature, and are sometimes renovated in older homes. The disadvantage of this type of floorboard is that they usually 'cup' after a period of time, causing the floor surface to become uneven with high spots along the raised joints. Premium-grade floorboards minimise the amount of cupping across the boards; this is because they are usually converted using the 'quarter sawn' method. The premium-grade boards would be narrower, with each board shrinking less. Although this system of converting trees is effective, it is also wasteful and costly.

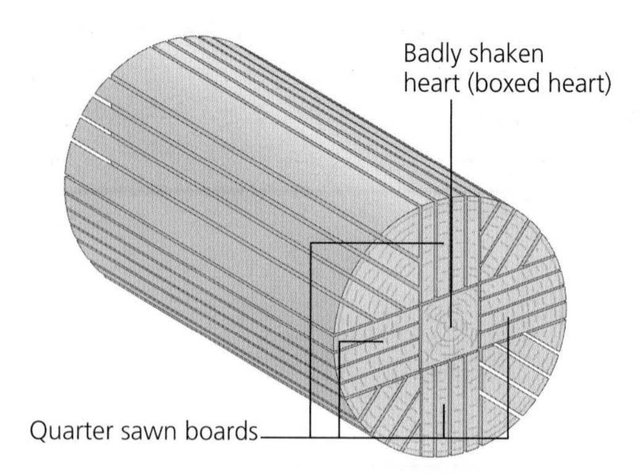

FIGURE 4.68 The quarter sawn method

Traditional floorboards would have had square edges that butted up against one another with *no* machined joints. The disadvantages of this style of floorboard were that it allowed cold draughts between the joints in the boards, and usually squeaked. The squeaking that occurred between the boards was due to the movement between the joints across the unsupported area between each floor joist.

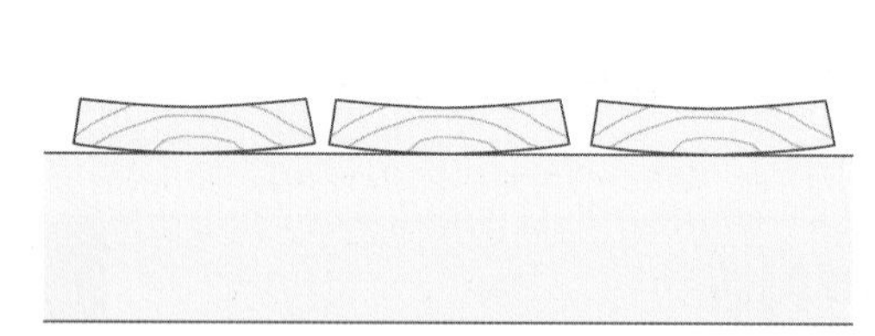

FIGURE 4.69 Cupping

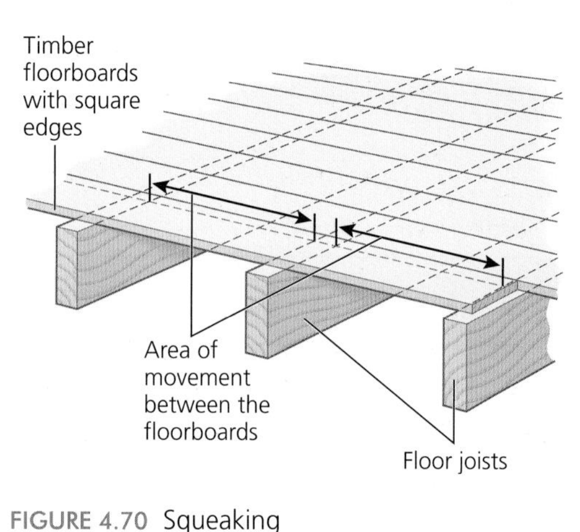

FIGURE 4.70 Squeaking

TONGUE AND GROOVE FLOORBOARDS

Solid timber flooring, with machined edges that interlock, reduces the likelihood of movement between the joints in the floor. The most common method of jointing the floorboards is with a tongue machined along one of the longest edges and a groove along the other. These mouldings are commonly referred to as 'tongue and groove'. The tongue and groove joints between floorboards should never be glued together, as this will affect the way the whole floor shrinks or contracts. If the boards are left 'dry' they will be allowed to move independently.

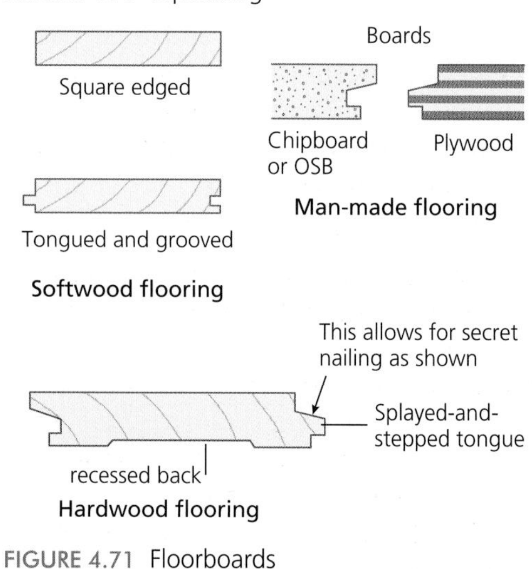

FIGURE 4.71 Floorboards

The length of solid timber floorboards will vary depending on the type used. The majority of solid timber or timber-based floorboards will need to be jointed at some stage; this joint is known as a 'heading joint'. Heading joints should always be made over a supporting joist and staggered across the whole floor to prevent a weakness. There are several methods of forming the lengthening joints; these include those pictured in Figure 4.72.

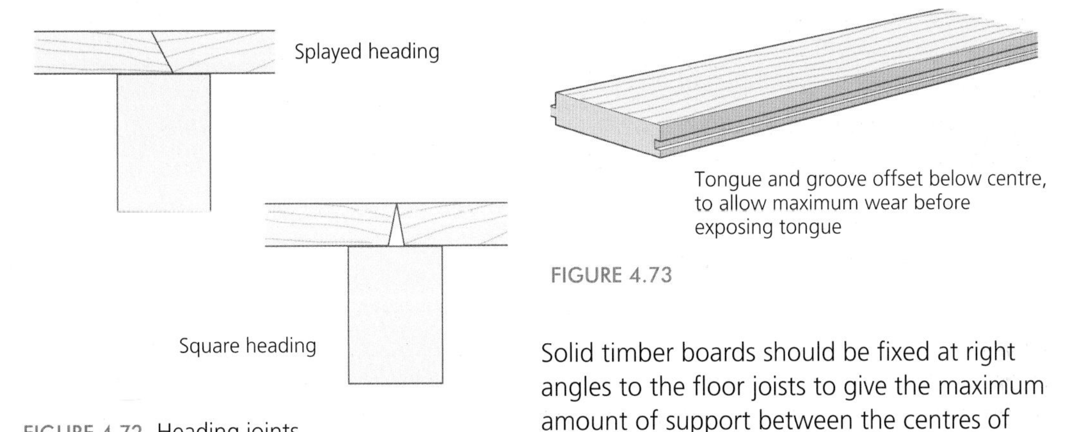

FIGURE 4.72 Heading joints

FIGURE 4.73

Solid timber boards should be fixed at right angles to the floor joists to give the maximum amount of support between the centres of the joists.

FREQUENTLY ASKED QUESTIONS

How are the floorboards fixed to the floor joists?

Softwood boards are normally fixed through their surface with floor brads, which are set below the surface with a nail punch. This will ensure that potentially raised nails do not damage sanding discs or belts, if the floor is sanded in the future. Expensive hardwood floors should be nailed through the tongues into the joists, using a 'floorboard nailer' and a large rubber mallet. A floorboard nailer is used to fire fixings through the tongues of the flooring at 45°.

SELECTING THE CORRECT FLOOR

Care must be taken to select the correct type of floor for a given situation; poorly selected materials could reduce the life span of the boards, or fail to suit the room size and appearance. Things to consider include:

- narrow or wide floorboards (wide boards in a small room give the appearance of a smaller floor area);
- type of material (hardwood, softwood, manufactured boards);
- species of timber (oak, beech, maple, etc.);
- finish (natural, oiled, waxed, etc.);
- wet or dry area (moisture-resistant boards for bathrooms and kitchens).

TRADE SECRETS

Special consideration should be given if softwood flooring is selected. Although softwood is one of the cheaper alternatives, it will damage very easily if heeled shoes are worn over it. Do not assume that all hardwoods are 'hard' wood; some have poor durability. Some manufactured floorboards are difficult to distinguish from solid timbers. High-quality manufactured boards are commonly used in wet areas and commercial buildings because they are less likely to swell or shrink, and they are extremely hard.

EQUILIBRIUM MOISTURE CONTENT

All natural timber contains a certain amount of water; the exact amount will depend on the type of timber used. It is important to use floorboards that have been dried to the correct moisture content to prevent rot, make the material easier to work with and to reduce the movement in the floor after fixing. Solid timber floorboards will expand or contract in their width and only a minimal amount in their length; this is due to the direction of the growth rings. The amount of movement that takes places in the timber will depend on the humidity and temperature in the surrounding air – for example:

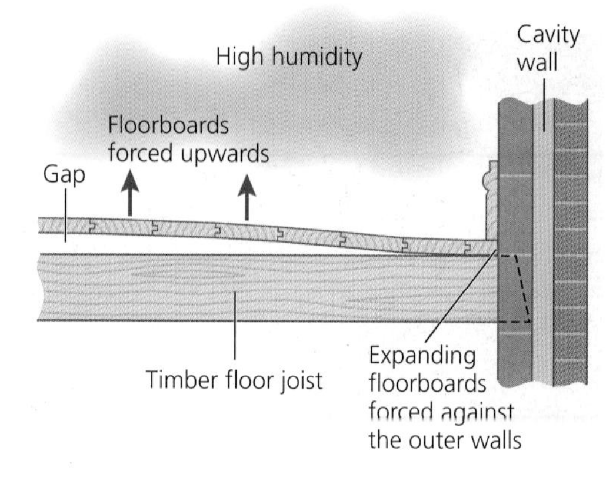

FIGURE 4.74 Moisture content

- timber left in a room with *high* humidity will absorb the water in the air and expand;
- timber left in a room with *low* humidity will dry and shrink in its width.

When timber reaches a balance in moisture content between itself and the surrounding air, the amount of shrinkage and expansion is reduced. The balance in the two moisture contents is known as the 'equilibrium' moisture content. Timber floorboards should be left in the area where they are to be fixed prior to installation for a period of time, a process known as 'second seasoning'. This will give the timber the opportunity to dry naturally until this balance is achieved. The exact moisture content can be measured with a moisture meter.

FREQUENTLY ASKED QUESTIONS

How does a 'moisture meter' work?

Although there are several methods of checking the moisture content of timber, a moisture meter is probably the most common. There are two types of moisture meter: 'pin' and 'pinless'. The pin-type model has two hardened steel pins protruding from one end of the meter. These are pressed into the timber and the unit will measure the moisture content as a percentage. The pinless model is less accurate but prevents damage to the timber being tested.

EXPANSION GAPS

Natural timber has the potential to expand and contract after it has been installed, even if it has the correct equilibrium moisture content, although this may be only a minimal amount. It is important to make provision for this movement around the perimeter of the floor; this is usually done by means of an 'expansion gap' concealed under the skirting board.

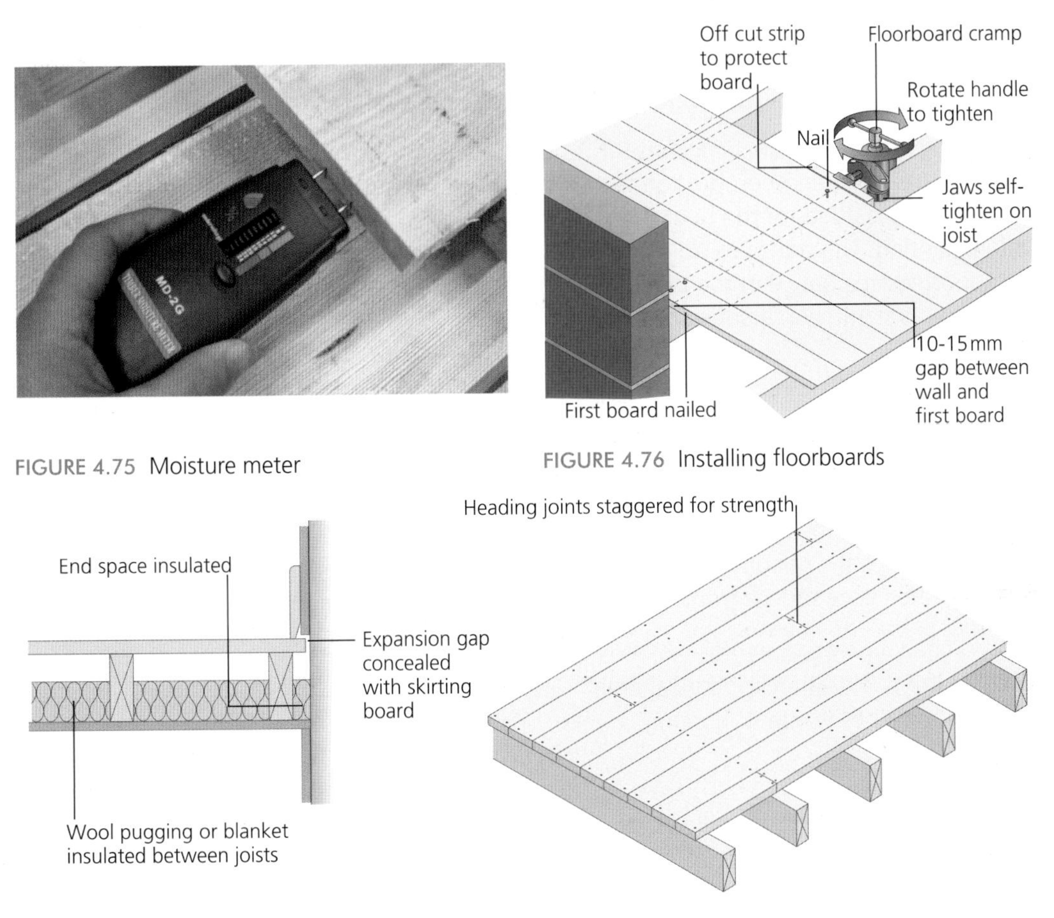

FIGURE 4.75 Moisture meter

FIGURE 4.76 Installing floorboards

FIGURE 4.77 Expansion gap

INSTALLING TONGUE AND GROOVE FLOORBOARDS

It is important to ensure that the joints between the boards in a solid timber floor are tight during installation prior to fixing; this will prevent the floor from squeaking and provide a neat, uniform finish. The floorboards can be squeezed together during installation, either with a floorboard clamp against the floor joists or with a ratchet clamp.

MAN-MADE FLOOR COVERINGS

Manufactured chipboard floorboards are the most common type of sheet material floor covering. It is available in a standard overall sheet size of 2400 × 600 mm and in different thicknesses, ranging between 18 and 22 mm through to 38 mm. Flooring-grade chipboard is more durable than the standard sheet materials; it also has machined tongue and groove edges along all four sides.

Flooring-grade chipboard is available in standard grade and moisture-resistant sheets

> **TRADE SECRETS**
>
> *Clamping against the tongue and groove floorboard can cause damage to the machined edge. Any damage that occurs could prevent the joints between the tongues and grooves fitting together correctly and may cause unsightly gaps. This could be prevented if an 'offcut' of the flooring is used as a packing piece between the floorboards and the clamp (see Figure 4.76).*

(used for wet areas, e.g. bathrooms and kitchens). It is common practice to use moisture-resistant boards throughout an entire floor if one or more rooms require it; this prevents the need to mix the different types, giving a better appearance.

Chipboard flooring should be laid at 90° to the floor joists, with the joints staggered across the floor area to strengthen the connections between the sheets. The tongue and groove joints along the edges of the boards should have a liberal amount of PVA glue applied before jointing. This adhesive will bond the boards together and prevent squeaking over the unsupported area between the joists. Each board should be fixed down to the joists using corrosive-resistant countersunk screws.

WEATHER-RESISTANT FLOORING

At the early stages of developing a dwelling, the floor joists on the ground and upper floors may become exposed to the elements until the roof has been constructed. The various floor levels within a building are normally used as working platforms to construct the upper parts of the development. Operating within the building increases the working area and productivity, and also reduces the risk of falls from a height. There are three methods of preventing damage to the finished floor surface from the elements and from construction materials:

1. lay a cheaper temporary floor (usually shuttering plywood);
2. cover the floor with hardboard sheets
3. install 'weather-resistant' flooring.

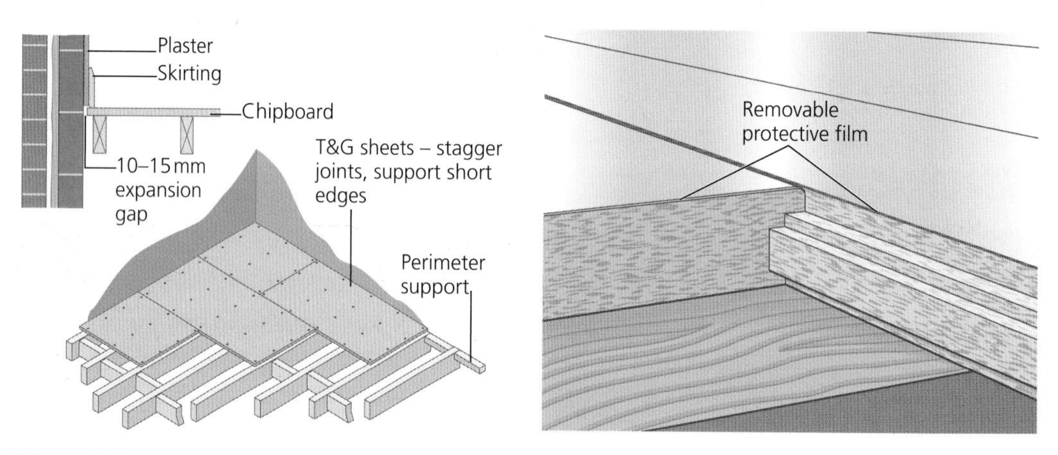

FIGURE 4.78

FIGURE 4.79 Weather-resistant flooring

For many years temporary floors have been the only option, but more recently alternative chipboard floors have been developed to resist water ingress. There are several different types and manufacturers of weather-resistant chipboard flooring, each one differing slightly in its design and installation methods. In general, the boards are either covered with permanent synthetic films over the exposed faces and sealed together at the joints, or are covered with removable heavy duty films and taped over at the joints. The removable films provide a non-slip and waterproof surface to work off during the construction phase, and when removed they reveal the undamaged floor surface.

FLOOR TRAPS

The term 'floor trap' refers to an access panel created in the floor covering. Floor traps are usually screwed in position to allow easy entry through the floor surface to hidden services below. Care must be taken when cutting through existing floorboards to ensure that the joists are not damaged and potentially weakened. Services such as water pipes and electrical cables could potentially be directly below an opening being formed. A thorough check should be made of the floor before any drilling or cutting takes place; this may be achieved with an electronic cable/wire/pipe detector. Floor traps must be adequately supported to prevent movement through use and a weak spot in the floor.

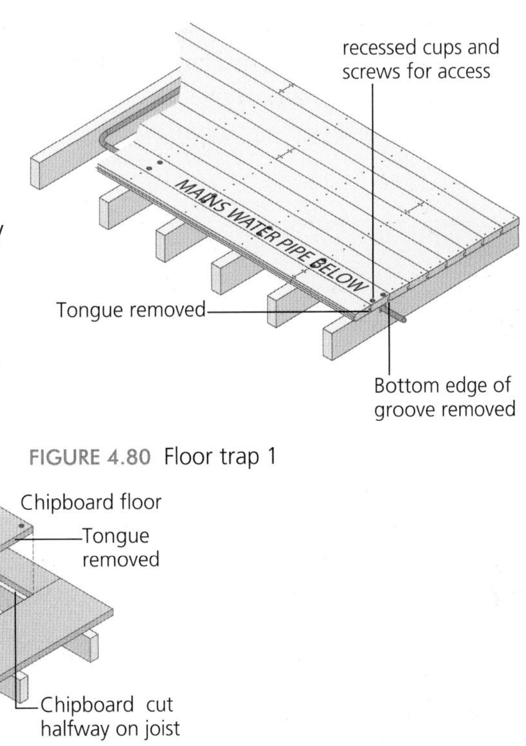

FIGURE 4.80 Floor trap 1

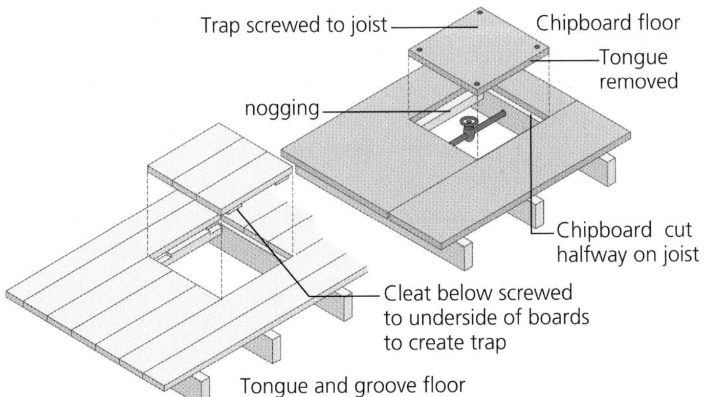

FIGURE 4.81 Floor trap 2

ACTIVITIES

Activity 17 – Fitting and fixing floor coverings and flat roof decking

Read through the following questions and answer them as fully as you can to help you develop your underpinning knowledge of this subject area.

1. Name two different types of floor covering.
2. Suggest possible methods of fixing the covering.
3. Explain the purpose of expansion gaps between floor coverings and walls.
4. What are the standard thicknesses of manufactured floor coverings?
5. Sketch a trap cut into a floor covering with adequate support between the joists.

PARTITION WALLS

The term 'partition walls' is the name generally given to the non-load-bearing, internal walls that are commonly used to divide open areas into smaller rooms. They can be constructed using a variety a different materials; these include:

- timber (carpenter);
- bricks or blockwork (bricklayer) (*note*: these are generally load-bearing partitions and used on the ground floor);
- metal stud (carpenter/shopfitter).

FIGURE 4.82 Timber stud

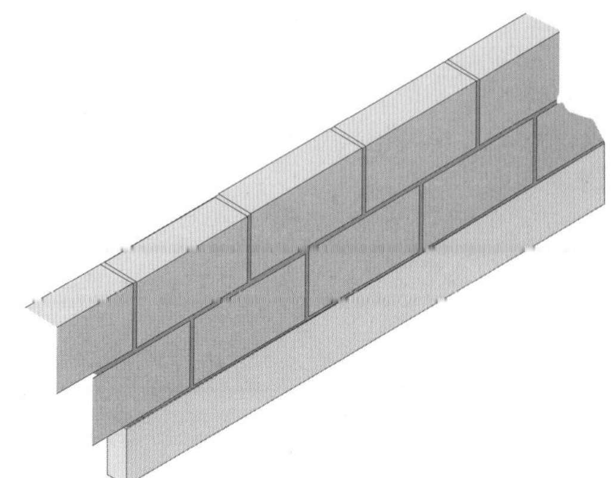

FIGURE 4.83 Light-weight blockwork

The methods used to build partition walls will usually depend on the location of the walls and their use (e.g. thermal insulating, soundproofing or a barrier to prevent the spread of fire). Generally, partition walls that are constructed using bricks and mortar are *less* cost-effective than other methods; this is because of the use of wet materials and the slow drying times. Bricklayers are also limited by the amount of bricks or blockwork courses that they can lay in one day; this restriction prevents the weight of the newly laid courses distorting the shape of the wall. As a rule, the less material that is used to construct a partition that requires 'drying out', the quicker the partition can be formed and completed.

Partition walls that are built using dense materials usually have good insulation and soundproofing qualities. The majority of partition walls built-in on upper floors in new constructions will be built using lightweight materials; this will reduce the amount of weight on the upper floor joists. Timber partition walls are used in many situations because the materials are readily available from suppliers, and are also easy to cut to length and fix. Hollow partition walls can be adapted to create resistance to fire and passage of sound, and to restrict the loss of heat through the wall (this is explained in greater detail later in this chapter).

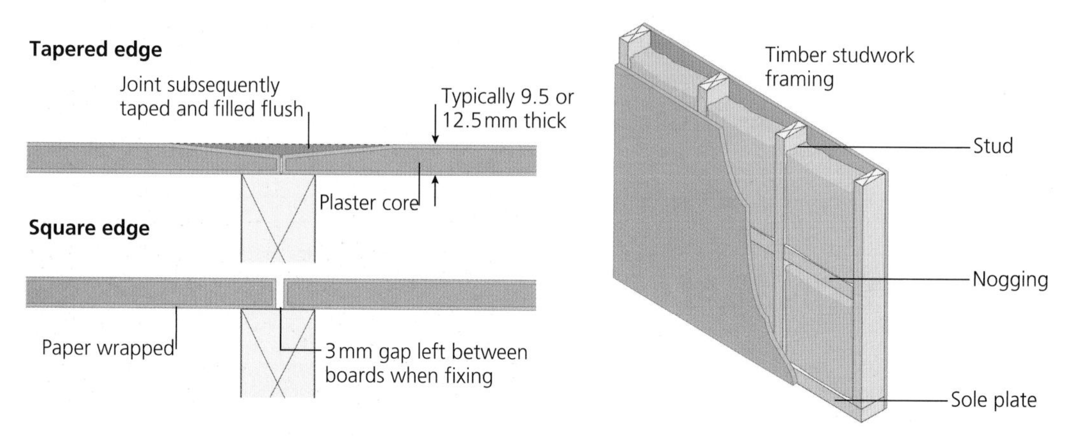

FIGURE 4.84 Partition walls

TIMBER PARTITIONS

Timber partition walls are one of the most common and simplest methods of constructing dividing walls. This method of building walls has been used in traditional house building for many years and is still used in new building today. There are two methods commonly used to construct timber partition walls: 'built in' or 'framed up'.

MATERIALS

Generally, fast grown, low-grade carcassing timber is used to build timber partition walls. This type of timber is commonly available in different forms, ranging between:

- sawn – all four sides have rough sawn edges from the saw mill; sawn carcassing timber may vary in width between ± 4 mm;
- Regularised – at least two edges will have machined edges to provide a uniform accurate width; the use of sawn timber for partitioning is considered bad practice because of the inconsistency between the sectional sizes;

FIGURE 4.85 Sawn carcassing timber

FIGURE 4.86 Regularised carcassing timber

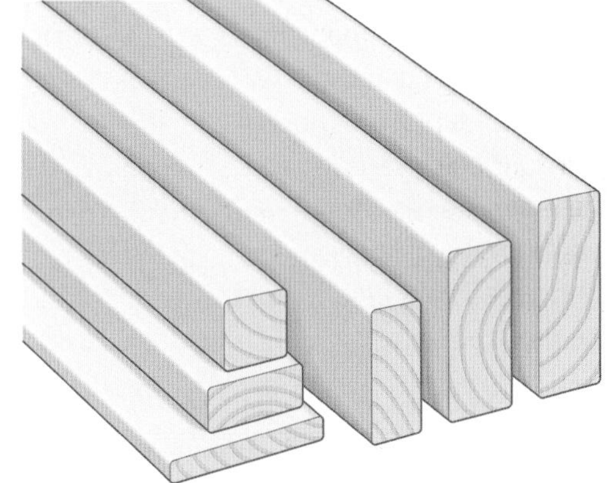

FIGURE 4.87 CLS

- Canadian lumber sizes (CLS) or American lumber sizes (ALS) – commonly used in timber framing, CLS and ALS have all four edges planed, with rounded edges; the use of these timbers originated in North America, but they are now also commonly used in European countries; its uniform section (at the machine sizing stage), rounded edges and relatively low cost make this partition material the most popular choice for most site carpenters;
- planed all round (PAR) – as the name suggests, this carcassing timber has all four

FIGURE 4.88 PAR carcassing timber

surfaces planed square. This is rarely used in house building, although the square edges give a greater bearing for the facing material.

JOINTS

Although the basic layout of the timbers within partitions has not altered over recent years, the method used to join them together has developed. Traditionally, the joints between timbers in partitions included housings, bridle joints, and mortise and tenons. Although these methods perform well to prevent the timbers within the partition wall from moving and twisting, they are complex and time consuming to form.

ALTERNATIVE FIXING METHODS

'Butt' joints are mostly used to connect the timbers within partition walls because they are simple to cut and fix, which increases productivity and reduces labour costs. The simplified butt joints rely heavily on the strength of the fixings holding the timbers together. An alternative method of securing the butt joints is with the use of mechanical 'framing anchors'.

Framing anchors are available in sections of folded galvanised mild steel, designed to sit over each butt joint and provide additional support to prevent the timber moving. Framing anchors have a series of 2 mm holes located evenly over each side; these provide easy fixing points for the galvanised round head nails used to hold them in place.

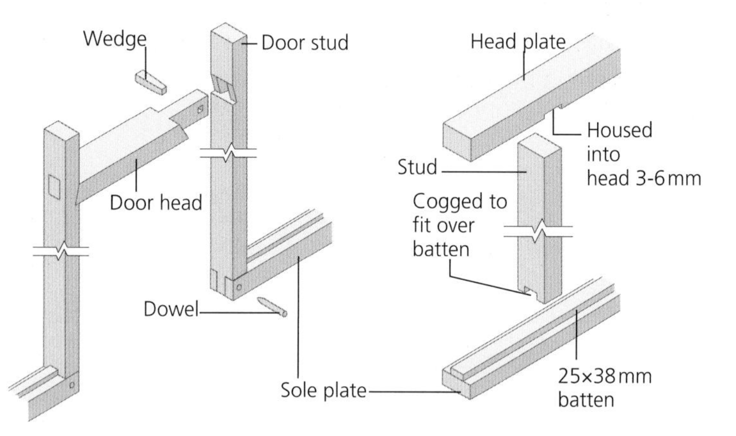

FIGURE 4.89 Doweled butt joints

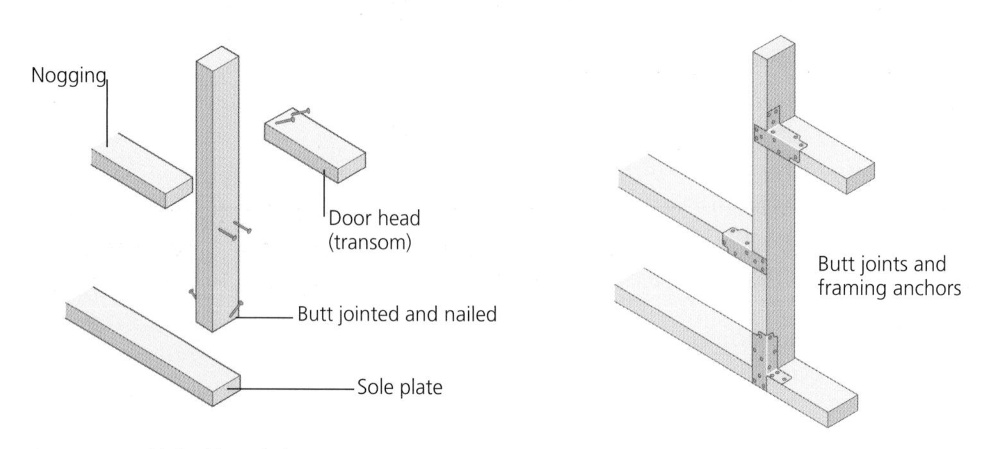

FIGURE 4.90 Nailed butt joints

FIGURE 4.91 Framing anchors

FREQUENTLY ASKED QUESTIONS

▶ How are the joints in traditional and modern timber walls secured?

The majority of the joints within a timber partition wall are securely fixed in position with round head nails. The most common lengths of round head nails used to fix the frames together are between 75 and 100 mm, depending on the section of the timbers used.

Doorways that are created within timber stud partition walls, particularly in the centre of a stud partition, are subject to the stresses of constant use. The continual opening and closing of doors within an opening may cause the studs to twist over a period of time. This movement in the timbers around the opening may also cause the door lining to twist. This may then reduce the margin between the lining and the door, preventing the door operating properly. Traditionally, any timbers fixed above a door opening would have been connected to the vertical studs using 'cogged shouldered tenon' joints. These joints would provide an extremely strong connection between the timbers and reduce the likelihood of movement.

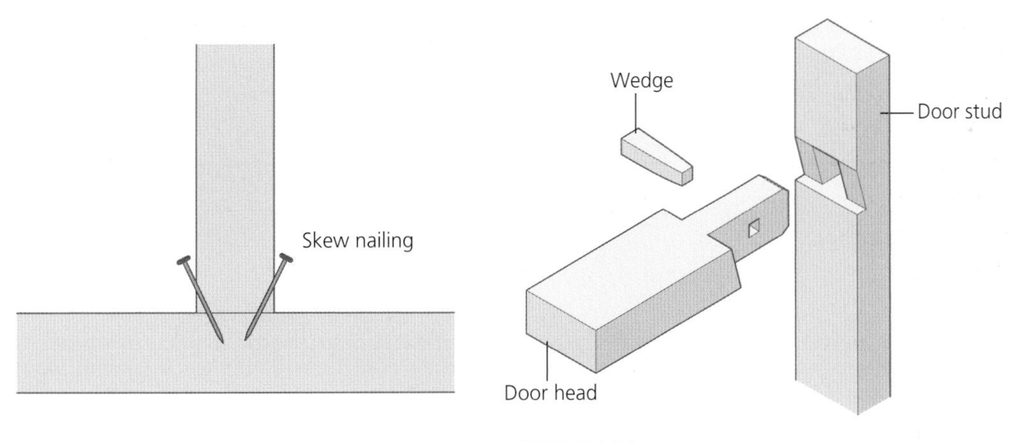

FIGURE 4.92

FIGURE 4.93

Timber partition walls are constructed with horizontal timbers along the top known as the 'head plates'; the timbers along the bottom are known as the 'sole plates'. The vertical intermediate timbers fixed at intervals between the head and sole plates are known as the 'common studs'.

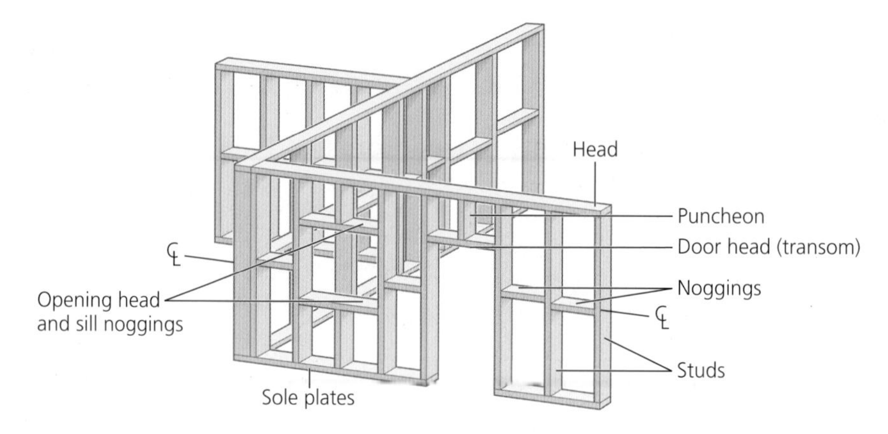

FIGURE 4.94

The positions of the vertical studs within a partition are determined by the covering materials. The most common material used to cover the internal framework of a partition wall is 2400 × 1200 mm plasterboard sheets. It is vital that the edge of each sheet is supported and has sufficient bearing for a sound fixing to the vertical stud, to prevent movement between the boards. The most common spacing for the studs is either 400 mm or 600 mm; this usually depends on the thickness of the covering sheet materials (see, for example, Figures 4.95–4.97).

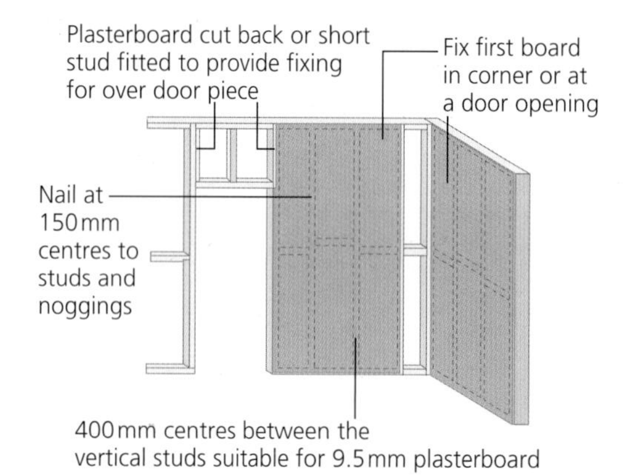

FIGURE 4.95

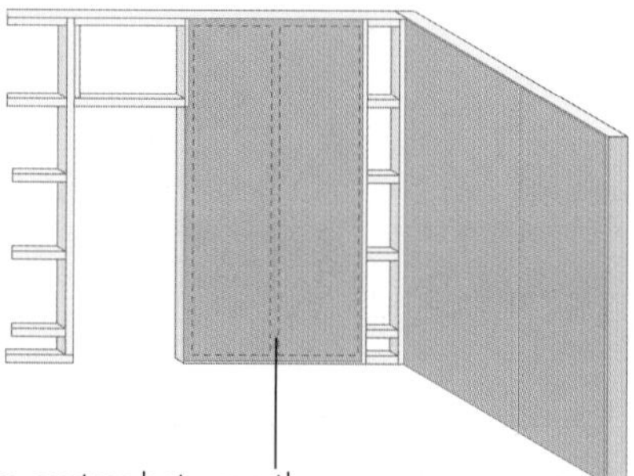

FIGURE 4.96

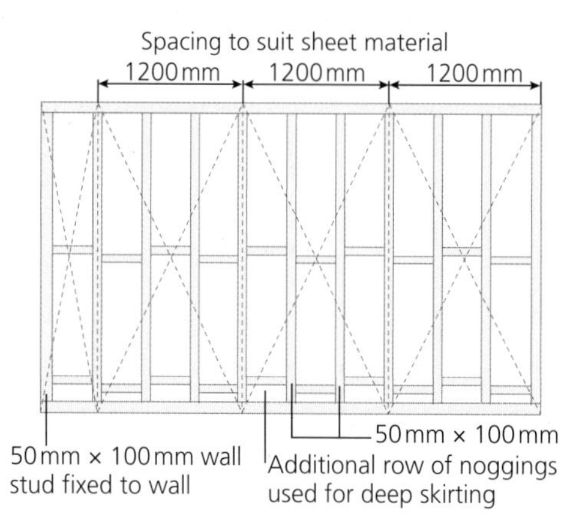

FIGURE 4.97

Remember – do not change the predetermined centres for the studs to suit the length of a wall or to create openings for doorways or hatches. These changes may not suit the covering sheet materials and may fail to provide an adequate fixing point.

FORMING CORNERS AND 'T' JUNCTIONS

The arrangement of the vertical studs will have to be altered where a timber partition wall has to be returned at 90° and also within the intersection of adjoining walls. Additional studs will have to be positioned within the corners and junctions between connecting walls so that suitable provision is made for fixing the covering sheet materials. Failure to build extra studs into the areas will result in a poor joint between the sheets, and possibly lead to cracks appearing along the joints after completion.

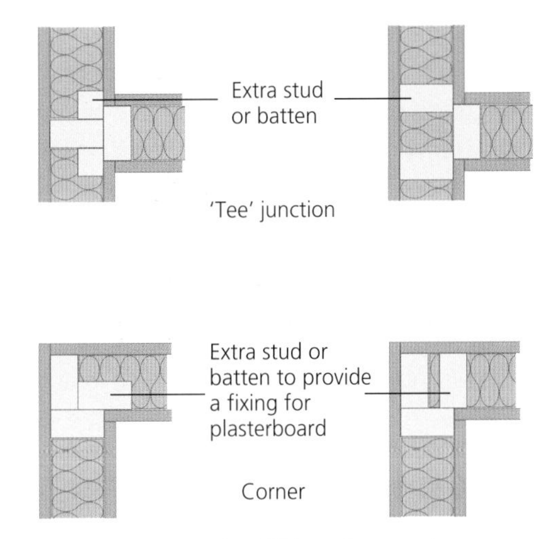

FIGURE 4.98 Forming 'T' junctions and corners

NOGGINGS

The full-length studs between the head and sole plates of a timber frame partition wall will distort and buckle if they are not adequately supported. Intermediate support should be provided with the use of 'noggings' fixed between each stud at approximately 1200 mm centres. This size will be reduced to allow even spacing of the noggings if the covering sheet material is over 2400 mm. Covering sheet materials can be laid either vertically or horizontally, with noggings installed accordingly to provide adequate support. Whichever method is used to lay the boards, the joints should be staggered to minimise the risk of weakness lines. Noggings are also installed between the studs just above the sole plates if deep skirting boards are used along the bottom edge of the wall. This will allow an additional fixing point for the top edge of the wide moulding, preventing the skirting from cupping away from the wall and revealing unsightly gaps.

Additional noggings are usually required within partition walls to provide a fixing point for heavy items to be fitted for example:

- kitchen and bedroom cupboards;
- fuse boards;
- sanitary ware (e.g. wall-hung sinks and WC).

INSTALLING NOGGINGS

There are several different methods of installing these intermediate supports between the vertical studs within a timber stud wall.

- In-line – the 'in-line' nogging is rarely used as an intermediate support because of the difficulty encountered when nailing the timbers into position between the studs. Noggings should be installed in this manner where edge support is required to adjoining boards if the full height of the covering sheet material requires support along the shortest or longest edge when boards are laid horizontally.
- Staggered – 'staggered' noggings are simply cut to length and positioned either side of a central line marked along the face of the wall. The staggered positioning of each nogging between the studs provides enough room to insert fixings square to the face of the timber.

- Herringbone (traditional method of nogging) – the main advantage of using 'herringbone' noggings over in-line or staggered is that they remain tight between the studs, even if the timber dries out after fixing; however, it is rarely used nowadays because it is the least efficient method. Noggings installed square to the studs will have gaps appear between the joints if the timber shrinks; this may lead to twisting of the studs and 'hairline' cracks appearing on the finished surface of the wall.

'FRAMED-UP' TIMBER PARTITIONS (PRE-CONSTRUCTED)

Timber partition walls are usually either built on site or manufactured in a factory as part of a timber frame house. There are many benefits from using factory-made partitioning, including the following:

- accuracy of construction under controlled factory conditions;
- speed of erection of the walls after delivery to site.

TRADE SECRETS

Measuring and marking the positions of the noggins in a timber stud wall can be very time consuming. A more efficient method is to use a packer, cut between the top of the sole plate and the underside of the nogging. During installation, each nogging is simply positioned in between the vertical studs, with the pack used underneath to govern its position, before securely fixing.

Alternatively, the heights of the noggings can be indicated each end of the partition wall, before a chalk line is used to mark a straight line between these two points.

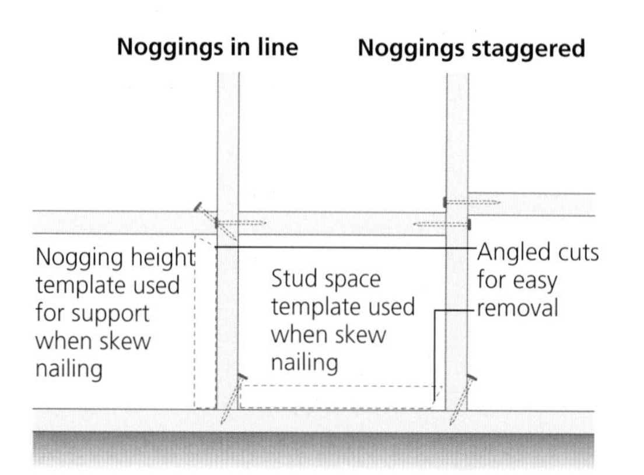

FIGURE 4.99 In-line and staggered noggings

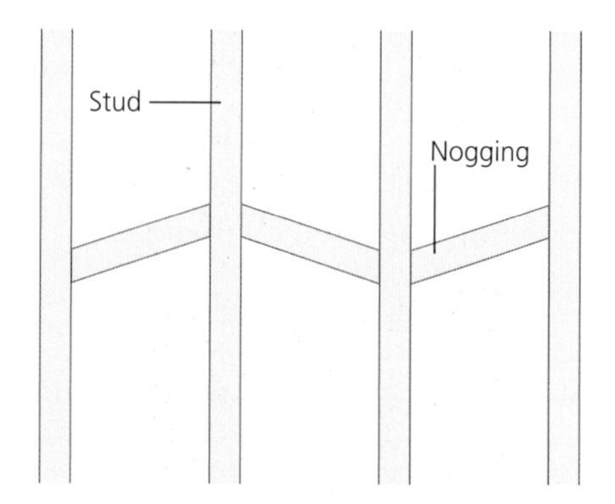

FIGURE 4.100 Herringbone noggings

Hollow timber partition walls can also be constructed on site, and wherever possible within the area where they are going to be used; this will prevent unnecessary manual handling of the frames across the building site.

The easiest method of constructing partition walls on site is to assemble them on a flat, level surface on the ground. The following steps should be used as guide to help produce a framed partition wall on site.

1. Refer to the architect's drawings to establish the correct position and size of the proposed new wall.
2. Measure the actual size of the space from the position that the wall is going to be located. Deduct approximately 25 mm from the overall measured height and width of the

proposed new wall sizes to allow a tolerance to manoeuvre the assembled frame into position. Alternatively, additional sole plates can be fixed at ground level prior to the timber frame being erected. This method creates 'double' sole plates at floor level, which allows the frame to be simply located and fixed in position. It also allows sufficient clearance between the floor and ceiling for framed-up partitions to easily be erected without wedging in their height.

3. Clear a floor space directly in front of the proposed new wall to ensure a flat, level assembly area.
4. Cut all the components for the wall to length and lay them out in position on the floor.
5. Securely fix all the joints together and check the frame is square. This is normally achieved by measuring diagonally across the frame to the opposite corners; these measurements should be equal. If these measurements are not equal, then light pressure needs to be applied across the frame between the longest diagonal measurements to correct the discrepancy. This method of squaring a frame is known as measuring from 'corner to corner'.

TRADE SECRETS

Remember – you must always fix either through the packers that you insert around the frame you are fitting or directly underneath. This will prevent the packers slipping out of position if there is any drying out of the timbers or movement occurs between the frames.

FIGURE 4.101 Measuring a stud wall frame

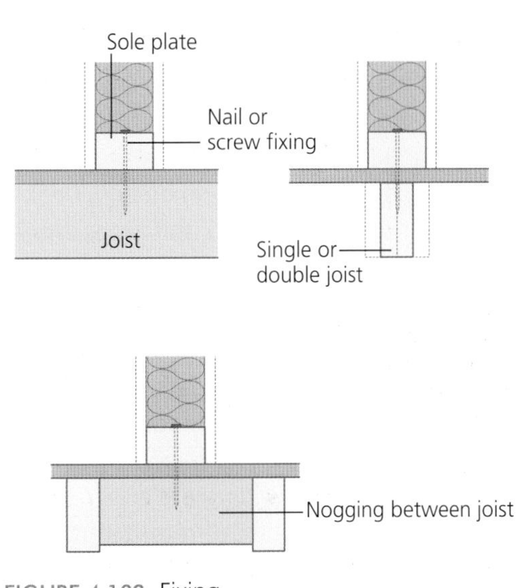

FIGURE 4.102 Fixing

6. The overall sizes of the frame should be checked one more time against the size of the opening in which it is going to be fitted. Any door openings in the frame should also be checked for square and braced; this will ensure that there will be no problems at a later stage when fitting the lining.
7. Carefully lift the frame into position and begin to insert 'folding wedges' between the top of the timber frame and the underside of the opening to temporarily hold the frame in place. Check the frame is vertical with a spirit level at several points across the wall.
8. Once the frame has been correctly aligned and further wedges have been inserted to hold the frame firmly in place, secure the partition with the appropriate fixings.

FREQUENTLY ASKED QUESTIONS

What are folding wedges?

'Folding wedges' are simply a couple of wooden wedges that work together as a pair by sliding against each other. They are particularly useful when timber frames require packing around the outer edges because they are easily adjusted during installation.

Alternatively, plywood of various thicknesses can be used, or standard purpose-made plastic packers. These are readily available from building suppliers and are commonly used by plastic window fitters. They are available in various sizes, ranging from 1 to 10 mm, and are identified by corresponding colours, e.g. yellow = 1 mm, blue = 2 mm.

'BUILT-IN' TIMBER PARTITIONS (CONSTRUCTED IN POSITION)

It is not always possible to pre-construct timber partition walls. This could be due to a lack of space or the complexity of the wall to be built; however, this can be overcome by building the frame in its position. This means that each piece of the frame will have to be cut, installed and fixed independently. Constructing frames in this manner can be time consuming but it ensures a perfectly fitted partition and, unlike the pre-constructed method, this process needs no allowance for fitting.

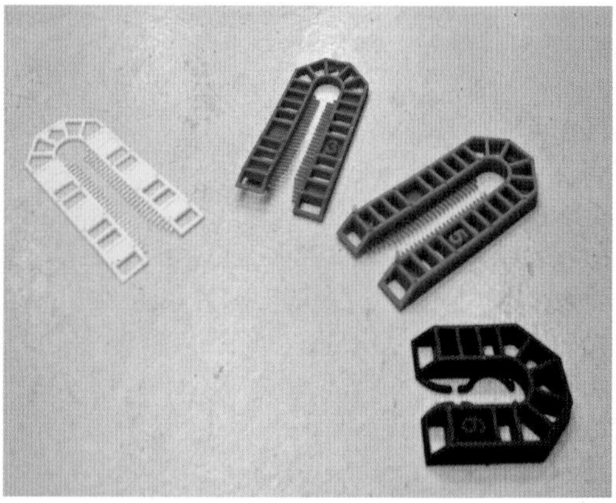

FIGURE 4.103 Standard plastic packers

SETTING OUT PARTITION WALLS

Before any partition wall is erected, it is important that it is accurately set out in its intended position. The vast majority of walls in residential house building are either straight or will have at least one 90° return in them at some point.

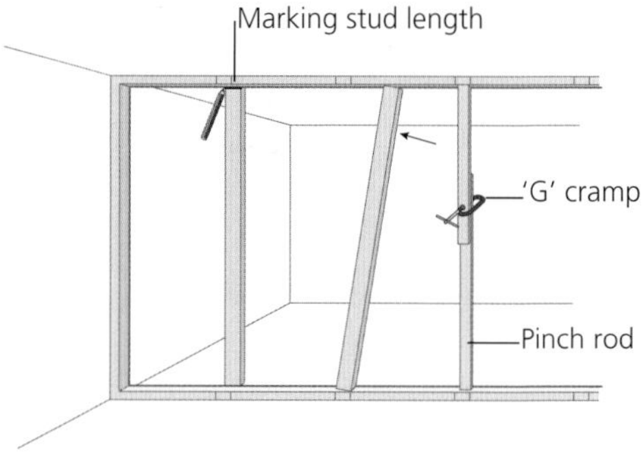

FIGURE 4.104 Setting out a partition wall

Complex partitioning of an entire area is normally marked out on the floor by the site carpenter; this is usually done before lifting the wall sections into position prior to fixing. The position of partition walls may be laid out on the ground with lengths of timber the same section as the studwork, prior to erecting the walls. As the lengths of timber (sole plates) are laid out on the floor to replicate the layout of the rooms to be created, they are permanently fixed to the ground. Although this method uses additional sole plates, it is easier to fix and build the frames up from the ground once the exact position has been established.

METHODS OF ACCURATELY FINDING A SQUARE LINE TO A WALL

FIGURE 4.105 A folding site square in use

Figure 4.105 demonstrates the basic principle of a right angle triangle and how this can be related to any partition wall with a 90° corner. In practice, the bigger the scale used to apply these principles, the more accurate the walls; for example, the scale demonstrated in Figure 4.107 is too small to accurately apply it to a large partition wall. This is because it will only check the wall is square up to 300 mm and 400 mm in each direction.

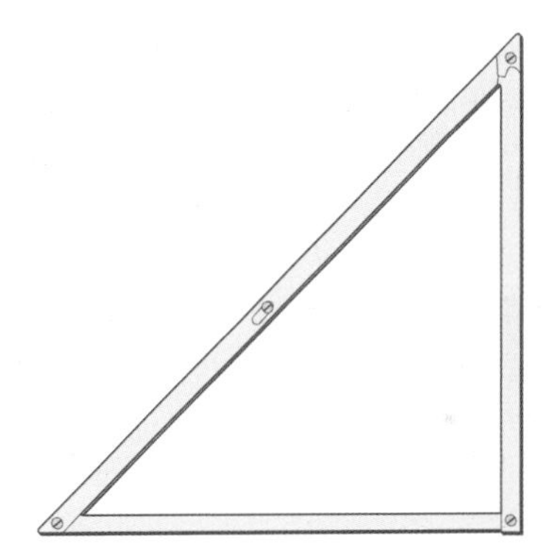

FIGURE 4.106 A folding site square

As a rule, the bigger the scale, the more accurate the corner will be. This scale works better for a partition wall. It is big enough to be able to set a wall out at 90°, but small enough to be able to measure it without additional labour.

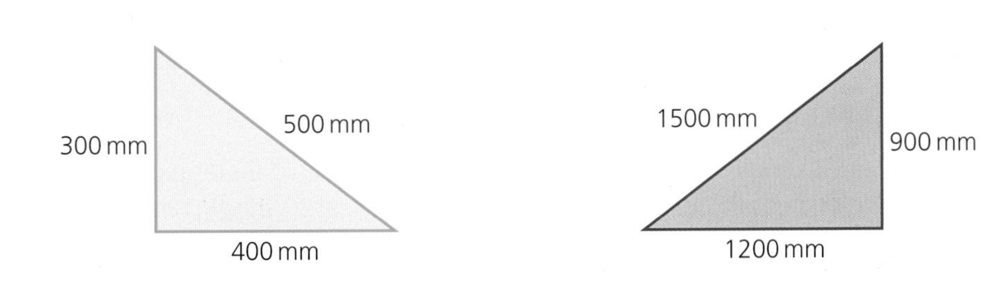

FIGURE 4.107

FIGURE 4.108

FREQUENTLY ASKED QUESTIONS

How do you make sure that the wall you are setting out is straight?

When you are working with carcassing grade timber for partitioning, it will be highly likely that some of the timber will be twisted or bent, especially if the material is sawn. Careful selection and use of this timber will correct the defect, and avoid having bent or poorly constructed walls. For example:

- check for bends and twists in the timber before use;
- use bent lengths of timber for noggings or short sole plates or heads, and the straightest for the vertical studs;
- start fixing sole plates at one end and work along their length, applying pressure to straighten them before inserting the next fixing.

Use a prepared (machined) piece of timber or length of a timber-based sheet material (e.g. plywood/MDF/blockboard) as a straight edge. This could be used to mark straight lines on the floor for short sections of walls. Longer runs of partition walls should be set out on the floor using a chalk or laser line.

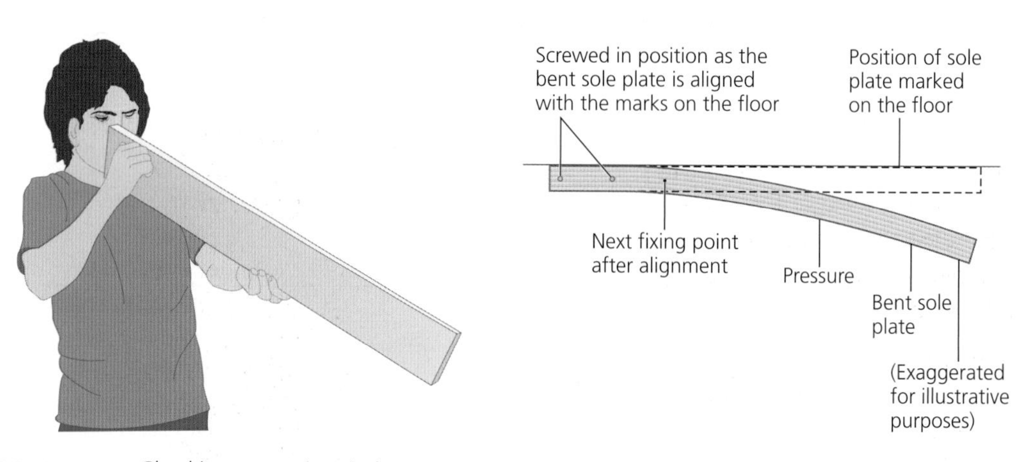

FIGURE 4.109 Checking carcassing timber

FIGURE 4.110 Aligning timber

DRILLING AND NOTCHING TIMBER STUDS

Hollow partition walls usually conceal services such as hot and cold water pipes, electrical wires or telephone and computer network cabling. In many cases, this will mean that the vertical studs will have to be either drilled or notched to accommodate these services. Too much drilling and notching through the studs in a partition wall could seriously weaken the structure. Building Regulations state that all holes should be drilled through the middle of the timber (the neutral stress line) up to 25 per cent of the stud's width, and positioned between 25 and 40 per cent from either end. Notching of timber studs is also permitted on either side, providing the notch is located no more than 20 per cent away from either end and no deeper than 15 per cent of the stud's width.

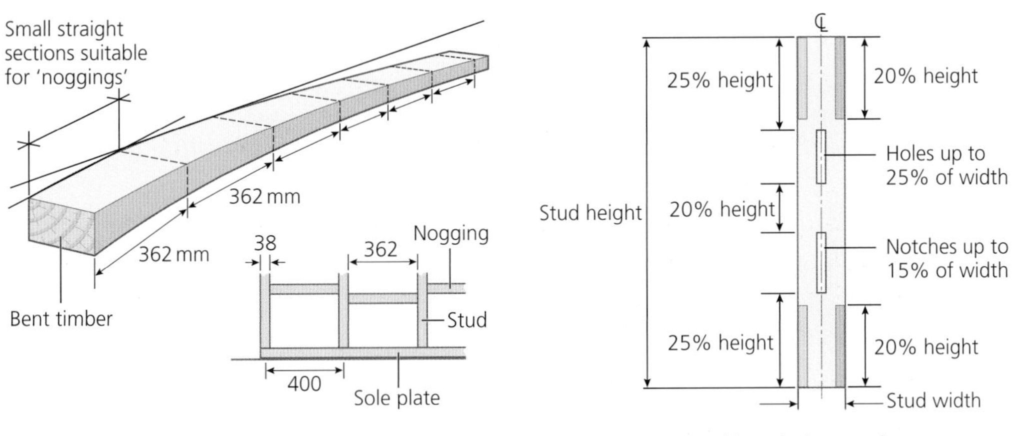

FIGURE 4.111

FIGURE 4.112 Notching timber studs

Particular care should be taken to ensure that any holes and notches that are being cut into a partition to accommodate wires, pipes, etc. are well away from fixing points. Coordinating with service trades such as plumbers and electricians will avoid them fixing in known vulnerable positions in the future. Some of these fixtures could include:

- fixings for dado rails (screws and nails);
- fixings for deep skirting boards;
- adjustable brackets or battens for kitchen wall units;
- fixing for brackets to support shelving, etc.

The most effective method of checking for a buried service is with the use of an electronic 'wall scanner'. These hand-held pocket devices have to be laid flat on the partition wall and repeatedly moved over the area where the fixing may be inserted; it will then pinpoint and identify the exact position of the following items:

- vertical studs and noggings (useful if a strong fixing point is required);
- electrical cables;
- water and gas pipes;
- metal objects.

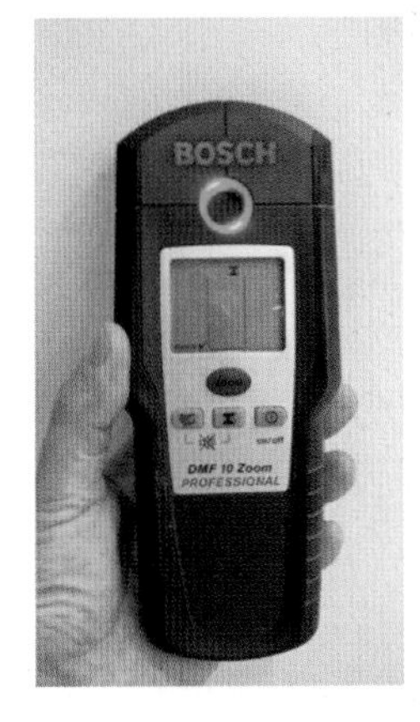

FIGURE 4.113 Wall scanner

FREQUENTLY ASKED QUESTIONS

What is a fixture?

A 'fixture' is a term given to an item that has been secured in position; it is often confused with the term 'fitting'. A fitting is an item that is 'free standing' and does not require securing or fixing in place. For example, an exterior door is a *fixture*; a chest of drawers would be considered a *fitting*.

METAL STUD PARTITION WALLS

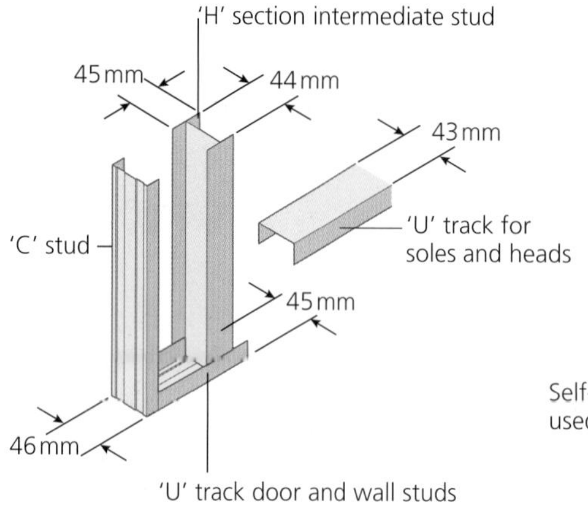

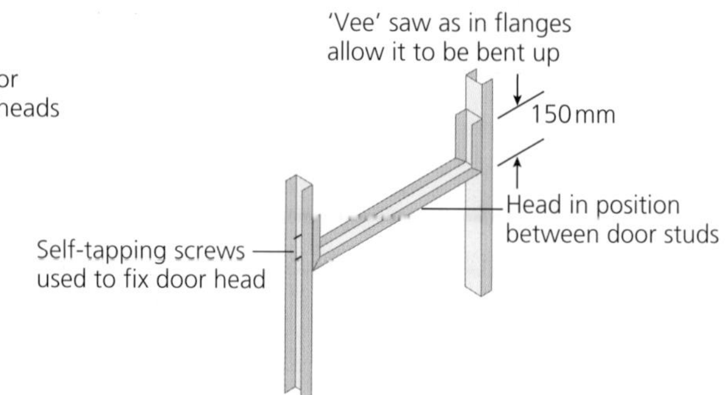

FIGURE 4.114 Metal stud partition walls

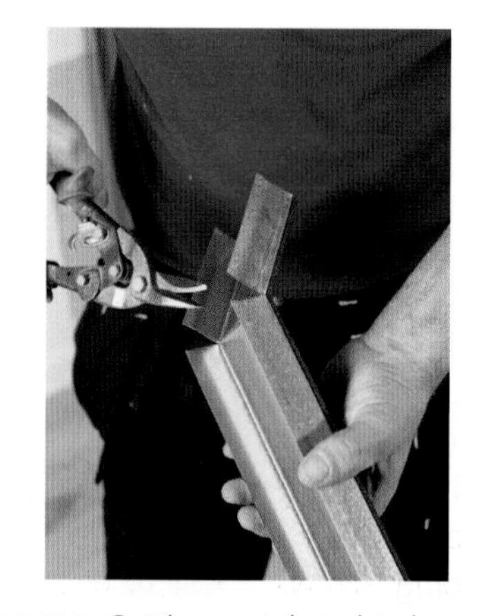

FIGURE 4.115 Cutting metal stud to length

An alternative material to traditional timber, metal stud walls offer a lightweight cheaper option that is quicker to construct and simple to use. Metal stud has been used to divide office spaces and in shopfitting for many years, and more recently in house building. The layout of the studs is very similar to that of timber partitioning: they both have head and sole plates (which are referred to as 'U' track) and vertical studs (also known as 'C' studs). Metal partitions do not require noggings between the studs to prevent them twisting because they are manufactured and not a natural material with characteristics such as grain and knots. The metal studs are factory made from 1 mm thick, folded galvanised steel into hollow sectional lengths. Metal studs do not require any drilling or notching for services to pass through because these holes are manufactured in the metal sections. (*Note*: care should be taken when wires are fed through the studs to prevent damage to the cable on the sharp edges.) Suppliers normally carry a wide range of standard lengths of both 'U' track and 'C' studs to avoid unnecessary cutting of the metal on site. Any metal studs or track that do require cutting to length are normally trimmed to size with large tin snips.

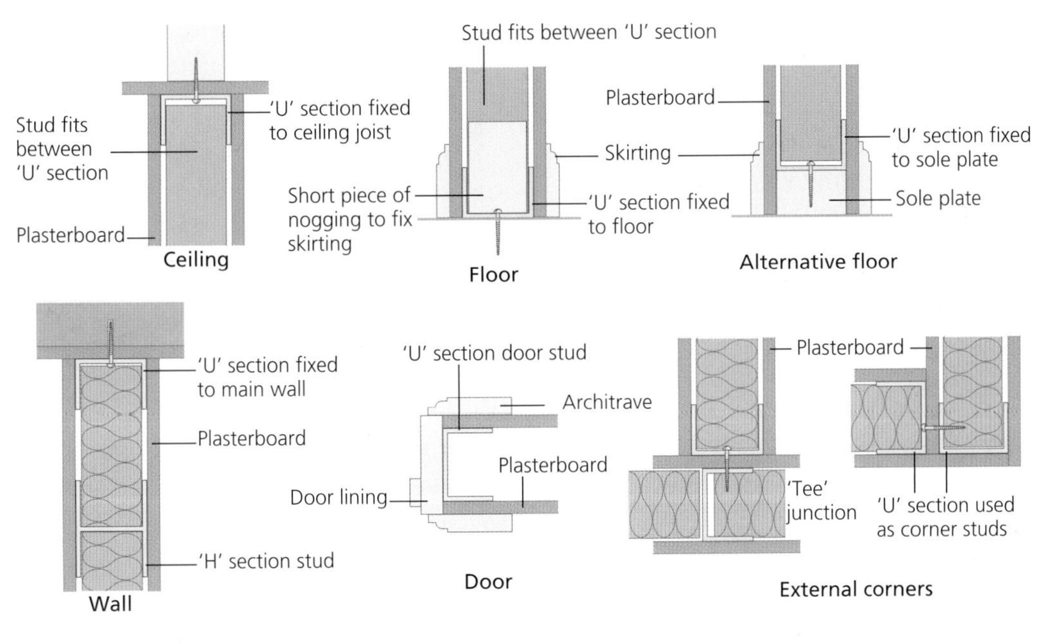

FIGURE 4.116

METAL STUD JOINTS

The vertical 'C' studs slot inside the 'U' track at the joints at the top and bottom of the frame, and are secured together with either a 'crimping' tool or 'self-tapping' screws through the sides of the joint. The frame is still relatively weak at this stage; it will gain its strength when it has been clad with plasterboard. The plasterboard is fixed to the metal sections with 'self-tapping' screws. These fixings do not require any pilot holes prior to use; the front of the screw has a cutting edge that cuts through the metal studs before the hard thread along the shaft of the screw bites into the joint.

The components of the metal stud are positioned exactly the same as with timber partition walls; they still have vertical studs and noggings to allow the covering sheet material to be fixed at their joints.

TRADE SECRETS

A long length of partition wall is sometimes referred to as a 'run'. Long runs of partition walls can sometimes lead to difficulties if the centre of the vertical stud positions do not correctly line up with the edges of the plasterboard. This may be due to small gaps between each joint in the plasterboard, leading to the incorrect alignment of the studs. This can be avoided if the studs towards the end of the run are loosely fitted in position and realigned before fixing; this is an advantage of metal stud partitions.

FIXING METAL STUD

Metal stud partition walls are usually constructed in situ unless the walls do not extend the full height of the room. This is because the frame is weak and prone to damage from twisting at the joints as it is being lifted from the floor into its position.

STAGES OF BUILDING IN SITU

1. Mark a straight line on the ceiling and use either a level or a 'plumb bob' to transfer these marks along the floor.
2. Fix the 'U' track to the ceiling and floor.
3. Cut the 'C' sections to length, insert one either end of the wall and fix them in position.
4. Clearly mark the centres of each stud along the top and bottom 'U' track and insert the 'C' studs.
5. Use the crimping tool to secure the studs in position and cover with plasterboard.

FIGURE 4.117 Crimping tool

(*Note*: insulation and services may have to be inserted before the plasterboard is fixed on both sides of the partition wall.)

FORMING OPENINGS IN THE METAL STUDS

Openings in metal stud partitions will have to be trimmed with lengths of 'U' track. These are only required between the studs and are connected together by cutting the 'U' track 300 mm longer than the required opening size. The 'U' track is then cut and bent 150 mm from each end and slotted over the 'C' section to form the opening. The sections of 'U' track used to trim the opening are then secured in place with the crimping tool.

Timber packers are used to fill the voids in the metal studs around the openings formed in the partition walls. This will help strengthen the wall and provide strong fixing points for the window and door linings.

PLASTERBOARD

Prior to the 1940s, all hollow partition walls would have been covered in 6 mm thick lengths of timber (known as 'laths'), with 5–6 mm gaps in between, as a foundation to allow plaster to be applied. This method of cladding a partition wall is known as 'lath and plaster'. This method is rarely used now, but can be found in older buildings or used in restoration work.

? FREQUENTLY ASKED QUESTIONS

▶ What is the difference between 'renovation' and 'restoration'?

A building that requires 'renovation' work usually has an upgrade; for example, the windows may be replaced or painted, and the interior may be modernised.

A building that requires 'restoration' cannot be modernised, but it can receive repairs and period or replica items installed to reinstate the building to its original condition.

Plasterboard is made from a gypsum core with lining paper on either side, and has been developed since the 1940s to replace the old method of lath and plaster. Its large sheet sizes make it quicker to cover the partition walls, and the flat surface reduces the labour time needed by the plasterer or dry liner. Plasterboard is available in thicknesses ranging between 9.5 and 12.5 mm through to 19 mm, and is manufactured in a variety of lengths and widths; some of these are listed below:

FIGURE 4.118 Lath and plaster partition wall

- 900 × 1800 mm;
- 1200 × 2400 mm;
- 1200 × 2700 mm;
- 1200 × 3000 mm.

Plasterboard is also available with square edges along all the sides; this is usually used for walls that require a skim coat of board finish plastering. The skim coat of plaster creates a seamless finish over the entire wall surface, masking the joints between the boards in preparation for a decorative finish to be applied. Plasterboard is also available with tapered edges manufactured on the two longest sides of the boards. After the sheets are installed, the joints between the tapered edges of the boards and the nail/screw fixing positions are then taped and filled with 'dry lining joint filler'. Once the filler has dried it is usually sanded flat, and then further layers of filler are applied until a perfectly flat surface is achieved. This method of wall boarding is known as 'dry lining'. Dry lining applied to partition walls avoids the need to plaster the whole surface.

INSTALLING PLASTERBOARD

Plasterboard is normally fixed to metal stud partition walls with self-tapping drywall screws, and fixed to timber partitions with either galvanised plasterboard nails or standard drywall screws. Drywall (bugle headed) screws are the preferred method of fixing because they can be installed with an auto-feed screwdriver and are less likely to withdraw through movement in the wall after fixing. All the fixings that are used to fix the plasterboard should be resistant to corrosion, preventing the likelihood of staining through the face of the plaster/filler or decorated surface.

TYPES OF PLASTERBOARD

The types of plasterboard available are:

- fire resistant (pink-coloured core);
- sound resistant;
- thermal resistant;
- vapour resistant/moisture resistant (green-coloured core);
- impact resistant (for use in hospitals, colleges, etc.)
- flexible board (used on walls with internal and external curves).

FREQUENTLY ASKED QUESTIONS

I have noticed that one side of the plasterboard is a different colour from the other. Does it matter which side I use?

Yes, one side of the plasterboard is ivory and the other a light brown. The ivory lining paper is specially designed to be plastered; under no circumstances should the light brown side be used.

ACTIVITIES

Activity 18 – Erecting timber stud partitions

Read through the following questions and answer them as fully as you can to help you develop your underpinning knowledge of this subject area.

1. Name two different joints used to construct timber stud partitioning.
2. What do the initials CLS stand for?
3. At what 'centres' are the vertical studs in partition walls normally positioned?
4. Explain the purpose of 'noggings'.
5. Name one method that could be used to square partitions of adjacent walls.

STAIRCASE IDENTIFICATION

Staircases are manufactured in a number of ways, depending on a number of factors. Their shape, size and style will be determined by the size of the stairwell opening and the economical use of space and architectural style (stairwell – the opening formed in the upper floor), entrance and exit points on and off the stairs, the headroom height and, most importantly, the finished floor level (FFL) to finished floor level. The finished floor level to finished floor level (or 'floor to floor') is a measurement taken from the lower floor level to the upper floor level. This measurement *must* take into account the floor finishings, including floor screeds, underfloor heating, insulation and wooden floor finishes.

Most new-build projects will already have the shape of the staircase determined at the planning stage by the architect, but in some cases, such as refurbishment or modernisation, the shape of the stairs will be determined by existing features such as doorways, landings and a suitable headroom height.

Before calculating the size and shape of the stairs it is important to be able to understand the components that make up the completed staircase. The names of these components will remain the same, regardless of the shape or size of the stairs; these components can be clearly identified in Figure 4.120.

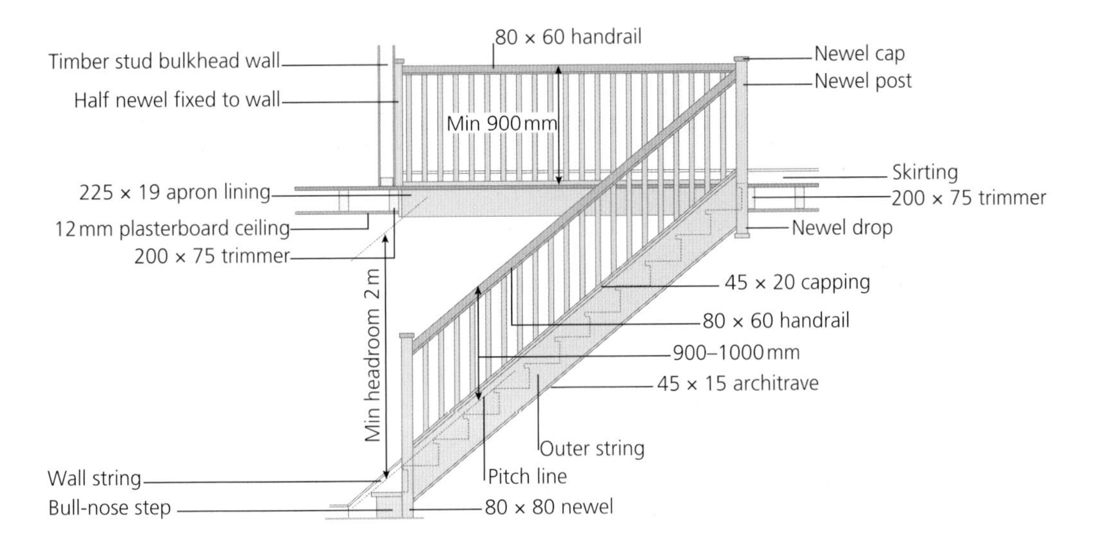

FIGURE 4.119

STAIRCASE DEFINITIONS

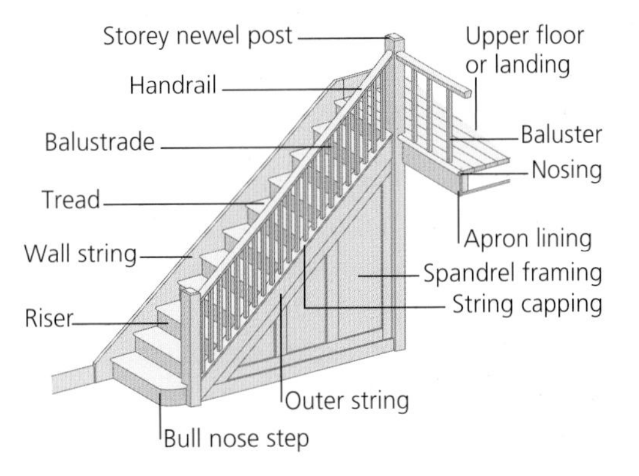

FIGURE 4.120 Components of a staircase

Apron lining – the component used to cover the rough face of the floor joists and provide a neat finish to the edge of the ceiling.

Baluster – the vertical component between the handrail and string capping, used to add strength to the handrail and form a part of the balustrade.

Balustrade – a combination of the handrail, baluster, string capping and string that forms the protection to the open side of a staircase.

Bull-nose step – a step usually found at the bottom of a staircase with a quarter-rounded corner on either one or two ends.

Flight – a continuous series of steps that form a staircase between landings.

Going – the horizontal dimension from the front to back of a tread, minus the overhang with the next tread above.

Handrail – the component supported by the balusters and usually running at the same pitch as the staircase. A requirement of building regulations that is used to support and steady the user.

Newel post – the heavy section vertical component used to support the handrails over long distances or where the stair changes direction.

Nosing – the section of a tread that overhangs the riser. The term is also used to refer to the narrow section of tread positioned at the top of a flight of stairs that sits over the trimmer joist.

Outer string – the structural component of the staircase that is used to support the treads, risers and balustrade on the open side.

Rise – the height between successive treads.

Riser – the component used between treads to form the vertical part of a step.

Spandrel frame – the panelling constructed directly under the outer string to enclose the underside of the staircase or form a cupboard.

Step – a combination of a tread and riser.

String capping – a length of timber positioned on the top edge of the outer string, used to locate the balusters.

Tapered treads (winders) – a shaped tread used to change the direction of a staircase where the nosing is not parallel to the tread or landing above it.

Tread – the horizontal component of a step.

Wall string (flier) – the structural component of a staircase used to support the steps against a wall.

FREQUENTLY ASKED QUESTIONS

What is the difference between a baluster and a spindle?

A baluster is the intermediate timber between the handrail and the string; a spindle is a baluster that has been shaped (turned) on a lathe.

Staircases can generally be identified by their shape on a plan, from above – an aerial view. The simplest and cheapest form of staircase is straight between walls, also known as 'cottage style'. This means that there is no need for a balustrade (although Building Regulations say there must be a continuous handrail fixed to the wall); if a staircase is not between walls then a balustrade must be installed to guard the space between the handrail, strings and newels on the open sides. The layout of the floors and room positioning will determine the shape of the staircase, and in some cases their shape can become quite complex. In simple terms, generally staircases can change direction by 90° (quarter turn) or 180° (half turn). A stair case may, however, change direction at any angle depending on the site layout/architectural plan. These twists and turns can be produced in a staircase with the use of quarter space landings, half space landings or winders (tapered steps); these are known as complex stairs.

Although staircases come in a variety of styles, their purpose remains the same – to allow pedestrian traffic (people) to move from one level to another with minimum effort.

Staircases can also be further categorised by their construction methods. If a staircase has the strings 'housed out' to accept the treads and possibly the risers, this is known as a 'closed string' staircase; if the strings follow the same profile as the treads and risers, this is known as a 'cut string' staircase. The closed string staircase is commonly used in the construction industry because of a number of factors; these include the fact that they are easier to manufacture and assemble and therefore less labour intensive, which enables costs to be kept down. The cut string staircase, although the more expensive option, offers a grander and more classical appearance, often with embellishments (trims and mouldings).

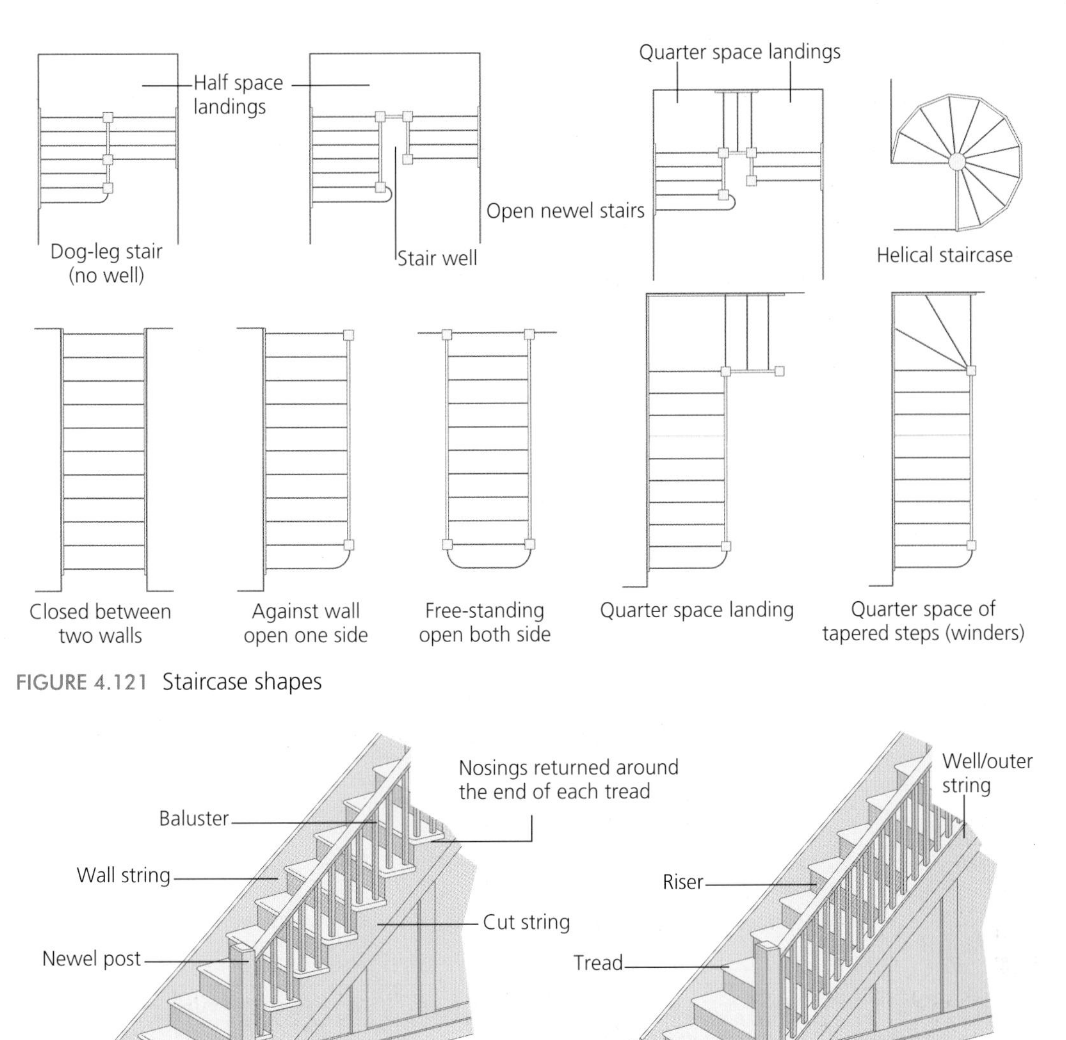

FIGURE 4.121 Staircase shapes

FIGURE 4.122 Examples of cut string and closed string staircases

Not all staircases need strings to support the treads and risers, 'spine back' being a good example. The stair has a central support that eliminates the need for the outer strings.

Figure 4.123 clearly demonstrates another example with the use of a central column into which all the steps are jointed on a helical staircase. The outer radius of the spiral staircase is supported from the balustrade to the next step; this pattern will continue on every step until the upper floor is reached. The balustrade in this case is jointed through the step and bolted into position; this process requires specially manufactured balusters.

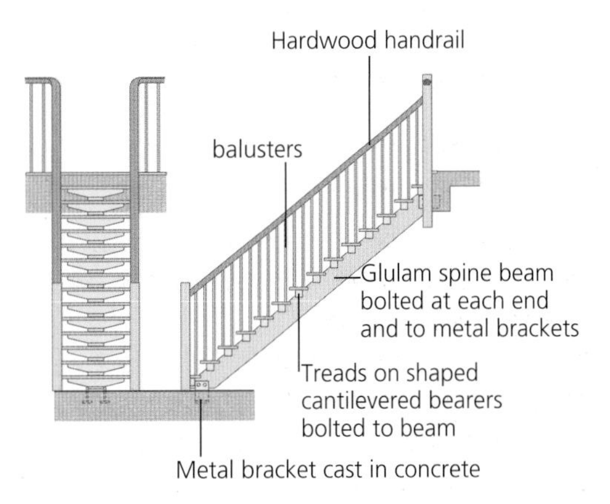

FIGURE 4.123

The combination of a tread and a riser is known as a step; a series of steps makes up a flight of stairs. A staircase can consist of just one or a number of individual flights. Although every flight of stairs has a series of steps, not all staircases have a riser; this type of staircase is known as 'open riser'. Building Regulations state that the opening between treads on an open riser staircase should not allow a 100 mm sphere to pass through. This can be overcome with the use of an intermediate rail, or a half riser, which can be jointed to the underside of the front of the tread or fixed at the rear of the step. Each of the methods mentioned meets with current Building Regulations, so the choice of guarding will depend upon the style and overall appearance of the staircase required.

The use of the half riser will provide additional support to the tread.

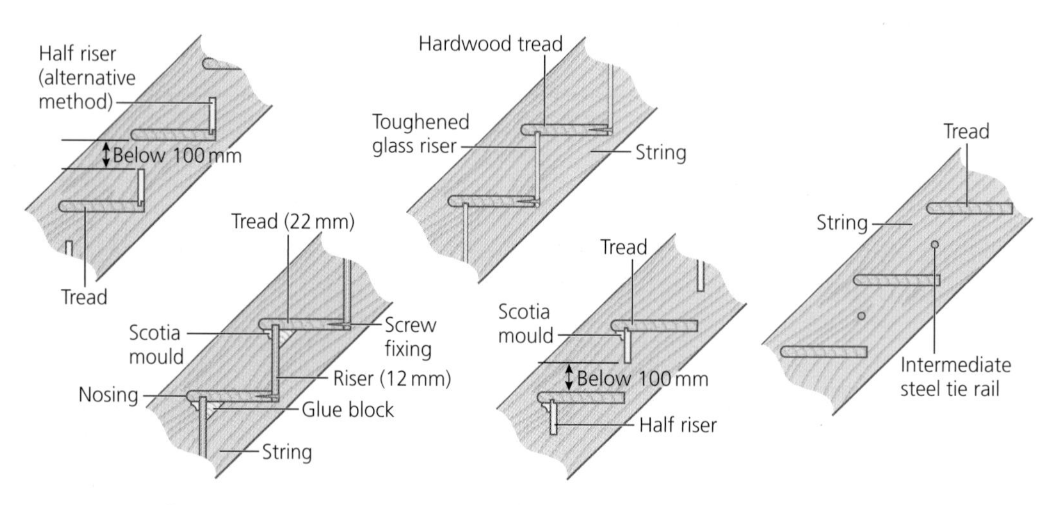

FIGURE 4.124 Steps

TAKING SITE MEASUREMENTS AND UNDERSTANDING ARCHITECTS' DRAWINGS

Before it is possible to produce or fit a flight of stairs, it is essential that the following information is established:

- the distance between the finished lower floor and finished upper floor level (or 'floor to floor');
- the width of the stairwell;
- the length of the stairwell;
- the length of the balustrade on the landing (if necessary);
- the depth of the upper floor joist;
- the thickness of the upper level flooring (to prepare the nosing so that it aligns flush with the upper floor level);
- the positioning of any doorways that may interfere with the proposed layout of the staircase;
- the height of the skirting board (to prepare the strings to align with the skirting).

The only accurate way of establishing this information is for the 'setter out' (a term used for the person who produces working drawings in a joinery workshop) to visit the site. Although the positioning, size and layout of the staircase may already have been designed by the architect or designers, the actual build may have slight differences from the initial design drawings. Although these differences may be minor to the builder, they are the difference between the staircase complying with Building Regulations and not. In extreme cases the intended layout may not fit due to unforeseen amendments that have had to be made to the build as it progresses. In a situation like this, it is important that the site carpenter or joiner has a strong understanding of both the Building Regulations and possible solutions as to layout.

A range of staircases are available from suppliers, but the choice of materials, sizes and layouts is limited. If a staircase is required to fit a non-standard opening, then it will have to be purpose made. These are normally made by joiners in a workshop, but it is not uncommon for site carpenters to adapt an 'off the shelf' staircase, providing it meets with Building Regulations.

FREQUENTLY ASKED QUESTIONS

What does the term 'off the shelf' mean?

The term 'off the shelf' broadly means that the item is common in shape, size, colour, etc., and is normally carried in stock by suppliers.

Although architects work to the same standards, the style of drawing will vary considerably according to the individual's approach. Essentially, the following items are obtainable from the drawings made by the architect:

- layout of the staircase – the plan is the clearest drawing;
- position of shaped treads (e.g. bull-nose steps);
- newel post positions;
- entrance and exit points on the stairs (this is normally indicated with an arrow pointing in the direction of the climb);
- the expected rise, going and handrail heights (this will be drawn to scale on the architect's drawings and also stated in the specification or written in the drawing panel);
- sections of component parts.

In addition to the views displayed on the architect's drawings, a specification supporting the design will be provided. The specification will enhance the architect's drawings and state the materials to be used, references to standard mouldings if applicable (e.g. handrail/carriage rail/newel posts) and the applied finishes to the materials used – in other words, information unable to be shown on the drawing without cluttering it with wordy explanations.

BUILDING REGULATIONS

In order for a staircase to allow people and materials to move from one floor to another in the safest and most comfortable manner, it must comply with current Building Regulations. It can be assumed that a staircase with an excessively large *rise* on each step will demand increased effort to reach the desired height and pose a danger, so the Building Regulations will always work to a maximum height restriction. When a staircase is designed it is not always possible to work with very small rises, because this means that the number of risers will have to be increased to reach the desired finished floor level. It also means that small rises will make the stair awkward and tiring to use. If the number of risers is increased, then the number of goings is increased, therefore increasing the overall going of the staircase and the space required for it. It is not always possible to accommodate a large staircase into a common dwelling. With these considerations in mind, the design of the staircase must comply not only with the Building Regulations but also with the space that it will demand.

Building Regulations also control the width of the treads on a staircase; these restrictions permit the *going* to be a *minimum* size of 220 mm. It would be considered that the wider a tread, the more comfortable the use of the stairs and, ultimately, the safer the stairs when in use. Building Regulations will also control the height of the balustrade, the width of the stairs, the pitch and the minimum headroom height (these regulations are explained in greater detail in the table on page 145).

FREQUENTLY ASKED QUESTIONS

What is the difference between a 'tread' and a 'going'?

The *going* is the measurement taken from the front of one riser to the front of the next riser; a *tread* is the component used to walk on as you advance on a stair. The width of the tread is measured from one riser to the next plus the overhang, or 'nosing'.

The restrictions to a staircase depend upon its intended use. Building Regulations split the intended uses of a staircase into three categories: 'private', 'institutional and assembly', and 'other or common'. An explanation of these categories is as follows:

- private – intended to be within *one* dwelling;
- institutional and assembly – serving a place where a substantial number of people will gather;
- other or common – in all other buildings.

Building Regulations 2000: Approved Document Part K

	'Private' stairs	'Other/common' stairs	'Institutional and assembly' stairs
Maximum rise	220 mm	190 mm	180 mm
Minimum going	220 mm	250 mm	280 mm
Maximum pitch	42°	Must comply with limitations stated below	Must comply with limitations stated below
Minimum handrail height – flight and landing	Minimum 900 mm Maximum 1 m	Minimum 900 mm Maximum 1 m	Minimum 900 mm Maximum 1 m
Minimum headroom height	2 m	2 m	2 m
Minimum width of stairs	1 m if used as a means of escape or used for disabled access; in other situations there is no minimum width	1 m if used as a means of escape or used for disabled access; in other situations there is no minimum width	1 m if used as a means of escape or used for disabled access; in other situations there is no minimum width

Regulations state that the relationship between the rise and the going should equate to the following formula: $2 \times R + G$ = between 550 and 700 mm (twice the rise plus the going equals between 550 and 700 mm).

Building Regulations do not state the maximum pitch for 'common' or 'institutional and assembly' staircases but they must fit within the following limits.

- Private staircase – any rise between 155 and 220 mm must have a going between 245 and 260 mm; any rise between 165 and 200 mm must have a going between 223 and 300 mm.
- Institutional and assembly staircase – any rise between 135 and 180 mm must have a going between 280 and 340 mm.
- Other/common staircase – any rise between 150 and 190 mm must have a going between 250 and 320 mm.

Stairs should have a handrail on at least one side if the width is 1 m or less; stairs over 1 m in width should have handrails on both sides. There is no requirement for handrails on the bottom two steps of a flight of stairs. Handrail heights should be measured between the nosing line and the top of the handrail.

The minimum headroom height can be relaxed for situations such as loft conversions. In these situations the headroom should be measured in the centre of the staircase and must be a minimum of 1.9 m. If the ceiling is sloping in the loft conversion, then the headroom may reduce further to 1.8 m measured to the sides of the centre line of the stairs.

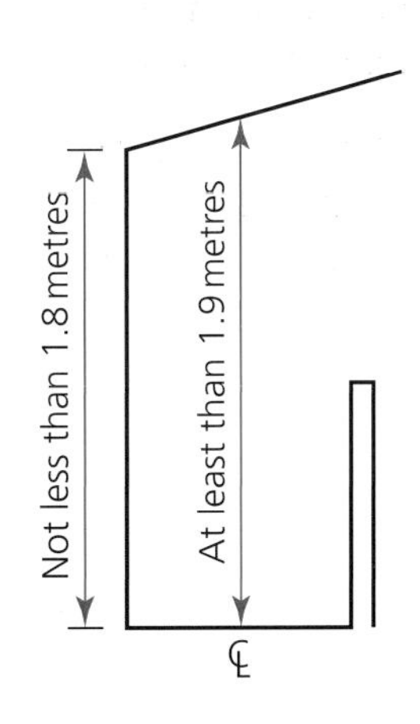

FIGURE 4.125

Part K of the Building Regulations states that if the building is likely to be used by children under the age of five years, any gap within the staircase must allow a 100 mm sphere to pass through it; this is to ensure that a child's head could not be trapped in it.

SETTING OUT STAIRS

CALCULATING THE RISE, GOING AND PITCH TO CONFORM TO THE REGULATIONS

The angle (more commonly known as the 'pitch') of a staircase is restricted by Building Regulations; these can be found in the table on page 145. Each staircase is designed to suit the environment for which it is intended, and considerations for public, private and disabled access are vital to meet with strict regulations. Space-saver staircases usually have an extremely steep pitch; this allows the benefit of less room used on the overall going. Although space-saver stairs serve their purpose, they are less comfortable to access due to the alternating arrangement of the treads and the severity of the pitch. Building Regulations will permit space-saver staircases to be used only in limited situations (e.g. access to one habitable room). These restrictions are because of the increased risk of injury when in use.

A suitable rise, going and pitch will be the basis for the layout of a staircase; these can be determined by working within the guidelines of the Building Regulations to calculate suitable measurements and angles. A simpler way to establish these sizes is to refer to Figure 4.126; this will provide the user with a simple and easy-to-use reference.

WORKSHOP RODS

To confirm the calculations of the staircase and to assist in the production of a cutting list, the stairs will need to be set out. It is normal practice for the 'setter out' to produce the drawings, either by using a drawing board and scaling tools in a small company or by using a computer-based program for larger companies. Once the setting out has been completed, this will then need to be transferred into a full-size workshop rod and accompanied by a cutting list.

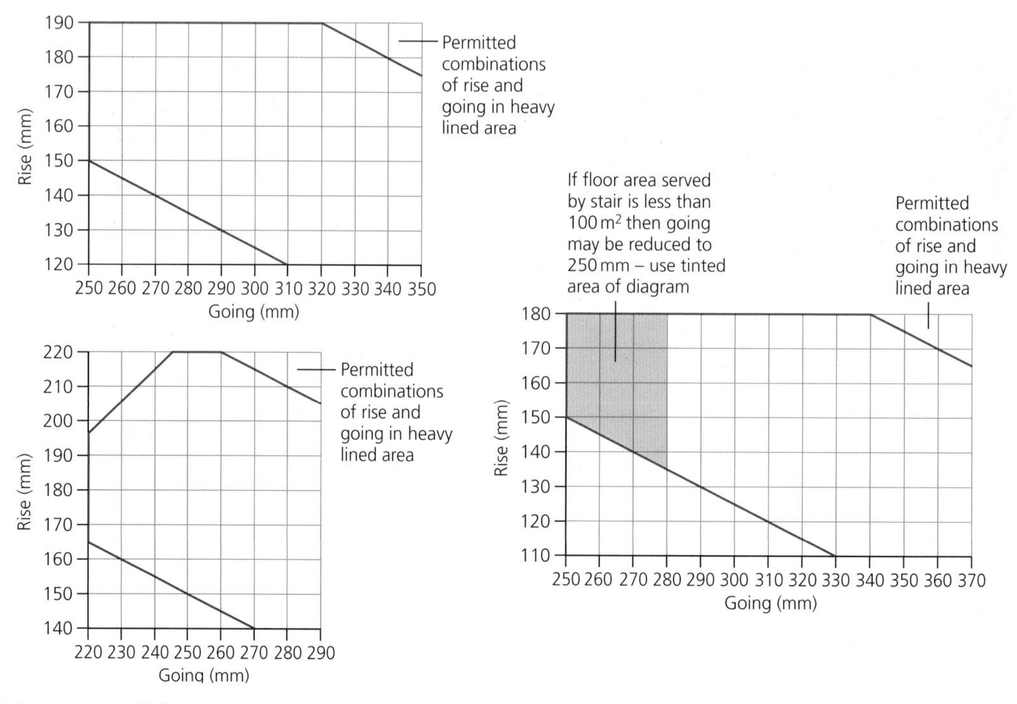

FIGURE 4.126 Tolerences

A staircase can be manufactured from a scaled elevation of the stair, from which newel, string and handrail lengths can be obtained; and a full-size section through a step, from which the pitch and string width can be obtained. Full-sized sections of moulded details will also be required. This is sufficient information for a cutting list to be produced and for the manufacture of the stair.

MARKING OUT

There are two common methods of marking out the strings and newel posts of a staircase:

1. pitch board and templates;
2. steel square.

In all cases a couple of items will have to be produced to assist with the marking out. First, the rise and going will have to be marked out on a 'pitch board' – a thin section of timber-based sheet material, normally plywood or MDF. This will have the rise and going marked on two edges.

The pitch board will be used to mark out the staircase in conjunction with a 'margin template'; this is used to position the risers and the treads consistently an equal distance from the front edge on the top of the string. Alternatively, the margin template and pitch board can be produced as one item, although
it is good practice to use the pitch board for further marking out on the newel post and more complex stairs.

Figure 4.127 demonstrates the first stage of marking out a 'closed string' staircase; this method of producing a staircase will require the strings of the stairs to be 'housed' or 'trenched' out 13 mm deep to receive the ends of the steps and for the wedging to secure their position. Each step is generally held in place with adhesive and wedges, depending on its method of construction. The process of marking out each step onto the strings is repeated along the length of the string until the correct number of steps is reached. As mentioned previously, the shape and design of the stairs can vary dramatically between jobs, but in most cases the newel post positions should be marked onto the well strings at this stage (Figure 4.128).

FIGURE 4.127 Marking out a 'closed string' staircase

FIGURE 4.128 Marking Newel post positions

To maintain the integrity (strength) of the newel posts when connecting the strings into them, it is important that the marking out of the haunched mortise and tenon joints are proportioned correctly.

The tenons cut on the string will have either one or two shoulders; this could depend on the thickness of the string (two gives additional strength and prevents twisting of the joint) or cost (one is less labour intensive). If using a tenon with one shoulder (or 'bare-faced tenon'), then consideration must be made for the mortise being 'offset'. This 'offset' will enable the strings to be lined up with the centre of the newel post, thus enabling the handrail over the strings, also centralised, to follow the same line. Where possible the housings should not reduce the thickness of the tenons.

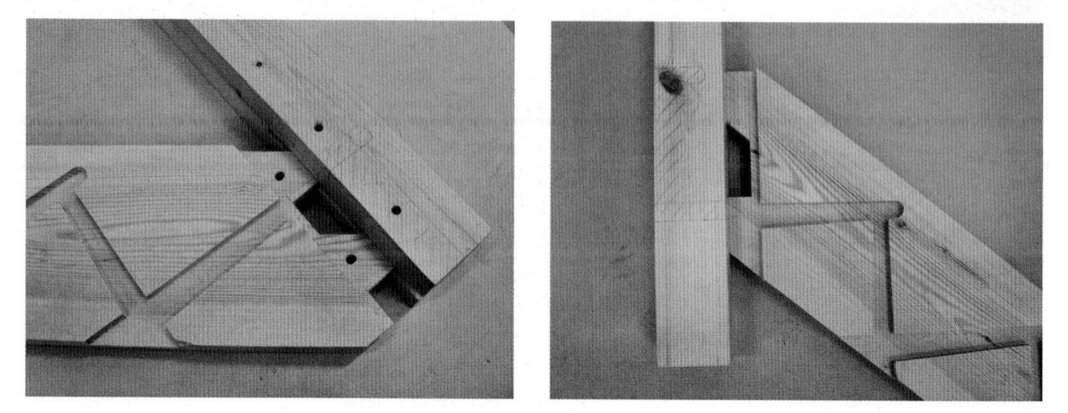

FIGURE 4.129

The height of a handrail on a staircase is determined by marking a pencil line along the margin line on the strings, and at the point this line hits the newel post the height can be established to the top edge of the handrail.

The purpose of the handrail is to offer support to the people using the stairs and to prevent falling from a height, so it is important that a strong joint is formed to the newel post and the handrail. This joint can be achieved in one of two ways: first, and the best option, is by using a mortise and tenon joint; second, it can be achieved by a mechanical fixing through the outer face of the newel post and into the handrail, with a pocket cover to hide the unsightly fixing. This method is usually used only for 'off the shelf' stairs.

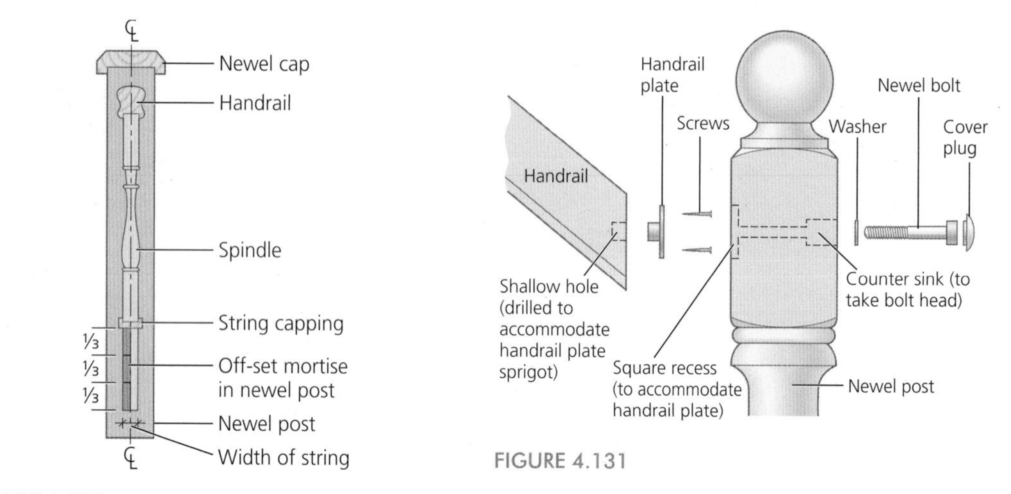

FIGURE 4.130

FIGURE 4.131

FIGURE 4.132 Marking out, drilling and assembling a draw dowel joint

The mortise and tenon joints on both the handrail and the string are held in place with a 'draw dowel'. The draw dowel is a section of timber driven into the newel post and into a hole in the tenon that is positioned 2–3 mm towards the shoulder of the tenon; this will allow the dowel to pass through the tenon and, at the same time, pull the joint tight.

Some more complex stairs may have continuous handrails formed over the top of the newel posts or over geometrical/continuous string stairs. (More often than not, these stairs do not incorporate newels.) In this case specially formed newel posts are required in order for the handrail to 'fly' over. There are also a variety of shaped handrail sections available 'off the shelf'; these short sections of handrail will offer solutions to overcome difficulties encountered when trying to keep a continuous handrail maintained at the correct height up a flight, on landings and around corners.

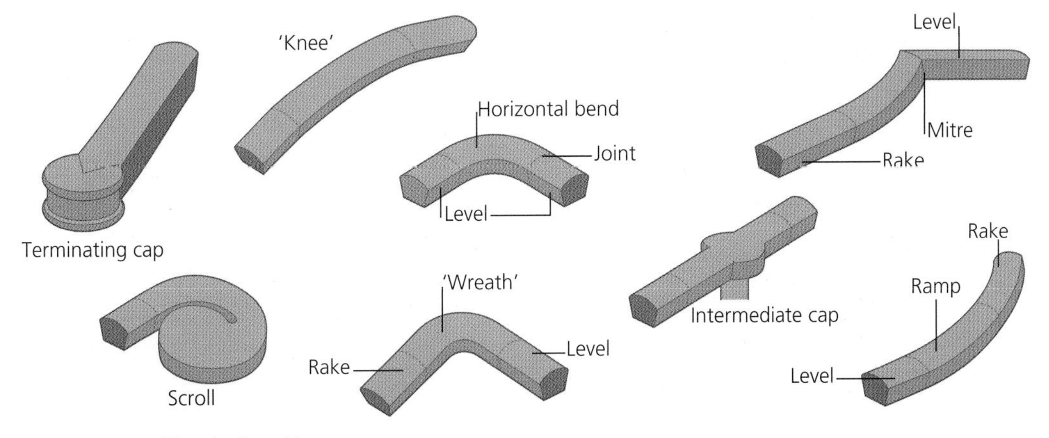

FIGURE 4.133 Handrail sections

To connect these example sections of handrail, or to extend a handrail in its length, a handrail bolt is needed in conjunction with a pair of dowels to prevent joint twisting. The handrail joint will provide a strong and secure fixing that is formed by drilling a series of location holes with the use of a jig; this will provide a consistently accurate joint.

STEPS

There are several methods of constructing steps within a staircase.

The advantage of the more complex jointing of the riser to the tread is that it provides added strength; this is due to the larger surface area to which the glue can adhere. The disadvantage of this method is that such steps are time consuming to manufacture and assemble, and will invariably increase the overall cost of the staircase. It is normal practice for the whole thickness of the riser to be recessed into a groove on the underside of the tread and the lower portion of the riser to be screwed through the back of the tread. This method will provide the minimum standard for a staircase manufactured from timber-based man-made sheet materials. A staircase manufactured from natural timber is better suited to complex joints; this will prevent any movement between the treads and risers and therefore eliminate any potential squeaks. (A noise heard when using the stairs caused by the movement between the treads and the risers or steps and strings.)

To provide additional strength between the treads and the risers, glue blocks are fitted. Generally, two to three glue blocks are fitted across the width of the staircase, depending on the overall width of the stairs. These are glued and rubbed (gently pushed back and forth until the suction holds them in place). Some joiners nail these into place, but this is counterproductive and when the nail is driven through the block a gap is formed where the grain is pushed out on the underside. This creates a gap, which later causes the glue joint to fail more quickly. The thin nature of modern risers also excludes this practice (for more detail consult BS 585).

MANUFACTURING STAIRS

One of the most important components on a staircase is the strings. The strings determine whether or not a staircase will reach its intended floor level, and provide the skeleton and strength for the treads, risers and newel posts to be constructed around.

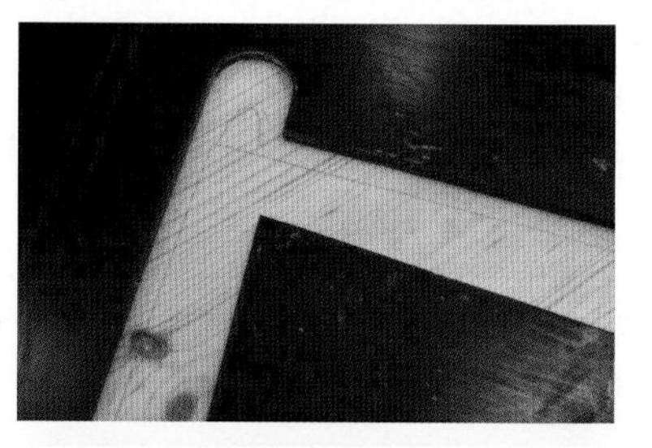
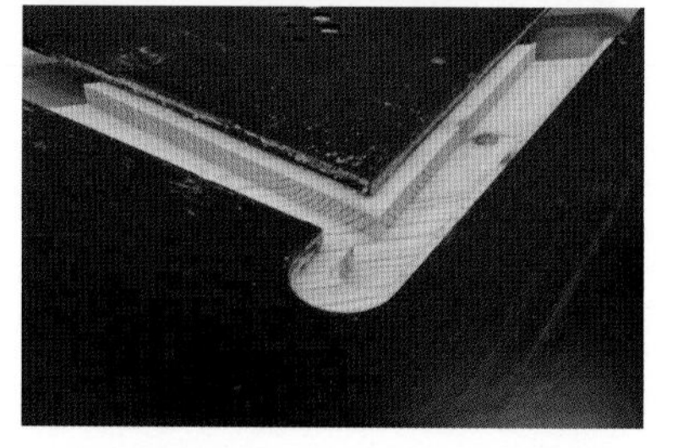

FIGURE 4.134 Methods of housing strings

These housings are generally produced with a portable router and a staircase jig; alternatively, the strings can be machined on a computer numerical control (CNC) machine.

Staircase jigs are usually made from plastic to withstand everyday use and are commercially available; alternatively, a cheaper option is to produce a jig from plywood or MDF, although these will wear with continued use.

Whichever method is used, it is important that the wedges holding the steps in place are positioned as in Figure 4.135. It is good practice to allow the wedge supporting the tread to run under the riser of the next step; this will prevent any movement between the riser and the tread and therefore prevent squeaks when it is under load.

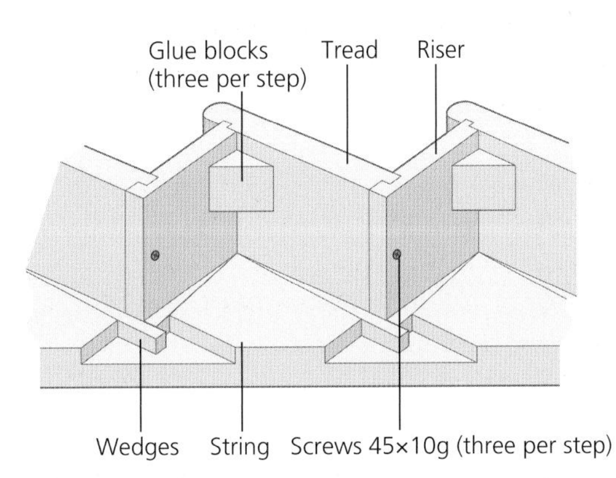

FIGURE 4.135 Correct positioning of wedges

Generally, the more work that is done on a staircase in the joinery shop the better; this will cut down on labour when installing, and reduce the potential for mistakes by the site carpenters when fitting. The amount each staircase can be assembled will depend on the availability of labour, transport arrangements to the site, clear walkways and the space available when positioning the stairs.

STAIRCASE ASSEMBLY

After dry fitting each step they are numbered to the string housing and set aside for assembly. All internal faces should be sanded prior to assembly. When assembling the staircase, it is important to ensure that the strings remain parallel to each other and that the overall width is maintained as originally designed. There are a number of ways that this can be achieved (see Figure 4.136).

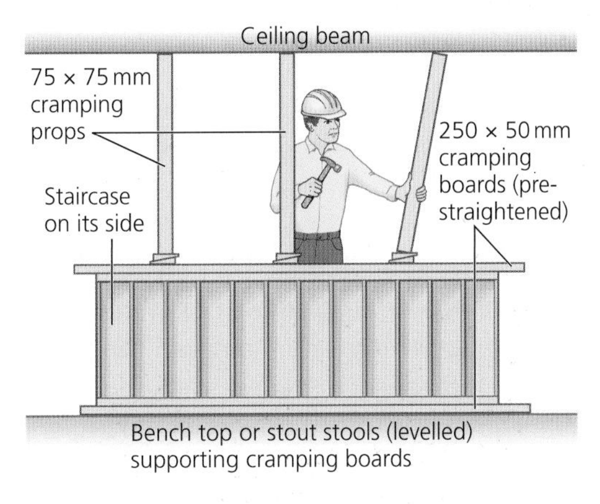

FIGURE 4.136 Staircase assembly

STAIRCASE INSTALLATION

When positioned, a staircase will achieve the majority of its strength from the notching of the newel posts and a 'bird's mouth' cut out of the string and over the trimmer joist.

The bird's mouth cut over the trimmer could be compared to similar cuts made on the rafters on a roof over the wall plate. The shaping of the top of the staircase over the trimmer joist will allow the distribution of the weight imposed upon it to be shared along the joist. In addition, any strings that are in contact with a wall will be secured with frame fixings or screws and plugs under every third step, and should have a batten fixed to the floor and riser underneath the bottom step. The fixing of the stairs to the batten will add strength to the unsupported bottom riser.

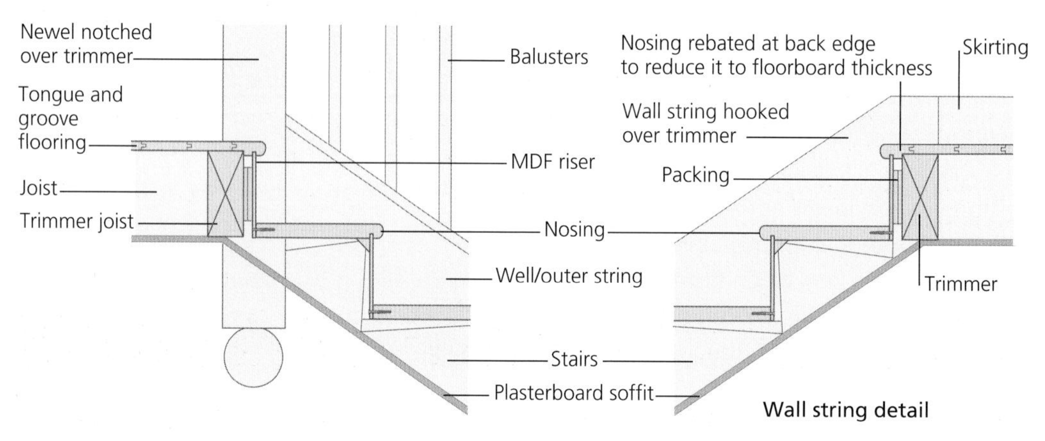

FIGURE 4.137 Staircase installation

INSTALLATION OF THE BALUSTRADE

As well as serving the purpose of guarding the open side(s) of a staircase, the balustrade can add decorative features with ornate newel posts and balusters. It is important that newel posts are securely fixed in position before the balusters (or spindles if turned) are installed. If the newel posts are turned, they are normally either in two sections (if the newel cap is formed on the newel turning) or in three sections (these include the newel base, newel turning (mid-section) and the newel cap). The newel turning formed from two sections of timber is a cheaper alternative to the three sections, although this option will provide a limited amount of alternative newel top finishes. The connections between each newel post can provide a weak spot in the balustrade if it is not firmly secured. The most effective method of connecting these components is by using a strong wood adhesive and wedging the bottom of the turned dowel into a blind hole drilled into the opposite section.

The number of balusters required for a flight of stairs is normally calculated by allowing two per step and just one each end of the balustrade where the string joins the newel posts. For example, a straight staircase with 12 steps would have 10 x 2 = 20 + 1+1 = 22 balusters. (*Note*: this is only a guide for costing and material ordering purposes. The exact calculation is normally workout at the time of installation following the Building Regulations Approved Document K (no openings to allow a 100 mm sphere to pass through).)

The balusters are cut to length and fitted vertically between the bottom of the groove in the underside of the handrail and the string capping. (*Note*: an alternative method of fixing the balusters is used if cut strings are used to construct the staircase.) In most cases, the balusters are secured vertically and equally spaced with small packers glued between each component along the length of the string capping and the underside of the handrail.

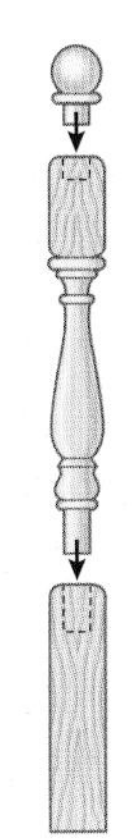
FIGURE 4.138 Baluster

PROTECTION AFTER FIXING

After the installation of a staircase is complete all the components should be covered to protect against moisture (staining), general dirt and damage. The treads are normally clad with strips of hardboard with a batten (down-stand) fixed along the front edge to protect the nosing. Handrails, newel posts, etc. are usually protected with bubble wrap, cardboard or thin timber-based sheet materials.

ACTIVITIES

Activity 19 – Assembling, erecting and fixing straight flights of stairs, including handrails

Read through the following questions and answer them as fully as you can to help you develop your underpinning knowledge of this subject area.

1. List the sequence of marking out and cutting a draw bored joint.
2. Explain how to measure and mark the handrail height on the top and bottom newel posts on a straight flight of stairs.
3. Explain the purpose of 'glue blocks'.
4. What is the maximum distance permitted between the balusters on a staircase?
5. According to current Building Regulations, which one of the following staircases is permitted to have the steepest pitch – private or common?

MULTIPLE-CHOICE QUESTIONS

1 Which **one** of the following activities is undertaken at the first fixing stage?
 a Fitting of cupboards and stairs
 b Fixing of partitions and stairs
 c Fixing skirting and architraves
 d Fitting of locks and letter plates

2 Which **one** of the following options lists stud partition component parts?
 a Stud, sole plate and nogging
 b Nogging, wind brace and stud
 c Transom, head plate and sub sill
 d Sole plate, skirting and head plate

3 Due to speed of erection and economy, stud partition members are usually jointed by
 a mortising and tenoning
 b bridling and stub tenoning
 c butt jointing and skew nailing
 d Dowelling and biscuit jointing

4 Solid timber floorboards are jointed in their length using which one of the following joints?
 a End
 b Heading
 c Footing
 d Dovetail

5 Which **one** of the following nail types is most suitable for fixing floorboards?
 a Panel pin
 b Floor brad
 c Flooring pin
 d Decking nail

6 Which **one** of the following joints is used to connect the head and jamb of a door lining?
 a Butt
 b Lapped dovetail
 c Tongued housing
 d Mortise and tenon

7 The component used to maintain the width of a door lining at the bottom is called a
 a tie
 b brace
 c stretcher
 d strutting

8 Stairs are covered by the Building Regulations within part
   a J
   b K
   c L
   d M

9 The maximum gap allowed between balustrades on a stair is
   a 80 mm
   b 90 mm
   c 100 mm
   d 110 mm

10 The maximum pitch allowed for a private stair is
   a 36°
   b 38°
   c 40°
   d 42°

SECOND FIXING

CHAPTER 5

LEARNING OUTCOMES

By the end of this chapter you should have developed a knowledge and understanding of:

- installing side-hung doors and ironmongery;
- installing mouldings;
- installing service encasements and cladding;
- installing wall and floor units and fitments.

INTRODUCTION

The aim of this chapter is for students to be able to recognise second fixing components and materials, and the methods used to install them. It also identifies and explains current Building Regulations applicable to this area of study, following all the relevant health and safety law and good working practices.

SECOND FIXING

Broadly speaking, site carpentry is broken down into two main areas of work; these are known as 'first fix' and 'second fix'. The terms 'first and second fix' are commonly used and recognised by construction workers in the building industry across the United Kingdom. First fixing building work is carried out during the initial construction phase of a project from the completion of the foundation through to the period of plastering the internal walls.

Examples of first fix carpentry:

- timber frame walls;
- floor joists;
- roofing;
- door frames and linings;
- windows;
- staircases.

(*Note*: installing first fixing components is covered in more detail in Chapter 4, First Fixing.)

Second fix carpentry involves the installation of manufactured joinery items, seasoned joinery-quality timber and ironmongery. Second fix carpentry items are non-structural, and can become vulnerable to damage from the weather and other trades working around them. It is essential that they are correctly stored and protected from the elements, accidental damage and theft prior to installation.

Qualified carpenters will have completed a period of training and instruction, which includes the study and practice of both first and second fixing. Some tradesmen prefer to specialise in either one element or the other, and usually purchase tools and equipment tailored to their chosen area.

Examples of second fixing:

- finishings, including skirtings, architraves and dado rails;
- units and fitments, including kitchens and built-in cupboards;
- internal and external doors;
- encasing services (constructing frames and panels to hide pipes, cables, etc.).

Other trades besides carpenters use the terms 'first and second fixing', as described below.

PLUMBER

- First fixing – installing hot/cold water supplies, waste pipes, etc.
- Second fixing – installing toilet suite, boiler and radiators, etc.

ELECTRICIAN

- First fixing – installing the main service cables and wires through the walls and floors, etc.
- Second fixing – connecting light fittings, switches and plug sockets, etc.

SECOND FIX CARPENTRY

DOOR REQUIREMENTS

There is a wide range of different internal and external doors available on the market, each one offering different performance criteria. Doors are capable of regulating the amount of light and ventilation through to other rooms or spaces within a building. Careful consideration should be given to the selection of doors to ensure that they suit their intended positions; for example, a door with a 'vision panel' (glazed section) would not be suitable for a bathroom. Although this seems to be an obvious mistake, it could easily be overlooked if the doors are ordered by referring to the architect's plans alone.

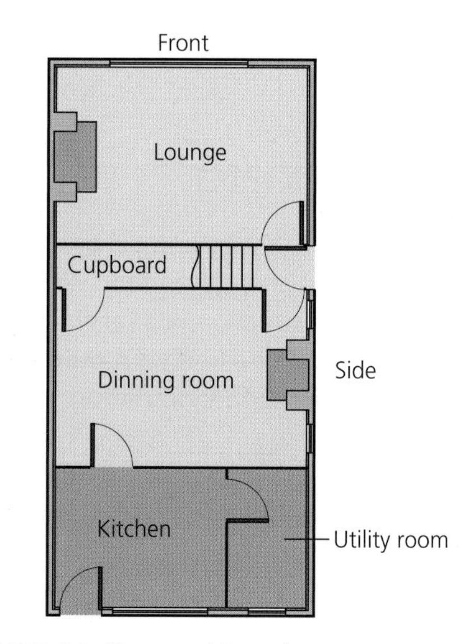

FIGURE 5.1 Door positions shown on architect's plan

DOOR CHARACTERISTICS

Security

Doors built of a sound construction and durable materials will offer a good level of security if they are installed correctly. They should also be fitted with good-quality locks and bolts. Some insurance companies insist that mortise deadlocks are fitted to all external doors leading into a domestic property before protective cover will be given. Internal doors in domestic buildings are rarely fitted with locks, with the exception of bathrooms, although they are commonly found in commercial buildings such as schools, colleges and offices.

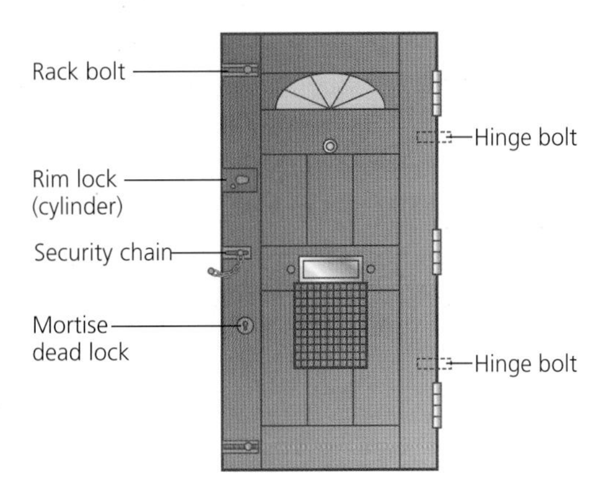

FIGURE 5.2 A typically secured external door

Weather protection

An exterior-grade door installed into a frame fitted with draught excluders or weather strips will prevent water penetrating into the building and improve the 'U' value (heat loss). It is important to check that any ironmongery to be fitted to the door can withstand the elements. Ironmongery not suited to an exterior position will tarnish, corrode, rust, and potentially, over a short period of time, break down and fail to function properly.

> **TRADE SECRETS**
>
> Fire escapes are considered to be one of the *least* secure points of a commercial or public building. Although many buildings are entered by intruders through fire escapes, they should *not* be secured with 'deadlocks' or any other type of fixing that requires the use of a key.
>
> In many situations fire doors are secured with chains and padlocks, and these remain in place during periods when the building is occupied. This is against factory regulations and may put lives at risk in the event of an emergency evacuation.

Privacy

Privacy is maintained in a room or space by regulating the position of the door. The pivot point or hinged side of a door can usually be established by referring to the architect's plan or by considering the following factors:

- make sure that all light switches are easily accessible as the door is opened;
- privacy is still maintained in the room after opening the door;
- wherever possible, the door should open into the room.

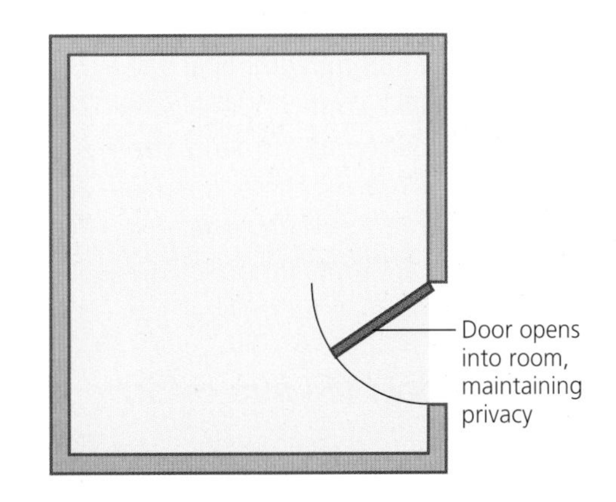

FIGURE 5.3 Maintaining privacy

Fire resistance

There are two types of fire door (also known as 'fire resisting' doors) commonly used in the construction industry; these are half an hour and one hour resistant. The spread of flames can be minimised from passing from one room to another if a fire door is correctly installed, with a suitable door frame and 'intumescent seals' (fire-retardant strips normally fitted to either the door or frame). (*Note*: fire doors are covered in more detail later in this chapter.)

Soundproofing

Most solid or dense doors will reduce the passage of sound through a door opening. In areas of excessive noise, such as theatres, music rooms or private offices where privacy must be maintained, acoustic doors must be fitted. The amount of sound that passes through a doorway can be reduced further by installing acoustic seals to both the door and the frame.

FREQUENTLY ASKED QUESTIONS

What are acoustic seals?

Acoustic seals come in a variety of different shapes and sizes, depending on their position on the door or frame. They work similarly to draught excluders, by sealing the joint between the door edge and the frame.

Access

Doors provide access in and out of buildings; their width will vary depending on their position. External doors are usually wider than internal doors because they have more traffic passing through them. Consideration should be given to the type of access and performance required by the doors; for instance, wheelchair or disabled access.

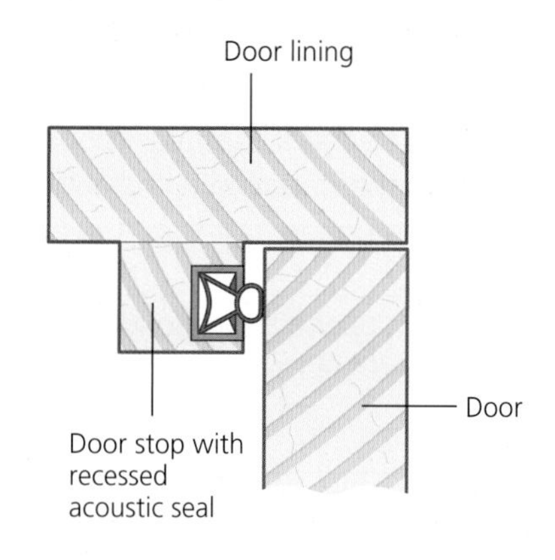

FIGURE 5.4 Acoustic seals

STANDARD DOOR SIZES

Interior doors (35–44 mm thick)

- 1981 mm × 457 mm (Imperial size – 18″ × 78″)
- 1981 mm × 533 mm (Imperial size – 21″ × 78″)
- 1981 mm × 610 mm (Imperial size – 24″ × 78″)

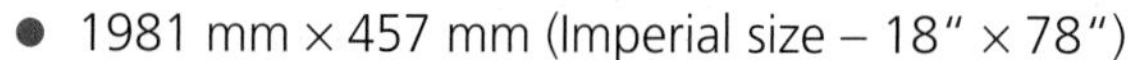

- 1981 mm × 686 mm (Imperial size – 27″ × 78″)
- 1981 mm × 711 mm (Imperial size – 28″ × 78″)
- 1981 mm × 762 mm (Imperial size – 30″ × 78″)
- 2032 mm × 813 mm (Imperial size – 32″ × 80″)
- 1981 mm × 838 mm (Imperial size – 33″ × 78″)

Exterior doors (44 mm thick)

- 1981 mm× 762 mm (Imperial size – 30″ × 78″)
- 2032 mm × 813 mm (Imperial size – 32″ × 80″)
- 2032 mm × 838 mm (Imperial size – 33″ × 78″)

- 2057 mm × 838 mm (Imperial size – 33″ × 81″)
- 2083 mm × 863 mm (Imperial size – 34″ × 82″)
- 2134 mm × 914 mm (Imperial size – 36″ × 84″)

DOOR SCHEDULES AND DRAWINGS

In the construction industry, a 'schedule' is a document that is used to reference all the details needed for a particular component. Schedules are usually supplied by the architect with the construction drawings; they provide repetitive detailed information for individual items such as doors, windows and ironmongery. Details are easily obtained from a schedule by using a referencing system between the architect's drawings. The relevant information is then used throughout the planning, building and completion stages of a project. Schedules are used by the main and subcontractors to provide the information needed to estimate the total costs of various items before ordering and programming deliveries. The following information could be obtained by referring to a detailed *door* schedule:

- the overall dimensions (height, width, thickness);
- design/type (flush, panelled, internal/external, etc.);
- glazed/non-glazed (pattern of glass, single/double glazed, safety glass, etc.);
- type of material (oak, beech, ash, redwood, etc.);
- type of ironmongery (stainless steel butt hinges, push plates, mortise latch, etc.);
- finish (colour of paint/stain/varnish, etc.).

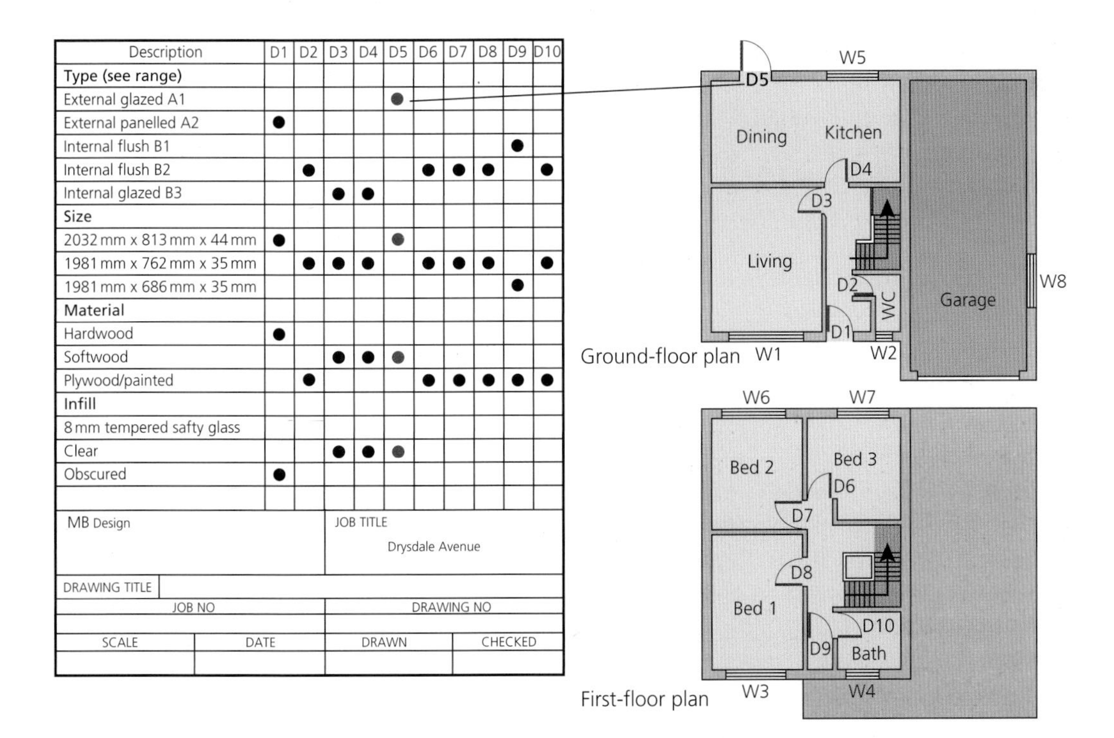

Description	D1	D2	D3	D4	D5	D6	D7	D8	D9	D10
Type (see range)										
External glazed A1					●					
External panelled A2	●									
Internal flush B1									●	
Internal flush B2		●				●	●	●		●
Internal glazed B3			●	●						
Size										
2032 mm x 813 mm x 44 mm	●				●					
1981 mm x 762 mm x 35 mm		●	●	●		●	●	●		●
1981 mm x 686 mm x 35 mm									●	
Material										
Hardwood	●									
Softwood			●	●	●					
Plywood/painted		●				●	●	●	●	●
Infill										
8 mm tempered safty glass										
Clear			●	●	●					
Obscured	●									

MB Design

JOB TITLE Drysdale Avenue

DRAWING TITLE

JOB NO | DRAWING NO

SCALE | DATE | DRAWN | CHECKED

FIGURE 5.5 A door schedule

DOOR TYPES

Internal and external doors

There are clear distinctions between internal and external doors, as follows.

Internal doors (35–44 mm thickness)	External doors (44 mm minimum thickness)
Sometimes faced in veneer or laminate	Seldom faced in veneer or laminate
Lightweight construction (except fire doors)	Mortise and tenon joints in framed doors
No preservative treatment	Treated to prevent against decay and insect attack
Interior adhesives used	Water-resistant adhesives used
Interior-grade timbers	Durable timbers or timber-based materials

Note: some cheaper, mass-produced external doors are constructed with dowel joints.

FRAMED/PANELLED DOORS

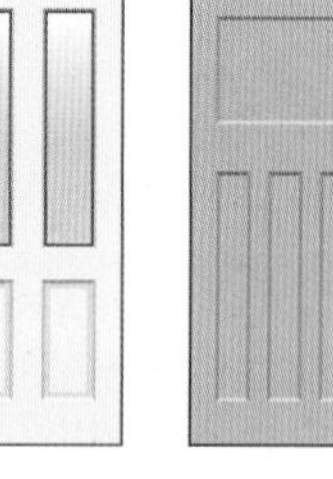

FIGURE 5.6 Timber panelled doors

Panelled doors are available in a variety of different shapes and styles to suit both modern and older traditional buildings. They are normally constructed with a series of solid timber rails, formed around either timber or glass panels. Panelled doors have been manufactured for many years with traditional mortise and tenon joints. Alternatively, the rails are 'scribed' over the profile or moulding around the inside of the door stiles and are dowelled together.

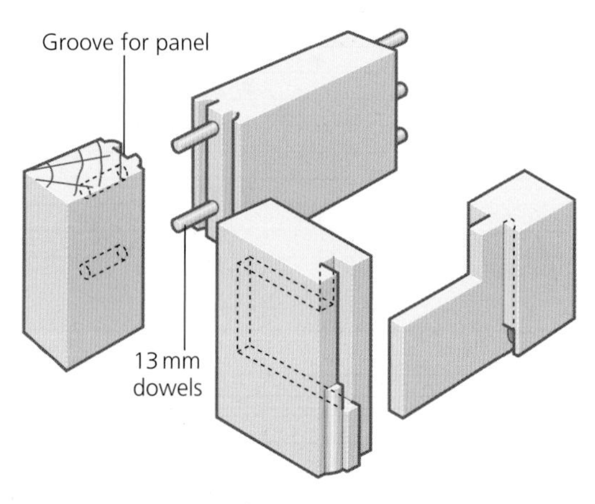

FIGURE 5.7

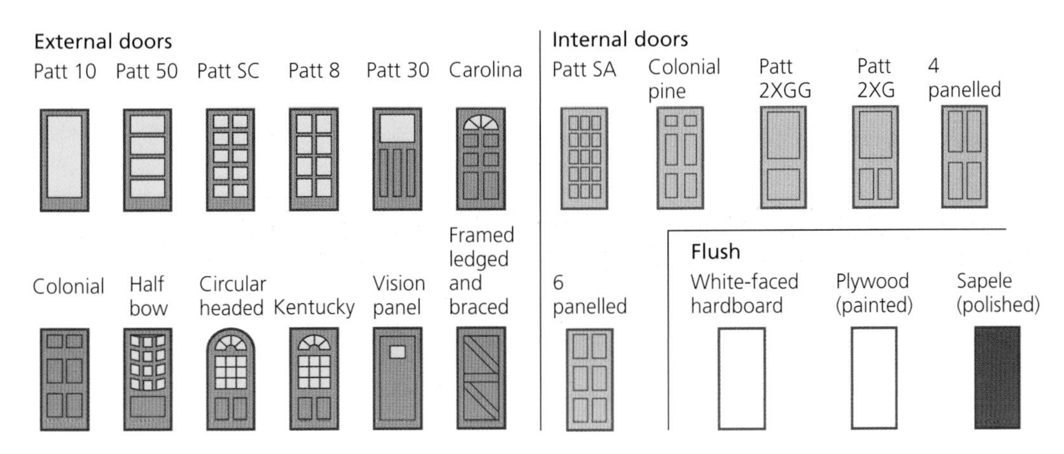

FIGURE 5.8 External and internal doors

The proportions of the mortise and tenon joints on a panelled door will vary depending on their position and the width of their component parts. The joints are held securely in place between timber wedges. Figure 5.9 highlights the sequence in which the wedges should be driven into position.

FREQUENTLY ASKED QUESTIONS

▶ Why is it important to install the wedges in this sequence?

Failure to follow these guidelines will result in the rails being wedged out of line. This movement would then cause the joints to 'open up' and reveal unsightly gaps. Gaps between the joints in the frame could result in a weakening in the structure and possibly cause the whole door to twist.

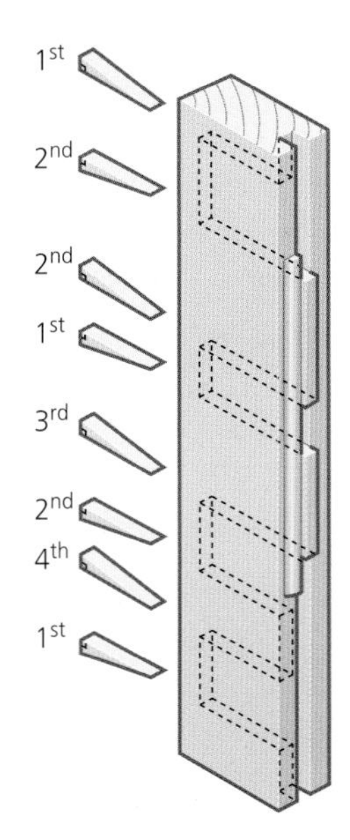

FIGURE 5.9 Hand scribed joints

GUNSTOCK STILES

Traditional panelled doors are designed to permit the maximum amount of light through

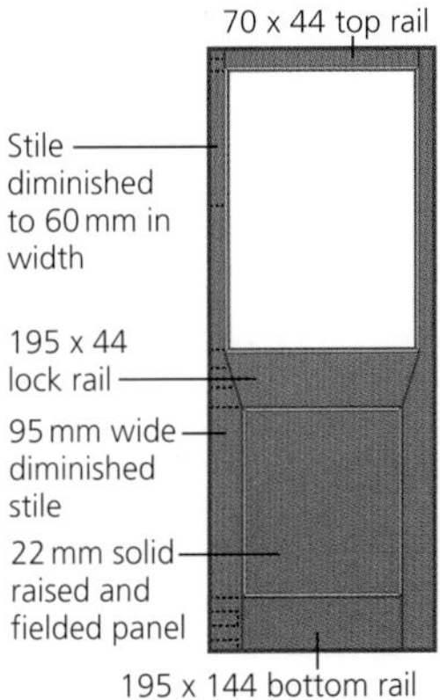

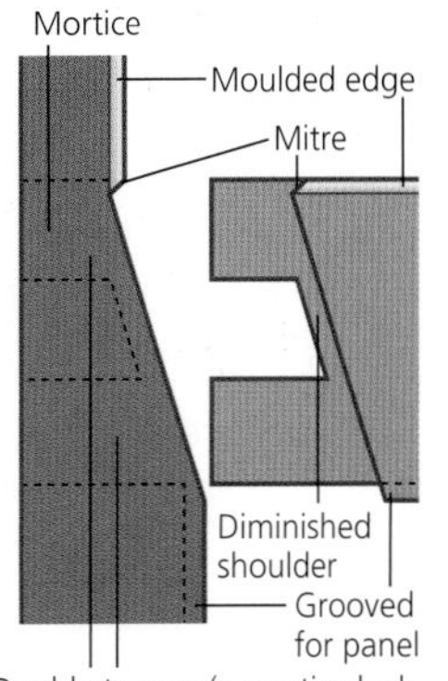

FIGURE 5.10

the glazed area without compromising the strength or integrity of the door. This is achieved by reducing the width of the stiles on either side of the door along the glazed area, yet maintaining the full width of the stiles where there are panels. Gunstock stiles are rarely used these days due to the complexity of the joints between the rails and the stiles.

PANEL MOULDINGS

The design of panelled doors may vary considerably, depending on the materials used and the shape of the panels/glazing and the mouldings (also known as profiles). In most cases the mouldings used on internal doors are purely cosmetic; however, consideration should be given to the outward facing mouldings on an external door. These profiles should be designed so that they prevent water from remaining on the large flat areas for long periods of time, as this may lead to wood decay and water penetration.

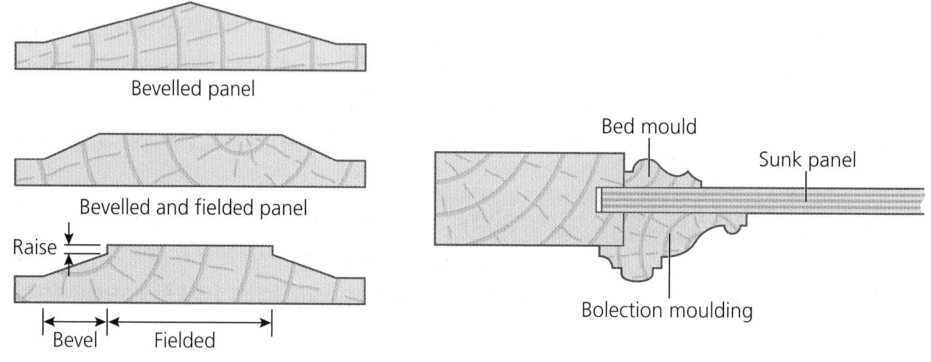

FIGURE 5.11 Panel mouldings

TRADE SECRETS

The panels within a timber-framed door will continue to move after they have been assembled because of the differences in moisture in the atmosphere. This movement in the panels will not affect the operation of the door if these guidelines are followed:

1. *Never* glue the panels in the frame of the door as this may lead to them splitting.
2. *Always* allow a gap of 2–3 mm between the outside edge of the panels and the grooves. This will permit the panels to shrink and expand freely, and prevent the joints on the door being forced apart.
3. *Always* prime or stain the edges of the panels with the same finish as the rest of the door before assembly. Shrinkage in the panels after completion may cause any untreated surfaces to be revealed.

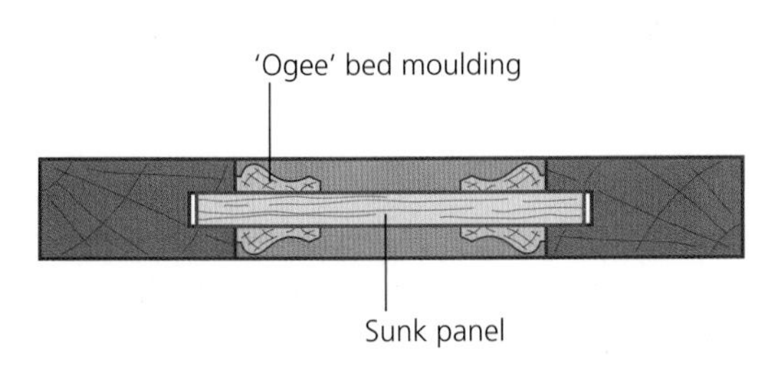

FIGURE 5.12

FIGURE 5.13 Fire doors

FIRE DOORS

A fire door's primary function is to delay the spread of flames, and dangerous smoke and gases onto fire escape routes for a set period of time. This will give the occupants of the building enough time to evacuate in a safe manner to their designated assembly points. Fire doors with increased resistance may be used to protect property or information contained in commercial buildings or offices.

Fire doors have to undergo vigorous testing to make sure that they conform to British Standards before being certified with a fire rating. Ratings and the certification of doors will vary depending on their ability to maintain their integrity and stability for a set period of time. The most common types of fire door are:

- FD30 = 30 minutes' integrity (44 mm thick);
- FD60 = 60 minutes' integrity (54 mm thick);
- FD90 = 90 minutes' integrity (54 mm thick);
- FD120 = 120 minutes' integrity (59 mm thick).

Notes:

- 'integrity' means the amount of time a door will resist cracking and fire breaking through the core of the door to the opposite face;
- 'stability' means the length of time a door will maintain its shape before collapsing.

TESTING

Fire doors must be tested in simulated conditions as a complete assembly to be approved; this includes:

- the door;
- door frame;
- intumescent seals;
- all the ironmongery, including hinges, locks and latches, and door closers.

(*Note*: all ironmongery must be made of metal that has a melting point above 850°C, e.g. steel or brass.)

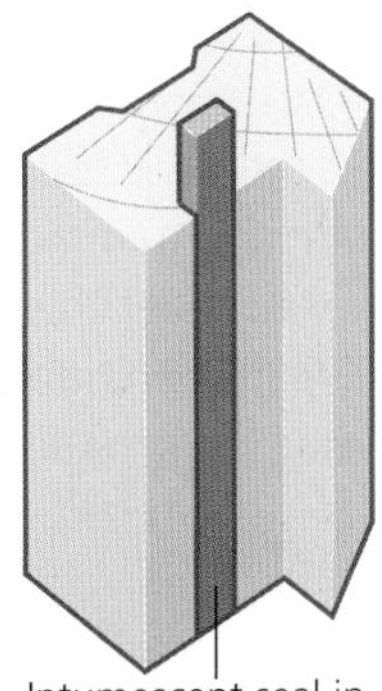

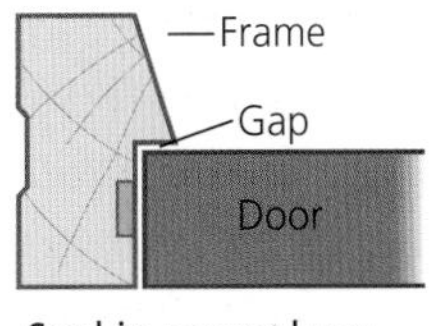

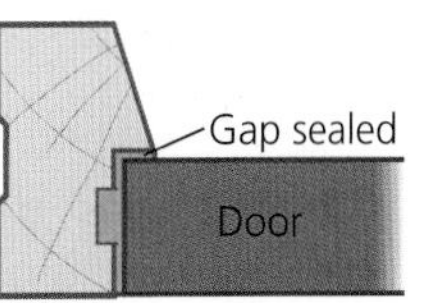

FIGURE 5.14

Fire doors that have undergone testing together with their frame etc. are usually sold by suppliers as a 'door set'. If a 'fire door' or 'fire door set' is not installed to the same specifications as the example that underwent testing, this will invalidate the fire certification and its rating. Any door or door set that has successfully completed testing to achieve certification must be supplied together with the installation instructions to ensure they are fitted to the same standards. An incorrectly installed fire door/door set may be the result of poor workmanship during the hanging or fitting – for example, large margins between the door edge and the frame (greater than 3 mm) or the incorrect amount of intumescent strips or smoke seals, etc.

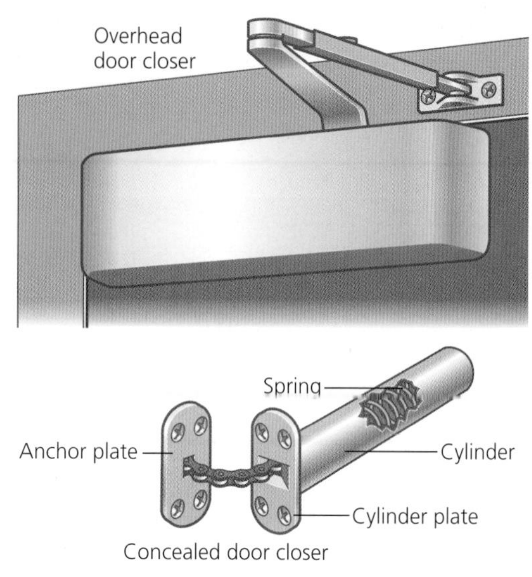

FIGURE 5.15 Door closer

FREQUENTLY ASKED QUESTIONS

My local builders' merchants stock 'intumescent paint'. How many coats of this do I need to apply to a timber door to achieve a half-an-hour fire rating?

All fire doors must conform to British Standards BS 476: Part 22: 1987. Intumescent paint applied to a timber door will not prevent the spread of flame for an acceptable period of time to meet with these minimum requirements.

FIRE DOOR IDENTIFICATION

It is important to be able to clearly identify fire-resistant doors and frames, to ensure that they meet with specifications and Building Regulations. Certified fire doors are sometimes difficult to identify through a visual inspection alone, so it may be necessary to request a copy of the test certificate for the exact door performance details. The British Woodworking Federation (BWF) runs a certification system known as 'BWF–CERTIFIRE Fire Door & Doorset Scheme'. The purpose of the scheme is for members to have their products rigorously tested to ensure that they comply with the minimum requirements to save lives and protect property. Doors and doorsets that meet the requirements of the scheme are identified with a unique labelling system. BWF-CERTIFIRE labels are normally attached to the top edge of fire doors, and contain the following information:

- member's name (manufacturer);
- member's phone number;
- certification number;
- unique serial number;
- fire rating (e.g. 30 or 60 minutes).

Alternatively, the Timber Research and Development Association (TRADA) uses a simple system of inserting a tree-shaped coloured-coded plastic plug into the edge of the door, which identifies the type of fire door/frame and intumescent seals. The following colours are used around the outside of the plugs to identify the rating of the door:

- yellow – 30 minutes;
- blue – 60 minutes;
- brown – 90 minutes;
- black – 120 minutes.

GLAZED FIRE DOORS

Most fire doors require vision panels or some form of glazed section; this can be achieved with the use of wired or fire-rated glass. The fire-retardant glazing must be secured in position with hardwood beads, either screwed or nailed in position, together with an approved intumescent glazing system. This will prevent fire and smoke permeating through the joint between the glazing and the recess in the door. The exact specification for the installation of glazing in fire doors should be obtained from the fire test data sheet.

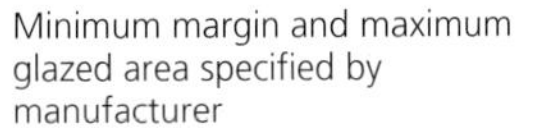

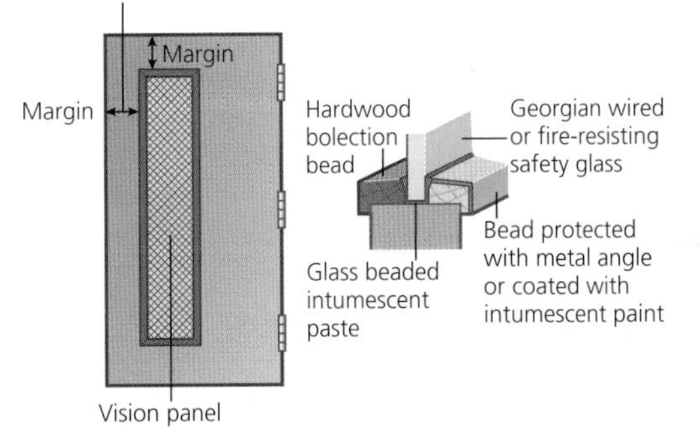

FIGURE 5.16 Glazed fire door

HOLLOW CORE DOORS

These are probably the most common type of doors used in the construction industry today because of their relatively low cost and light weight. Although the term 'hollow core' suggests that the door has been constructed without any structure in the middle, this is not entirely true. There are several different methods of building hollow core doors; the most common modern method is with the use of a timber frame infilled with a honeycomb pattern of corrugated cardboard. This arrangement is then covered with 3–4 mm plywood on either side or, alternatively, hardboard. In most cases hardboard is used because it is cheaper, smoother and can easily be pressed into panel shapes with wood grain effects to give the appearance of a panelled framed door, constructed with solid materials.

LOCK BLOCKS

'Lock blocks' are sections of solid timber, approximately 500 × 127 mm, sandwiched into the core of hollow doors. They are usually found on only one side of each door to enable locks, latches and handles to be securely fixed, although they can be found on both sides on some better-quality doors. It is important that the sides containing the lock blocks are clearly identified to prevent the doors being hinged on the wrong side, leading to difficulties in fixing the ironmongery. Manufacturers normally label the top edge of the side containing the lock block with either a key symbol or the lettering 'LOCK'.

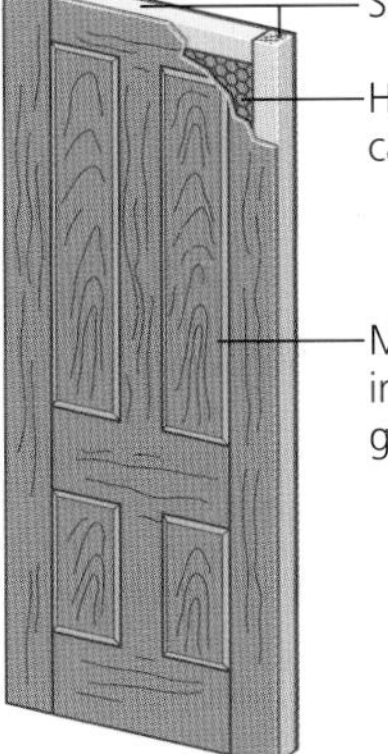

FIGURE 5.17 Hollow core door

MATCHBOARDED DOORS

'Matchboarding' is a term used for a series of timber boards jointed together in their width. The most common way of connecting the boards is with the use of 'tongue and groove' mouldings. Matchboarded doors are probably the simplest method of door construction, and are commonly used internally and externally for garden gates, sheds and older buildings such as cottages. Tongue and groove boards are used in the construction of different types of matchboarded doors; these include:

- ledged doors;
- ledged and braced doors;
- framed ledged and braced doors.

V-jointed on two sides

Beaded matching

V-jointed on one side

Recessed matching

Centre matched tongue and groove

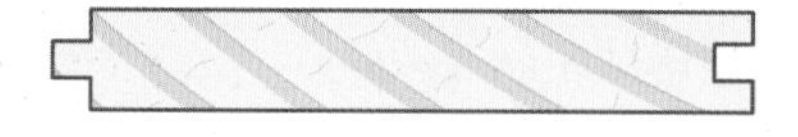

FIGURE 5.18

LEDGES

Ledges are the horizontal timbers running across the back of matchboarded doors. They are used as fixing points for the vertical timbers on the face of the door and help to increase the thickness of the door and its stability.

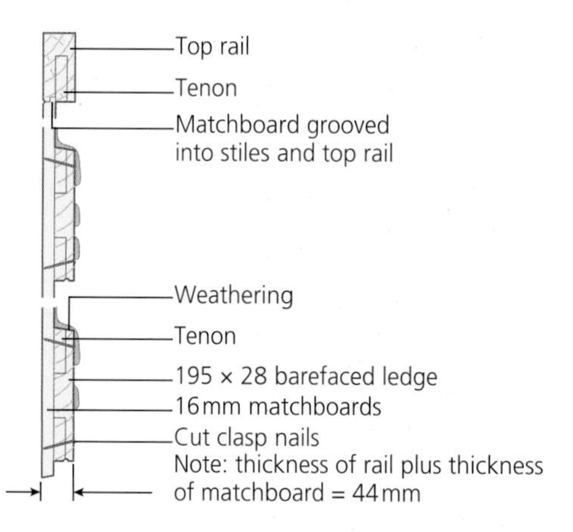

FIGURE 5.19 Ledged matchboard door

TRADE SECRETS

To prolong the life expectancy of matchboarded doors installed in exterior positions, all the top edges of the ledges should be 'weathered'. This will prevent water and moisture holding on the level surfaces for long periods of time and potentially instigating the onset of rot. The weathering will also direct water away from the joint between the matchboarding and the back of the ledges.

Painting the tongues and grooves, and the back faces of the ledges and braces, before assembly will protect the hidden areas from trapped moisture. This will also ensure that any movement in the timber will not reveal any bare or untreated timber.

BRACE

Braces are used diagonally across the back of matchboarded doors to maintain their shape and reduce the risk of 'dropping' under their own weight. 'Dropping' is a term used to describe the side of a door opposite the hinges that has failed to retain its shape due to poor supporting. Braces should also be positioned so that they support the side opposite to the hinges; an incorrectly braced door may result in the door failing to function properly. Braces should be fitted at 45° or more to the ledges; if they are fitted below this angle they will not provide adequate support to the door. Commercially produced matchboarded doors usually have the braces left loose so that they can be fitted by the carpenter once the hanging side of the door has been established. Some suppliers manufacture matchboarded doors with the bracing suited to both left and right handing. Although this method will provide some support to the door, it is considered to be a compromise and will not offer the same strength as alternative methods.

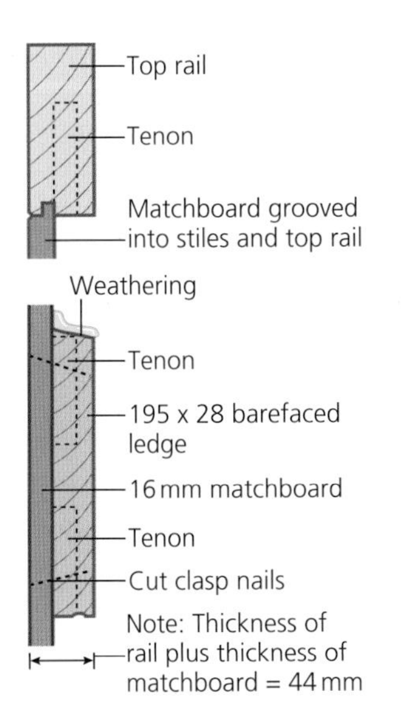

FIGURE 5.20 Matchboarding

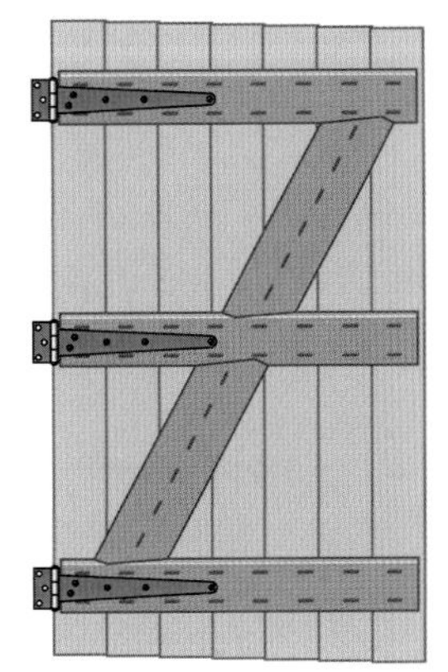

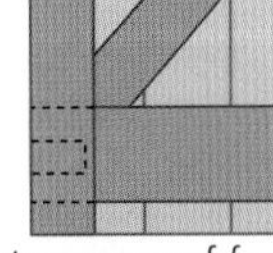

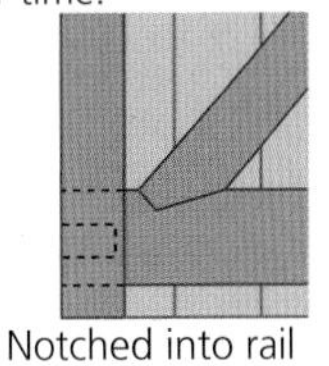

FIGURE 5.21 Braces

LEDGED DOORS

Ledged matchboarded doors are probably the simplest form of doors to manufacture, and for this reason they are sometimes manufactured on site by the second-fix carpenters. This eliminates transport and expensive joinery manufacturing costs; it also prevents delay between ordering, manufacture and delivery. These doors consist of a set of three ledges screwed at either end to the back of a series of tongue and groove boards; the remaining boards are fixed in position using a traditional method known as 'clench nailing' (Figure 5.23). The ledges on hardwood matchboarded doors are usually fixed to the back of the tongue and groove boards with screws, counter-bored below the surface of the timber and filled with wooden plugs to conceal the fixing.

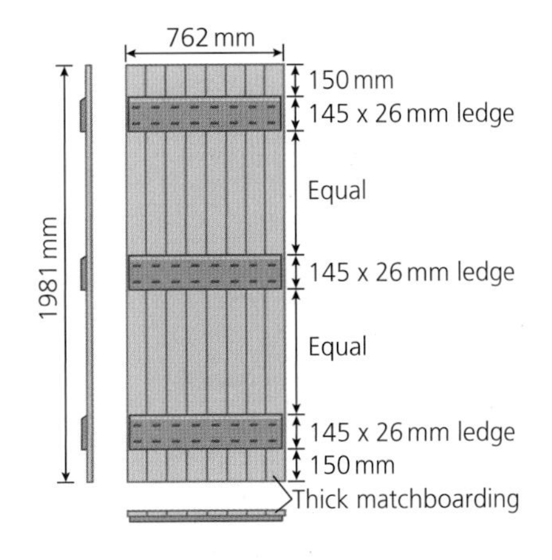

FIGURE 5.22 Ledged door

FIGURE 5.23 Clench nailing

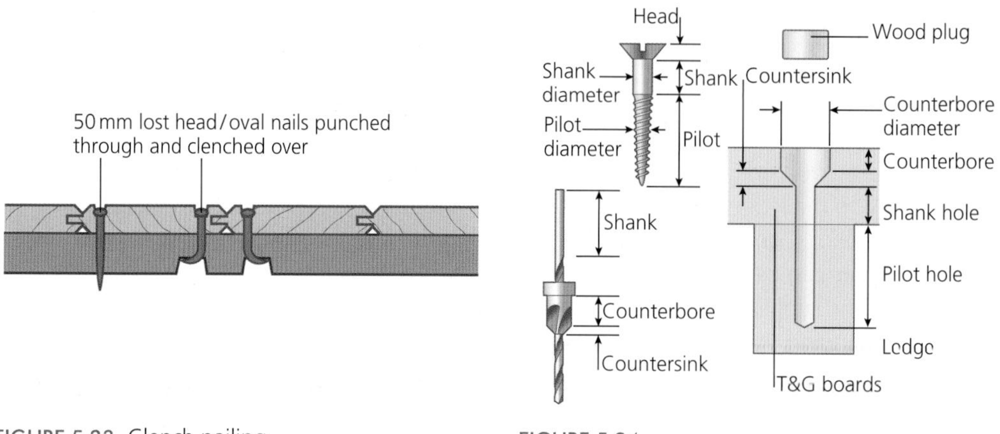

FIGURE 5.24

Ledged matchboarded doors are rarely used these days because they have a tendency to 'drop' on their lock side. The risk of 'dropping' can be reduced if ledged matchboarded doors are used only for narrower doorways. If wider tongue and groove boards are also used, this will help because the amount of joints in the width of the door has been reduced.

IRONMONGERY

Ledged matchboarded doors are relatively thin because, unlike many other doors, they do not have a substantial frame around their perimeter. This means that any ironmongery fitted on the door should be aligned with the ledges (Figure 5.26); this will ensure a secure fixing point when screwing or bolting into the timber. 'Black japanned', 'zinc plated' or galvanised 'T' hinges are normally used to pivot side-hung doors because they are relatively easy to install and are capable of supporting the light weight of the door.

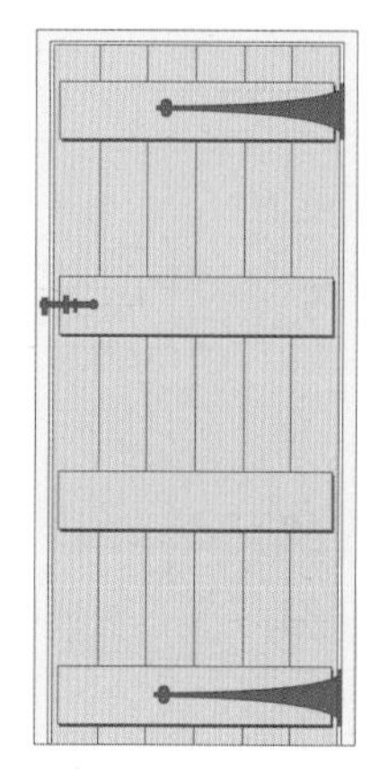

FIGURE 5.25

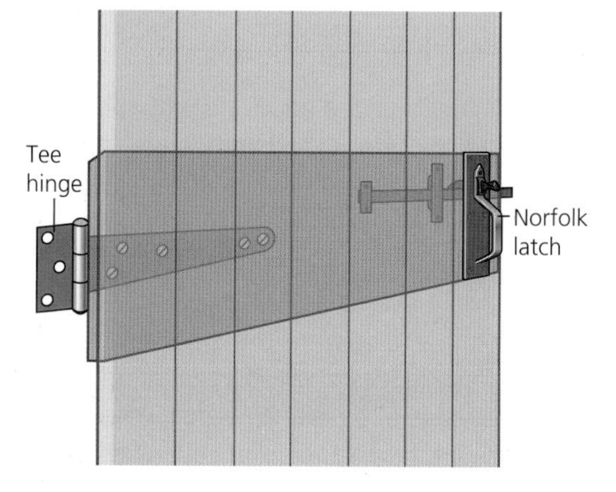

FIGURE 5.26 Ironmongery should be aligned with ledges

FREQUENTLY ASKED QUESTIONS

What are 'black japanned' 'T' hinges?

Items that are 'black japanned' have gone through a process of having a black weatherproof coating baked on.

Black japanned ironmongery is commonly used for garden gates, sheds and 'rustic' items of internal and external joinery.

LEDGED AND BRACED MATCHBOARDED DOORS

These are constructed in the same way as ledged matchboarded doors but have the addition of diagonal bracing to prevent 'dropping'. The bracing across the back of a ledged and braced matchboarded door increases its rigidity and minimises the potential of a twist developing in the door.

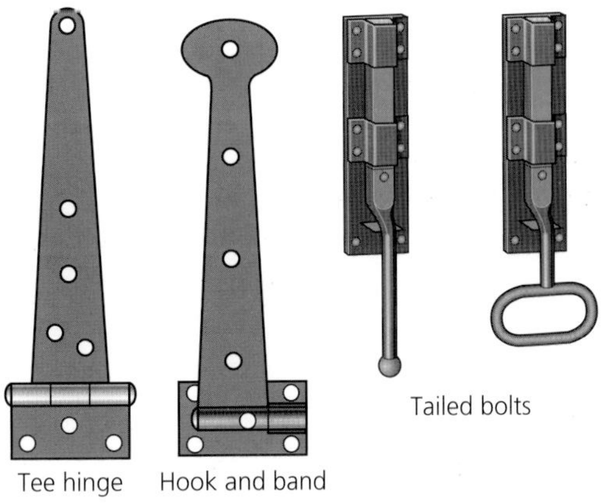

Tee hinge

Hook and band

Tailed bolts

FIGURE 5.27 Black jappaned ironmongery

FRAMED, LEDGED AND BRACED MATCHBOARDED DOORS

As the name suggests, these types of door are similar to ledged and braced doors, but, unlike them, they have a framed method of construction to increase their overall strength and performance. The frame and mortise and tenon joints could be comparable to those of panelled doors. In general, both doors use the same section timber for the stiles and head, and the same width for the middle and bottom ledges. Both the middle and bottom ledges should be reduced in their thickness to allow the matchboard covering to pass over their face without increasing the total thickness of the door.

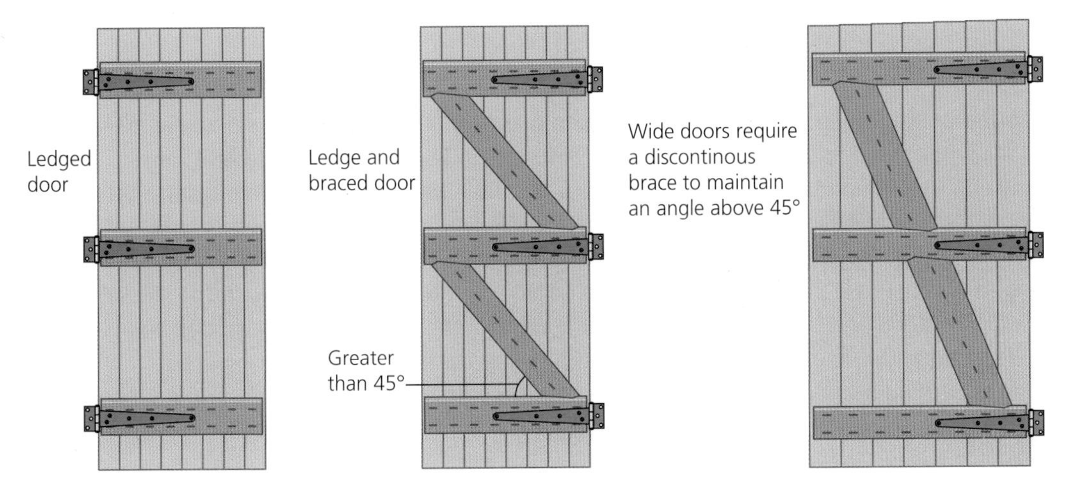

FIGURE 5.28

The reductions in the thickness of the middle and bottom rails mean that standard mortise and tenons with two shoulders can no longer be used. In order to maintain the mortise and tenon position along the length of the stiles, 'bare-faced tenons' should be used; this simply means that there is only one shoulder on one face of the joint.

There are two ways of jointing the matchboarding into framed, ledged and braced doors:

1. tongue and grooved;
2. rebated.

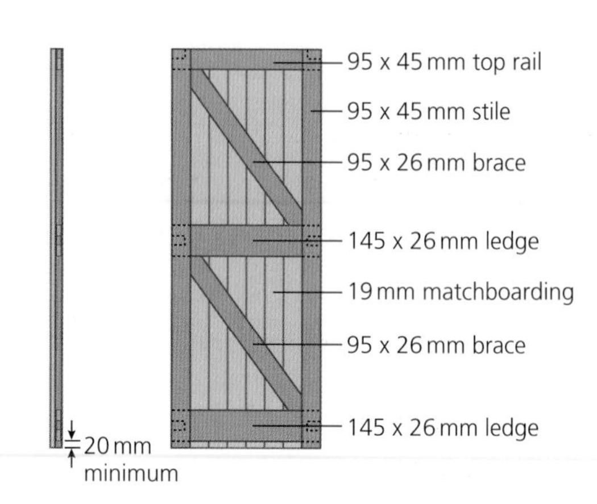

FIGURE 5.29 Framed, ledged and braced door

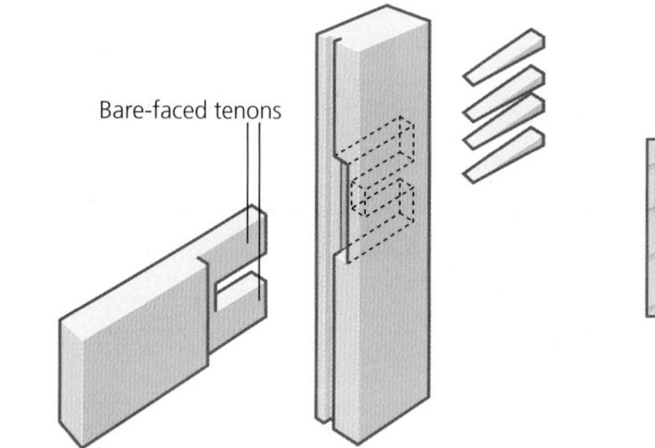

FIGURE 5.30

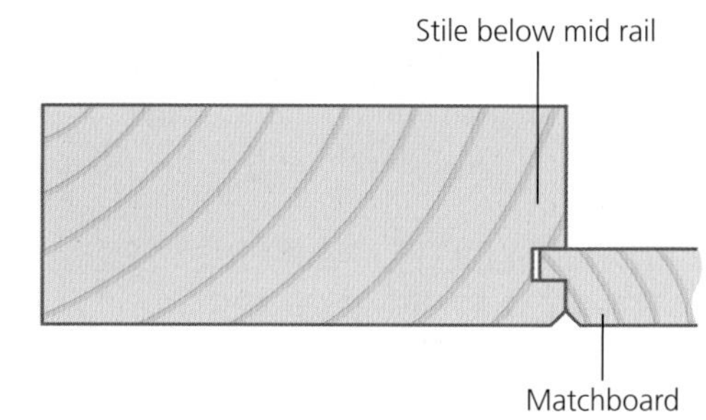

FIGURE 5.31 A section through the stile of a framed ledged and braced door

TIMBER SELECTION

Timber is a natural resource that will shrink and expand with changes in humidity and temperatures in the climate. This movement should be considered when calculating the number of boards required for matchboarded doors, to avoid distortion. The amount of movement will vary between different timbers or species, and will depend on the initial moisture content of the timber, the sectional size of the timber and the direction of the grain.

The seasoned tongue and groove boards used on internal matchboarded doors are likely to shrink when the doors are fitted. This is likely to continue until the balance is achieved between the timber and the atmosphere; this is known as the 'equilibrium moisture content'. Doors that are hung in exterior positions are also likely to experience movement, with the reverse results. The effects of any potential movement damaging or distorting internal doors can be eliminated if the tongue and groove boards are fitted tightly between the stiles or the width of ledged and braced doors. External doors require the tongue and groove boards to have a gap of 2–3 mm between each board to allow for expansion. Poorly spaced matchboarding across the width of a framed door could result in expansion in the timber, forcing the stiles apart and opening the joints along the shoulders. As a result, the structure of the door will be weakened and will probably cause it to 'drop'. It may also lead to an increase in the overall width of the door, causing it to 'bind' in the opening or fail to function properly.

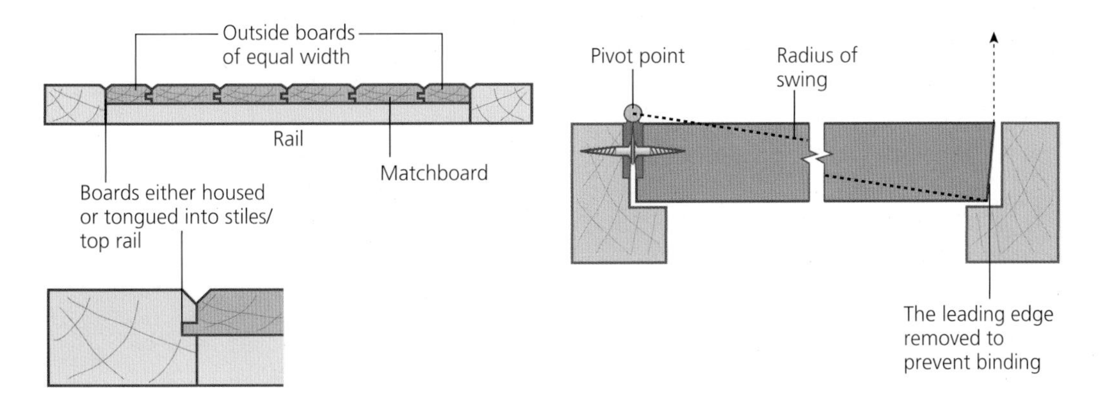

FIGURE 5.32 Ensuring an equal margin on the outer boards

FIGURE 5.33 Removing the leading edge

FREQUENTLY ASKED QUESTIONS

What does the term 'bind' mean?

In relation to doors, the term refers to the leading edge of the door (the edge opposite the hinges) rubbing against the frame as it operates. In the worst cases, doors may fail to close fully within the frame or the operator may have difficulty opening the door. This can be prevented and overcome by reducing the width of the door, so that a clearance gap of 3 mm is achieved between the edge of the door and the frame on all four sides.

DOOR OPERATIONS

There are many different ways that a single door or combination of doors can be arranged within a frame to suit different purposes. These purposes may include a pair of glazed doors to borrow extra light from one room to another, or sliding doors to minimise the amount of space needed for the doors to function properly. When an opening has more than one operational door, they are then known as 'leaves'. The operation of such doors will fall into one of the following categories:

- single action side-hung doors (single and double);
- double action side-hung doors (single and double);
- sliding doors;
- folding doors;
- revolving doors.

The term 'action' refers to the operation of a door; for example, a single 'action' door will swing only in one direction, while a double 'action' door will swing both inwards and outwards through the lining.

SINGLE ACTION SIDE-HUNG DOORS (SINGLE AND DOUBLE)

This is probably the most common way of hanging any type of internal or external door. In simple terms, a single action door is hung between the jamb on a door frame or lining and one side of a door, along its edge, and usually with butt hinges. Single action doors used in conjunction with butt hinges are designed to pivot through at least 90° in one direction.

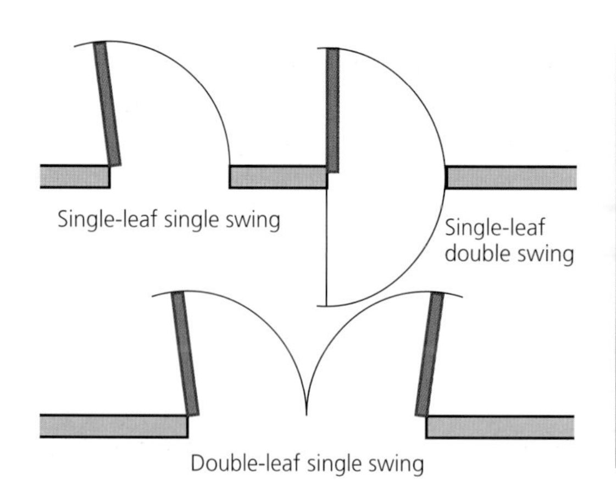

FIGURE 5.34 Single action side-hung doors

FIGURE 5.35 Butt hinges

DOUBLE ACTION SIDE-HUNG DOORS (SINGLE AND DOUBLE)

Double action doors swing through at least 180°. They are commonly found in areas of heavy traffic such as schools, offices and hospitals. Double action doors could potentially become a hazard unless vision panels are positioned at eye level to avoid the door(s) being used at the same time from both sides. The potentially heavy use that the door(s) could endure should be adequately protected against. This is normally achieved by using rigidly constructed doors and suitable ironmongery, including:

- push plates;
- kick plates;
- bump protectors.

Double action doors are hung using either 'helical butt hinges' (commonly referred to as 'bomber' hinges) or 'floor /transom springs'. Both methods provide pivot points for the door to operate in both directions, while a spring mechanism concealed in the ironmongery allows the door to return to the closed position after use.

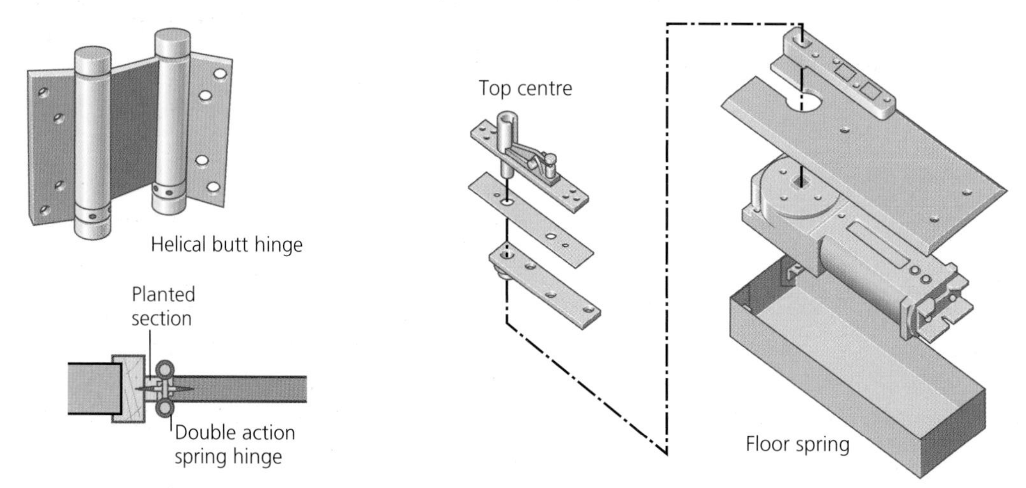

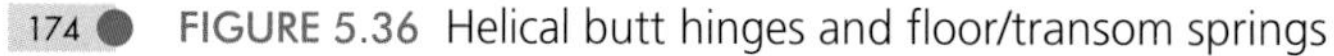

FIGURE 5.36 Helical butt hinges and floor/transom springs

SLIDING DOORS

Sliding doors are commonly used on wardrobes, kitchen units and areas where there is a lack of space. They are beneficial because they do not need the full width of the door(s) to swing or pivot into the room, therefore saving space.

FOLDING DOORS

The term 'folding' refers to door openings that are constructed in two or more parts. They are sometimes preferred to sliding doors, because they retract the full width of the door frame, allowing good access through the opening. There are two methods of pivoting folding doors:

1. side folding;
2. centre folding.

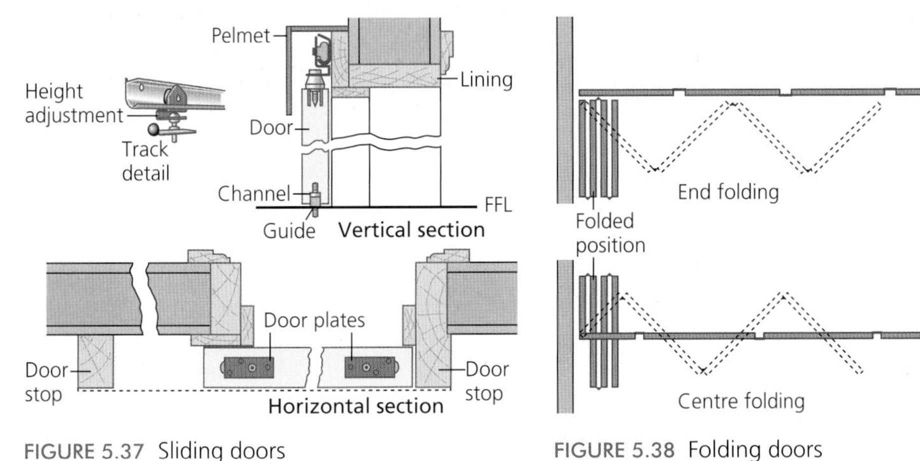

FIGURE 5.37 Sliding doors

FIGURE 5.38 Folding doors

REVOLVING DOORS

These are usually found in the lobbies of large hotels, shops or offices. They permit a steady flow of traffic both in and out of a building without ever creating a direct draught. Revolving doors are rarely fitted by carpenters because of the complexity of the assembly; they are normally installed by specialist companies.

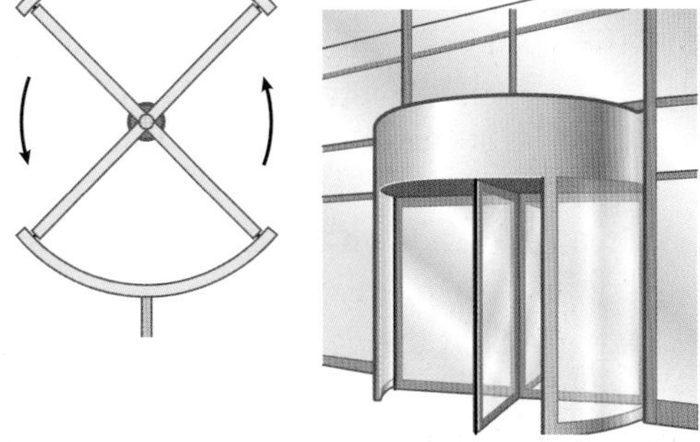

FIGURE 5.39 Four-compartment revolving doors

INSTALLING SINGLE ACTION SIDE-HUNG DOORS

The process of fitting either an internal or an external single action side-hung door will follow the same methods of installation. Generally, doors should be suited to fit their intended lining or frame, and heavy planing and cutting down of the door should be avoided. Reducing doors in their width by cutting along the length of the stiles could seriously weaken the structure of some doors and may result in the door twisting, dropping or falling apart at the joints. Heavily reducing the width of framed doors may lead to the stiles being under size, which could lead to some locks/latches or items of ironmongery being unsuitable.

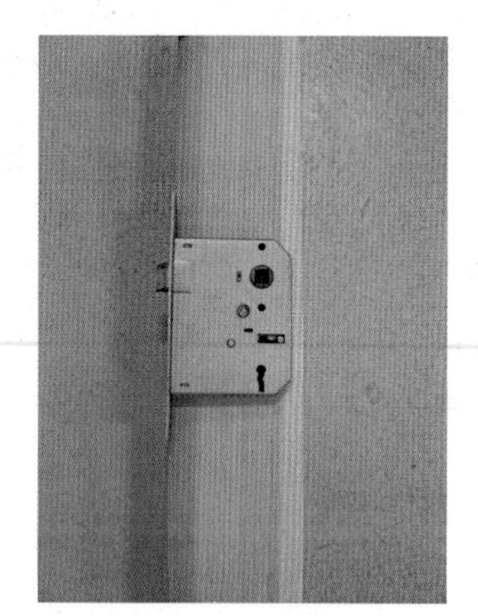

FIGURE 5.40 A reduced stile is unsuitable for a mortise lock

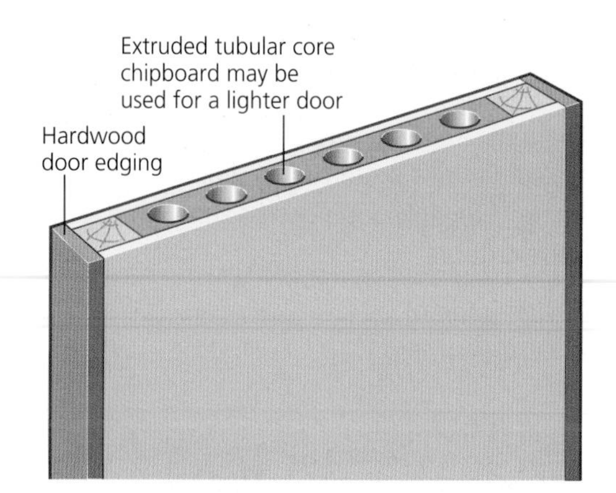

FIGURE 5.41 Re-edging a hollow core door

Care should be taken when reducing hollow core doors in both their width and length to ensure that they still maintain their stability after cutting. If the inner lining of a hollow core door is removed through the reduction of a door, then it must be replaced to strengthen the structure and allow strong fixing points for ironmongery.

STAGES OF INSTALLATION

- Stage 1 – Refer to the architect's drawings and door schedule to establish the door's specifications, its intended position within the building and the hinged side.
- Stage 2 – Check the dimensions of the door against the door opening and clearly identify the lock block on hollow core doors. (*Note*: The door should ideally be slightly under the opening size to prevent excessive planing and make the installation easier.)
- Stage 3 – Framed doors will require the projection of the stiles above and below the top and bottom rails to be removed with a handsaw; these areas of waste timber are known as the 'horns'. The horns should remain on the door until it is ready to be fitted; this will prevent potential damage to the top and bottom rails while they are stored in a vertical position.
- Stage 4 – Offer the door into the frame, reducing the width and height where necessary with a 'jack plane'. If the door requires more than a few millimetres to be removed, then a portable electric planer could be used to speed up this process.
- Stage 5 – Shift the door over to the hanging side of the frame, tight up against the head. Check along the joint between the edge of the door and the frame to make sure that there are no visible gaps. Any gaps at this stage should be reduced by planning the points where the door edge comes into contact with the frame.
- Stage 6 – The door should now be reduced along the other two edges (lock/latch side and the bottom) so that a gap is equal to twice the intended margin around the door once completed. It is common practice to leave a gap of 2 mm around the outside edge of a door, although in some situations this may be marginally increased to allow for thicker paint finishes in exterior positions. A bigger margin will also be allowed where

doors are subject to continued expansion and contraction due to changes in the weather (e.g. front/back or shed doors).

- Stage 7 – Remove the leading edge from the door to prevent binding as it closes into the frame. Remove all the sharp edges (the 'arrises') with a plane or piece of abrasive paper wrapped around a sanding block. This will improve the adhesion of the applied paint finish on the corners, and give a better feel to the door when in use.

FIGURE 5.42 Removing horns

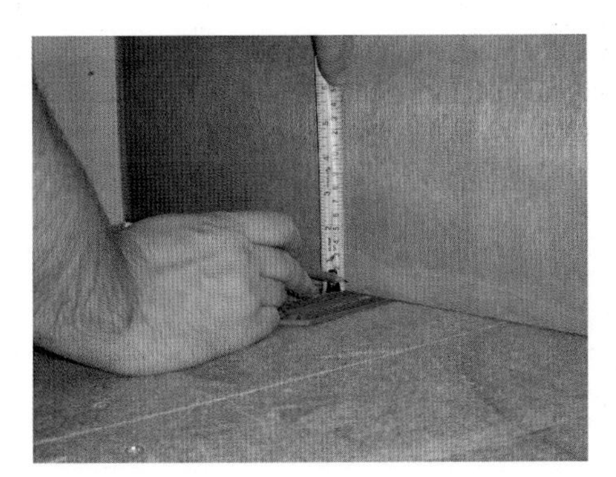

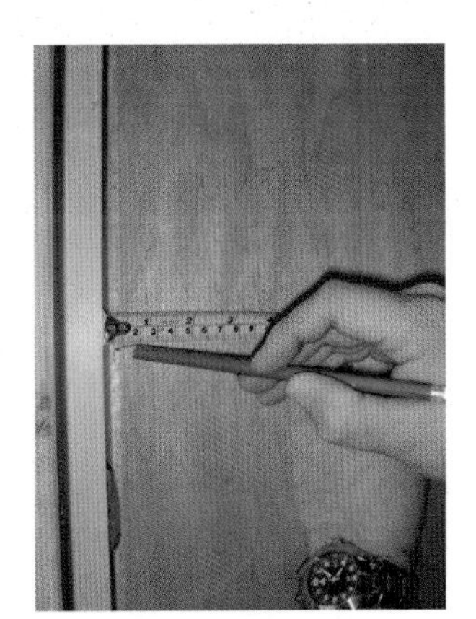

FIGURE 5.43 Marking twice the margin along the bottom and latch side of the door

- Stage 8 – Mark the position of the top butt hinge 150 mm from the top of the door and the bottom hinge 225 mm up to the underside of the bottom butt hinge (further hinges may be marked out at this stage for heavier doors, fire doors and exterior doors, etc.). Use one butt hinge to mark its length against the hinge positions indicated along the edge of the door. Try to avoid measuring the length of the hinge and transferring this measurement as this could lead to mistakes and is less accurate. The depth and width of the recesses needed to house the butt hinge leaves need to be determined by the following steps.

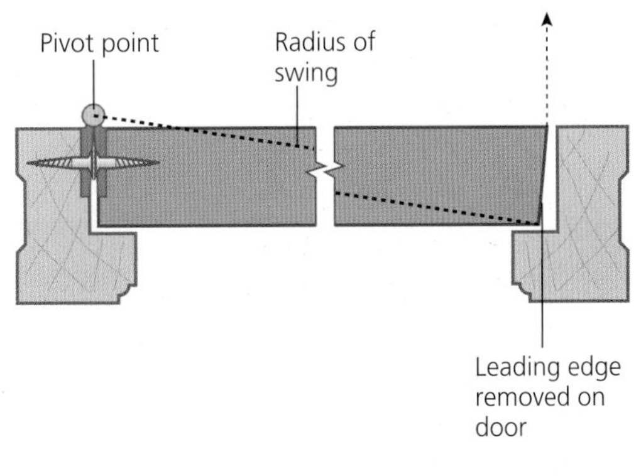

FIGURE 5.44 Removal of the leading edge

1. Adjust a marking gauge to the thickness of one hinge leaf and transfer this setting onto the face of the door, between the markings on the door edge. Recessing the hinge leaves at this depth will determine the gap remaining between the door and the frame once it is in its closed position. Closing the two leaves of the butt hinge together

so that they are parallel to each other will indicate the finished joint between the edge of the door and the frame if the leaves are recessed flush. As this may show that the joint will be too big when completed, the marking gauge will have to be set slightly deeper so that the hinge leaves will sit just below the surface of the door edge.

2. Adjust a second marking gauge to the width of the butt hinge leaf. This should then be transferred onto the door edge, between the marking out towards the top and bottom of the hinged side. This setting will determine the amount of the hinge knuckle projecting beyond the face of the door and frame when the door is in a closed position. Care should be taken when transferring these settings onto the door and frame to ensure that the face of the door sits flush within the frame. Careless marking out at this stage could result in either the door projecting beyond the face of the door frame or being set too deep within the frame, which could lead to binding.

FIGURE 5.45 Adjusting a marking gauge to one hinge leaf thickness

FIGURE 5.46 Adjusting a second marking gauge to the width of the butt hinge leaf

- Stage 9 – Use a bevel-edged chisel to remove the waste material from the recesses marked along the door edge. Check the hinge leaves fit neatly to the correct depth within each of the recesses. Once this has been achieved, drill a small pilot hole through each hole in the hinge leaf to allow the fixing screws to locate centrally within the recessed area, and then screw the hinges into position.

FREQUENTLY ASKED QUESTIONS

I find it difficult to see the lines when I use a marking gauge. I find it better to run my pencil through the gauge lines so that I can clearly see my markings. Do you think this is a good idea?

No, running a pencil through the lines will result in thicker marking-out lines and reduce the accuracy of the marking out. If you cannot see the gauge lines on the timber keep repeating the process of marking the timber with the marking gauge with slightly more weight. *Remember* – the marking gauge lines are only for marking out, not for cutting the timber.

- Stage 10 – Lift the door back into the frame. Use packers or wedges to raise the door along the bottom edge until a gap of 2 mm is achieved between the head of the frame and the top of the door. Use a pencil to transfer the positions of the top and bottom hinges across onto the door frame. Remove the door from the frame and repeat the process described in Stage 8, points (1) and (2), to complete the marking out of the butt hinges and recess the hinges below the surface of the frame. (*Note*: the marking gauges previously set for the marking out of the door edge should be reused at this stage.)
- Stage 11 – Locate the door in front of the frame in an open position. Align the hinges on the door with the recesses on the frame and secure the hinges with the appropriate screws. Check the door has installed correctly by opening and closing it in its frame. (*Note*: the underside of the door should be approximately 6 mm clear of the finished floor covering (e.g. carpet, laminate). This margin should be considered when cutting the door to the correct height to avoid having to refit the door at a later stage when the floor finish has been installed.)

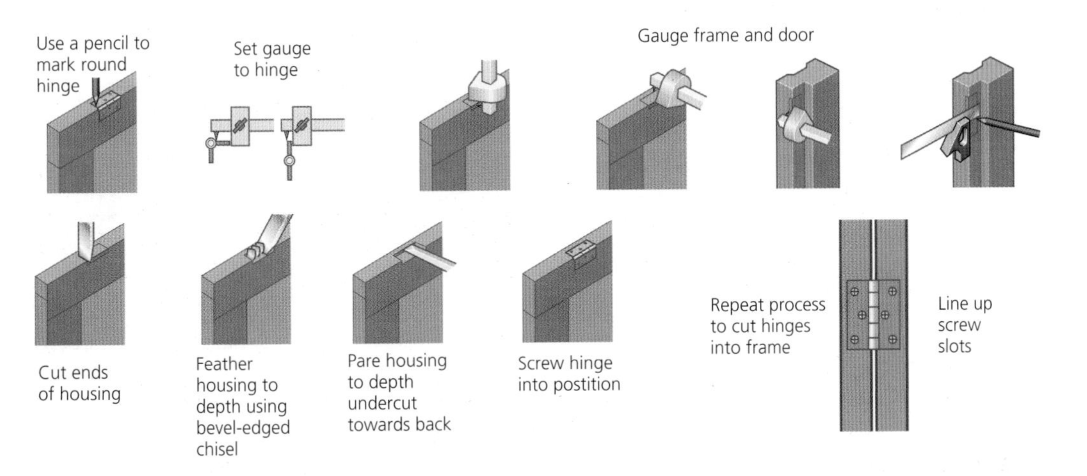

FIGURE 5.47 Cutting recesses for butt hinges

FIGURE 5.48 Ensure the bottom of the hinge recess is flat

TRADE SECRETS

When recessing the hinges into the edge of the door and frame, use the edge of the hinge leaf to check that the recess is flat.

It can sometimes be difficult to hold a door steady to plane its edge and cut the recesses for the hinges while on site. Some purpose-made 'saw horses' have a 'V'-shaped cut out on one end. This will allow the edge of a door to wedge into the saw horse and hold it temporarily. Alternatively, a block of timber (known as a 'chock') with a housing cut into it to suit the thickness of the door, plus the space for a loose wedge, can be used.

FIGURE 5.49 V-shaped cut out in a saw horse

FIGURE 5.50 Timber chock and wedge

ALTERNATIVE METHODS OF RECESSING BUTT HINGES

Side-hung doors are usually installed using a variety of hand tools, including a jack plane/electric power planer, and an assortment of chisels and marking gauges. Alternatively, purpose-made jigs can be used in conjunction with a router to remove the hinge recesses. Although it will take time to produce any jigs required, this time is not wasted and will be caught up as the use of the jig increases productivity. As the router is guided by the shape of the jig, it will produce a rounded internal corner in the recess. This can be squared out with either a sharp wood chisel or a 'corner chisel' (corner chisels are specialist tools designed to square out the shallow recesses remaining after the use of a router and jig). Some manufacturers have designed various hinges with radius corners to eliminate the need for squaring out after routing.

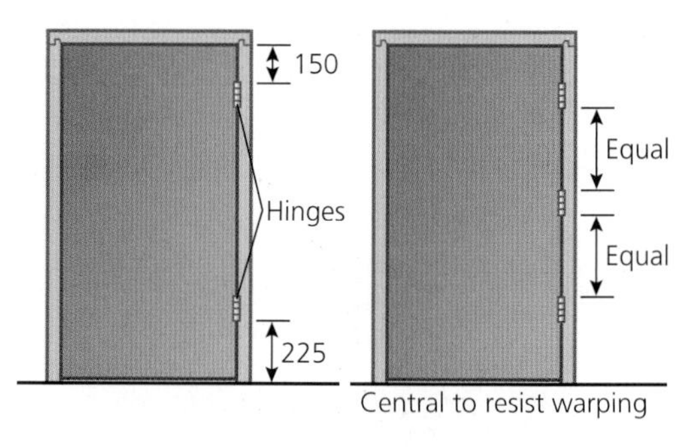

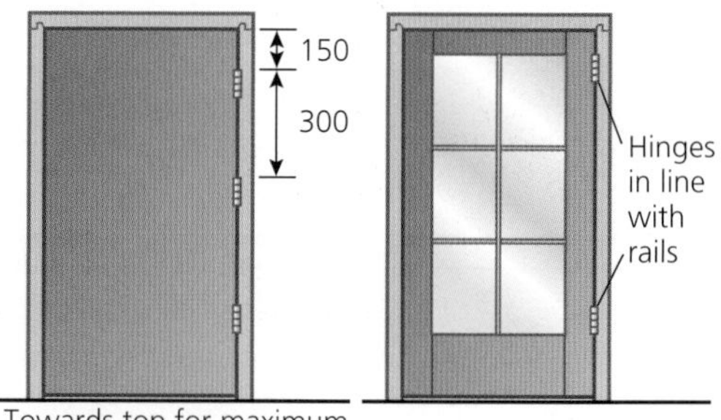

FIGURE 5.51 A completed side-hung door

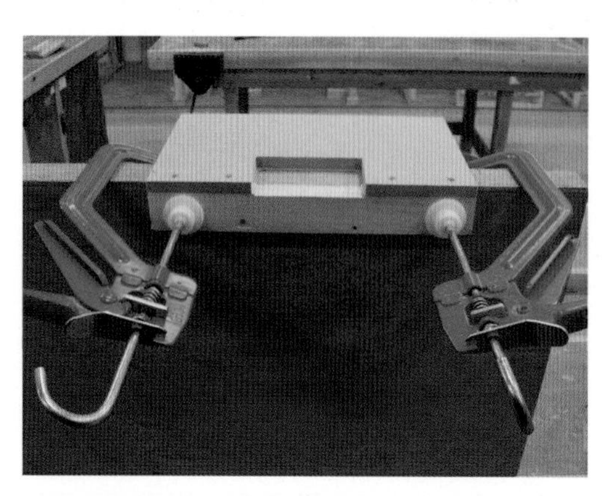

FIGURE 5.52

FIGURE 5.53

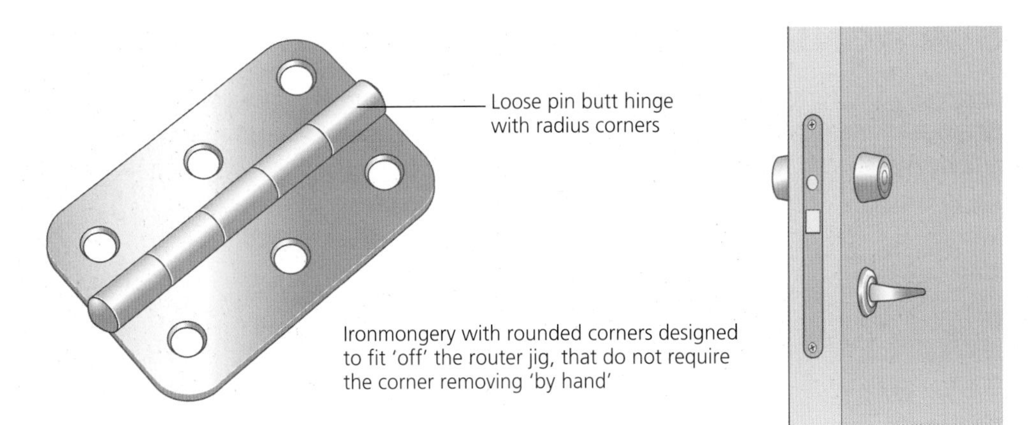

FIGURE 5.54 Door iron mongery with radius corners

Manufactured, lightweight, high-pressure laminate jigs are becoming increasingly popular with carpenters to aid the installation of hinges, locks and latches in timber doors and frames. Each jig is easily adapted to suit a wide range of different standard-size hinges, locks and latches.

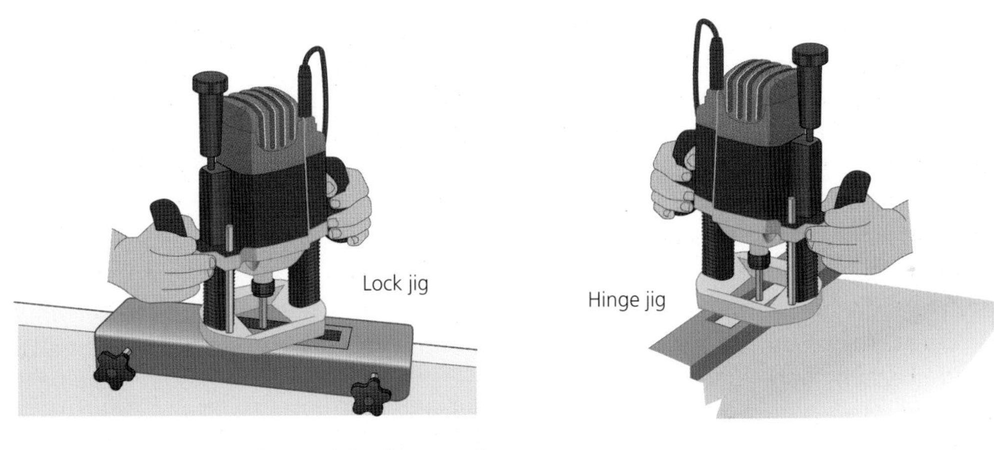

FIGURE 5.55 Using manufactured jigs for recessing

A correctly fitted door should:

- be free from twist;
- be flush with the surface of the door frame;
- not bind as it closes into its frame;
- not be 'hinge bound';
- have parallel joints not exceeding 3 mm between the door and the edge of the frame;
- have all the arrises removed.

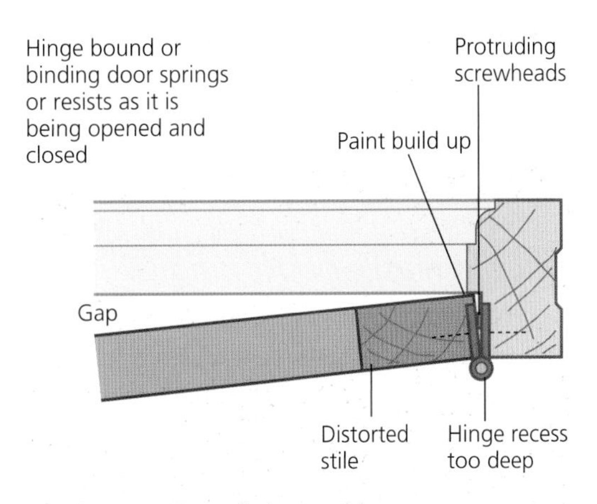

FIGURE 5.56 Door fitting problems

DOOR IRONMONGERY

Ironmongery and furniture are readily available for internal and external side-hung doors in a wide range of materials and finishes. Careful consideration should be given to the type of ironmongery selected for doors in damp and exterior positions to ensure they continue to function correctly without deteriorating prematurely when exposed to moisture.

TYPES OF DOOR HINGE

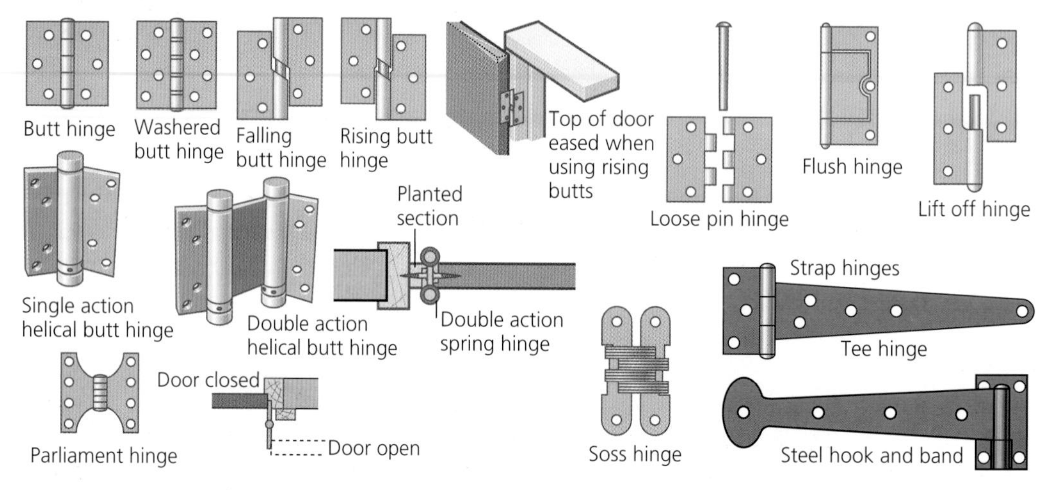

FIGURE 5.57 Types of door hinge

Ball bearing butt hinges

These hinges are generally used for heavier doors such as half-an-hour and one hour fire check doors. They consist of a set of small ball bearings encased in the knuckle of the butt hinge. Ball bearing butt hinges will operate almost silently and allow the smooth movement of the door, which should prolong the life of the hinges.

Butt hinges

Butt hinges are available in sizes ranging from 25 to 100 mm in length, and in a variety of different widths to suit different situations. The leaves on butt hinges are recessed into the door and frame equally, with a margin between to allow the door to operate without binding. Standard butt hinges are generally used for lightweight doors. Heavier doors may lead to the hinges grinding and wearing, which may affect the margin along from the hinge side.

Double washer butt hinges

Double washer butt hinges are similar to standard butt hinges, the main difference being the double washers between each joint in the knuckle of the hinge. The unique design of the washer butt hinges allows the smooth operation of the door and protects against knuckle wear.

Falling butt hinges

Falling butt hinges are used where the door will automatically fall open when not secured. These are commonly used on public toilet doors to inform users that the cubicle is not in use. They are purchased handed to suit.

Flush hinges

Flush hinges are generally used for hollow core or lightweight doors. They are simple and quick to install because no recessing is required into the door or frame. They are accurately set out by transferring the centre point of the hinge from the door onto the frame, and aligning the knuckle of the hinge against the face of the door.

Half-and-half method used on high quality work

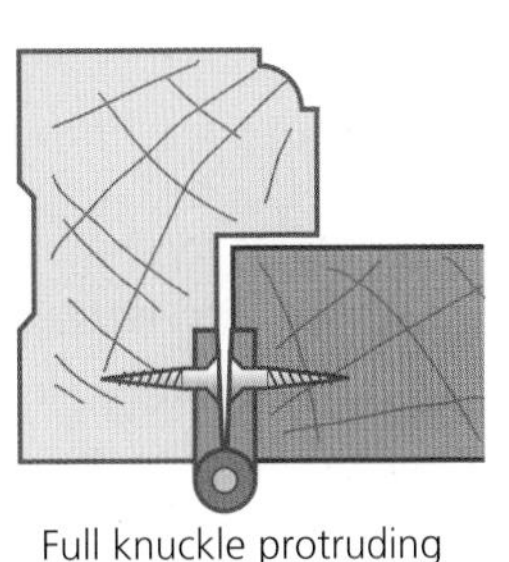

FIGURE 5.58 Flush hinge

Lift-off hinges

These types of hinge are normally used by shopfitters and partitioning manufacturers. The hinges are split into two parts. The lower section has a fixed pin that is usually fitted to the door frame to allow the upper part of the hinge that is fitted to the door to be lifted over during the installation stage. 'Lift-off' hinges are 'handed' to suit right- and left-side-hung doors.

Loose pin hinges

Removal of the loose pin driven through the 'knuckle' of the hinge allows the disconnection of the two parts of the hinge and the removal of the door without the need to unscrew the fixings. These lightweight hinges are used internally for the quick installation and replacement of doors.

Parliament hinges

The projection of the pivot point (the 'knuckle') allows doors to swing 180° without binding on large architectural mouldings, such as architraves and plinth blocks. Parliament hinges are available in a range of different sizes, offering different projections from the face of the door. Door closers should be avoided with doors hung with this type of hinge because the large margin created behind between the door and frame as the door opens could be a potential hazard.

Rising/butt hinges

These hinges are used to clear the surface of deep-pile carpets and uneven floor surfaces. They are handed to suit both right- and left-side-hung doors. The rising action of the hinge as it opens will result in the door 'self-closing' under its own weight, which will maintain privacy in the room and prevent draughts. The top corner on the hinged side of the door will have to be removed to prevent the door binding against the head of the frame, as the door rises as it swings open.

Single and double action hinges

These types of hinge are sprung loaded, which means that they will return to their original closed position after opening. Double action hinges are sometimes preferred to other types of door closer, including floor springs and transom door closers. Doors hung with double action hinges require the leading edge to be removed from both sides of the door to prevent binding as it swings inwards and outwards. This is achieved with rounded surfaces along the leading edge and the hinged side of the door. Double action hinges are commonly used in commercial kitchens, restaurants and public houses.

Soss hinges

Soss hinges are high-quality hinges commonly used by cabinet makers and shopfitters. They are secured into housings located between the edge of the door and frame to conceal their appearance once the door is in the closed position.

Strap hinges

There are several different types of hinge that fall into this category; they include 'T' hinges and 'hook and band hinges'. 'T' hinges are generally used for ledged and braced shed doors and garden gates. Hook and band hinges are usually made from thicker-gauge metal so they are capable of supporting heavier timber garage doors, framed ledged and braced doors, and large gates.

LOCKS AND LATCHES

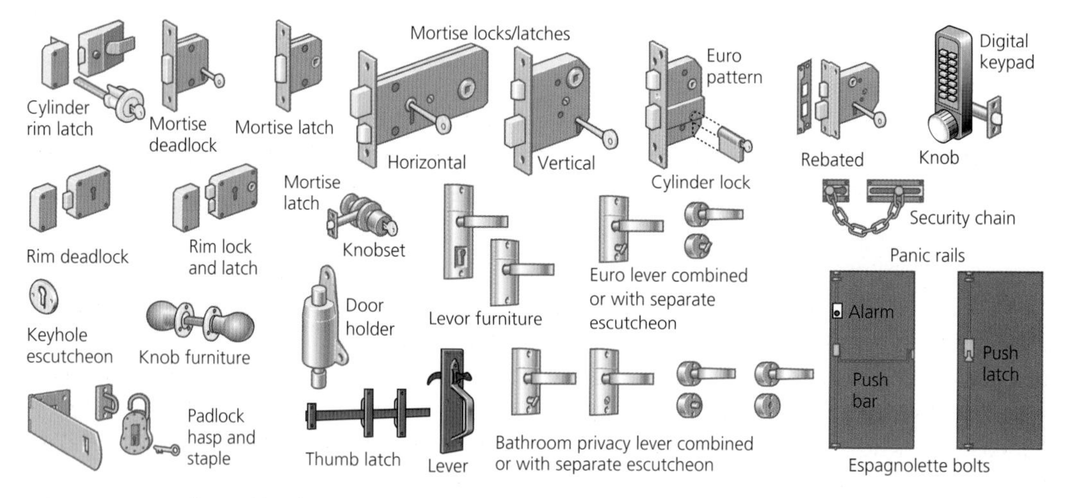

FIGURE 5.59 Locks and latches

Cylinder night latches

Cylinder night latches are commonly used to secure inward-opening front doors. Some insurance companies insist that the latch alone does not provide a suitable level of security, and require an additional mortise deadlock. Some cylinder night latches incorporate a deadlock to satisfy the standards required by insurance companies and the recommendations of the police.

Digital key pads

Digital key pads are used in conjunction with mortise latches to allow keyless entry through doorways. A number combination is required to allow the latch mechanism to release, permitting access by turning either a knob or handle on the body of the pad. The combination of numbers can be reprogrammed into the unit at any time to maintain security.

Escutcheons

Escutcheons are used to cover keyholes cut into the face of timber doors and prevent damage from the key. Some key escutcheons are very discreet with a loose cover that swings either side of the keyhole to give the opening some weather protection and a decorative feature.

Euro locks

Euro locks provide high security with hardened anti-drill steel along the length of a cylinder that operates the deadlock within the mortise lock.

Mortise deadlocks

Mortise deadlocks are available with between two and seven levers, depending on the level of security required. Five levers are normally recommended as the minimum requirement by insurance companies and the police for exterior doors. Any locks fitted on exterior doors must be certified to meet with British Standard BS3621-2004.

Mortise latches

Mortise latches are located in the edge of internal doors to hold them in a closed position as they engage with the striking plate; they do not offer any security. Mortise latches are normally operated with a pair of lever handles or a pair of door knobs fitted on either side of the door.

Mortise lock/latch

A mortise lock/latch is a combination of a mortise deadlock and a mortise latch. They are used to retain side-hung doors in the closed position as well as securing them. The latch part of the lock is shaped with a 'lead' on one side; this allows it to ease into the striking plate as it closes. The location of the lead on the latch will determine the 'handing' of the lock. The handing of the lock can be changed to suit both right- and left-handed doors, although the methods used to achieve this will vary between different manufacturers.

Rebated mortise lock/latch

These are used on pairs of doors where the meeting stiles/edges are rebated together. Rebated door kits are also available to adapt conventional mortise lock/latches to enable them fit over the rebates along the door's edges.

Roller catches

Roller catches are commonly used on lightweight internal doors, such as wardrobe doors. The roller ball projecting from the face of the catch can normally be adjusted to suit the margin between the door and the frame, and allow the door to glide over the striking plate as it opens and closes.

Thumb latches

Thumb latches (also known as 'Norfolk' or 'Suffolk' latches) are surface fixed on the faces of ledged and braced matchboarded doors, such as garden gates, and are also used on internal doors found in traditional-style cottages.

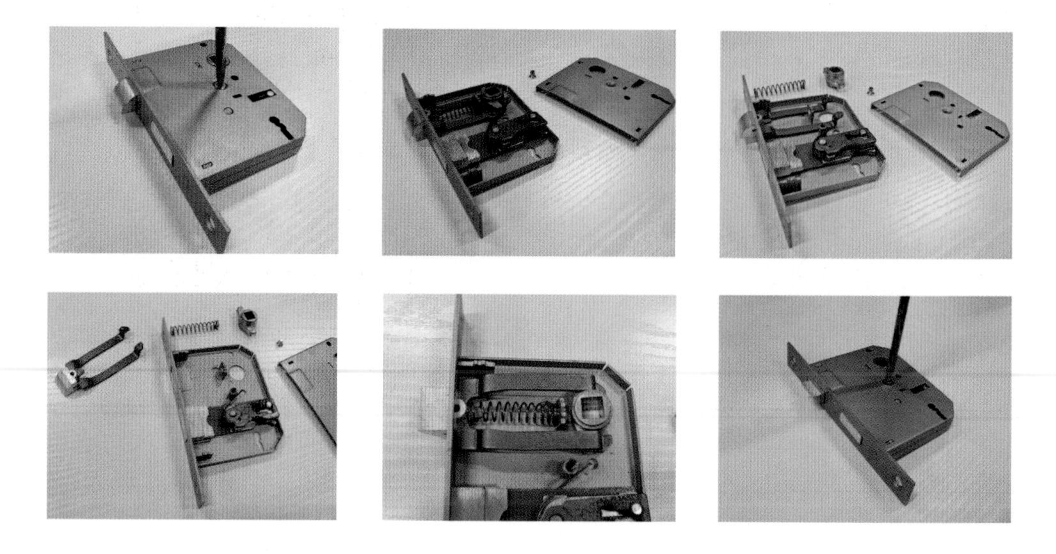

FIGURE 5.60 Mortise locks/latches

Traditional rim locks

Traditional rim locks are available as lock/latches and deadlocks. They are traditionally used to secure matchboard ledged and braced doors because the thickness of these doors prevents the use of other types of mortise lock/latches.

BOLTS

Barrel and tower bolts

Barrel and tower bolts are used for external gates and garage doors, either as a secondary form of security or to secure one half of a pair of doors. If the bolts are used to secure full-height double doors, they should be used as a pair, one at the top and one at the bottom.

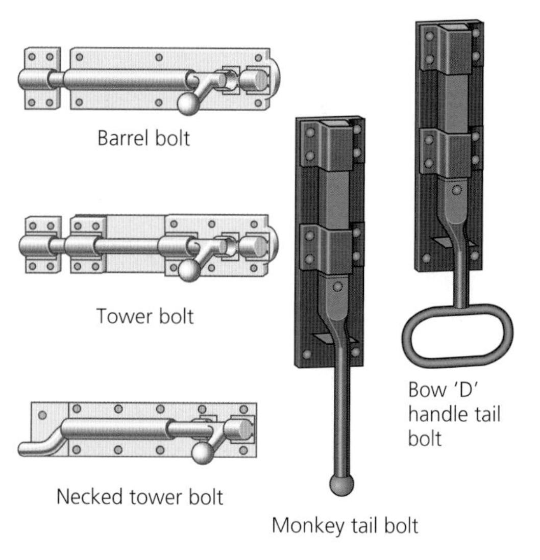

FIGURE 5.61 Bolts

Bow 'D' handle bolts

These are commonly used to secure garage doors. The 'D'-shaped handle allows easy operation of the heavyweight bolts.

Flush bolts

Flush bolts are fitted flush to the surface of side-hung doors (on a single door), or on the top and bottom edge of one half of a pair of doors. They provide a discreet appearance that is pleasing to the eye.

Hinge bolts

Hinge bolts are located on the hinged side of the door, in the joint between its edge and the frame. They are installed to improve security and prevent the door being levered and forced away from its hinges.

Monkey tail bolts

Monkey tail bolts are very similar to bow 'D' handle bolts except that they have a monkey tail-shaped handle. The extended length of the monkey tail bolts makes them ideal for securing the top half of taller gates and exterior doors.

FIGURE 5.62 Flush bolt

Necked bolts

Necked bolts are used to secure one half of a pair of double doors. The end of the bolt has a cranked section to allow it to be fitted on the rear face of doors located in rebated frames.

Panic bolts and latches (espangolette)

These are used to secure single and double fire doors. Panic bolts have a double-point locking system, one at the top and one at the bottom of the door, providing increased security. Once closed, the bolts are released by depressing the panic bar running across the middle of the fire door, which allows easy exit from the building in the event of an emergency. 'Single-point panic latches' are also available, providing a cheaper, lower-level security device, with reduced installation time. These are operated in exactly the same way as the double-point locking bolts.

Rack bolts

Rack bolts are used for a number of different applications, including additional security for single side-hung, sliding and double doors. They are operated either with a fluted key or with the use of a thumb turn mounted on the face of the door. Rack bolts are normally used in pairs at the top and bottom of the door leaves to provide the best results.

DOOR CLOSERS

Door closers are available in a number of different forms, ranging from a simple 'gate spring' to 'overhead closers'. Their purpose is to return doors to their original closed position. This is especially important on fire doors, because they are installed to prevent the spread of fire and smoke through doorways. Some door closers are fitted with heat sensors that automatically close the door in the event of a fire. The most commonly used door closers are those described below.

- Gate springs – as the name suggests, these are generally used on garden gates to provide security for young children and pets.

- Overhead door closers – these are usually used for heavier doors to control the speed at which the doors close. There are three different types of overhead door closers; each one is designed so the body fits either on the frame or the top of an inward-opening door. Alternative closers are available to be fixed to the head of the frame, allowing the doors to swing outwards. These are used to prevent the door closer being exposed to the elements and potential damage. Overhead door closers are available in a range of different strength capacities to suit different door sizes, although most of them can be finely adjusted. British Standard BS EN 1154 classifies door closers into seven power sizes and grades them 1–7, 7 being the largest capacity.
- Concealed spring closer – this is simply a cylinder containing a spring, which is housed into the door edge with a chain attached to an anchor plate, which is then fixed to the frame. These types of closer are particularly useful to conceal the appearance of any door control. The tension on the spring can be adjusted to suit different-size doors and allow the doors to engage the lock or latch correctly. This can be achieved by sliding the metal holding plate through the chain close to the cylinder and unscrewing the anchor plate from the edge of the door. The tension on the spring can be increased by twisting the anchor plate in a clockwise direction; this will result in the door closing quicker. Turning it in an anticlockwise direction will have the opposite effect.
- Roller arm closers – these are face fixed to the frame of side-hung interior doors up to approximately 30 kg. They are easily fixed and adjusted, and provide a cheaper alternative to other door closers mentioned previously. Roller arm closers are normally fitted in the middle of the hinged side of the door; if there is a risk of children using the door, it should be positioned at a higher level.

FINGER PLATES (ALSO KNOWN AS 'PUSH PLATES') AND KICKING PLATES

Finger and kicking plates are fitted to doors in vulnerable locations such as schools, hotels and hospitals. Kicking plates are installed to protect the lower faces of the doors from general wear and tear caused at floor level. Finger plates are usually positioned either directly behind pull handles or adjacent to this height.

INSTALLING DOOR FURNITURE

It is strongly recommended that, before installing any ironmongery or door furniture, the following documents are referred to:

- the door schedule – will provide the exact details of all the items needed for each individual door;
- the specification – will give the correct fixing positions for the door ironmongery;
- the manufacturer's fitting instructions – the guidance provided in these instructions will prevent damage to the furniture during fitting and allow the items to be correctly fitted; incorrectly installed ironmongery will invalidate the guarantee on the items and potentially cause damage to the door and frame.

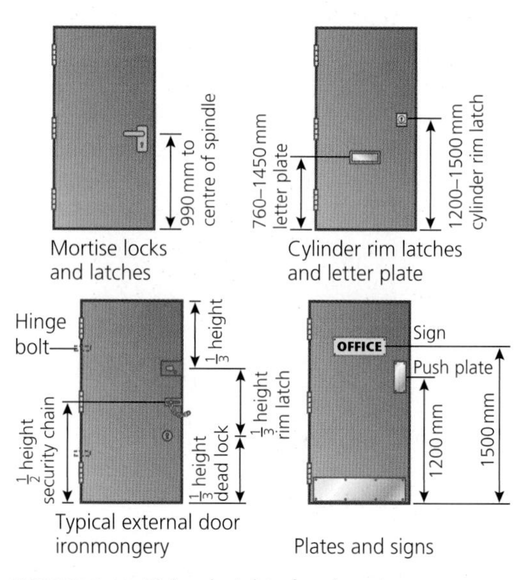

FIGURE 5.63 Fixing heights for door ironmongery

INSTALLING A MORTISE SASH LOCK/LATCH

- Stage 1 – Open the side-hung door and secure it temporarily by sliding a pair of timber folding wedges underneath its bottom edge.
- Stage 2 – Mark the height of the mortise lock on the edge of the door. Generally mortise locks are positioned 990 mm from the bottom edge of the door to the centre of the spindle. Align the lock with the centre point and mark the width of the body of the lock. This will be the width of the mortise required to house the lock, so a few millimetres should be allowed either side of these marks to allow the lock to slide easily into position.

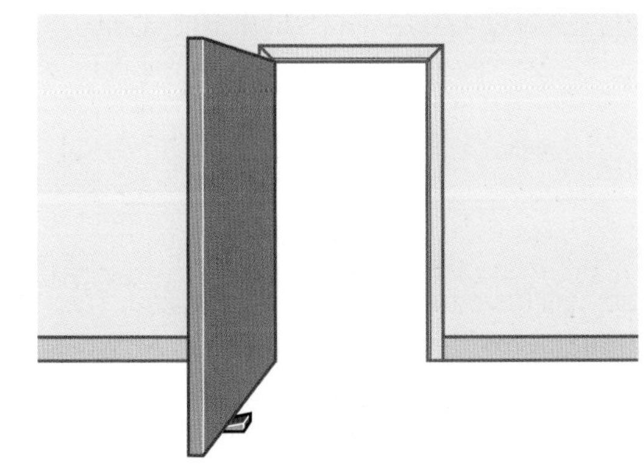

FIGURE 5.64 Stage 1 – Wedge door in place, secure with one screw per hinge

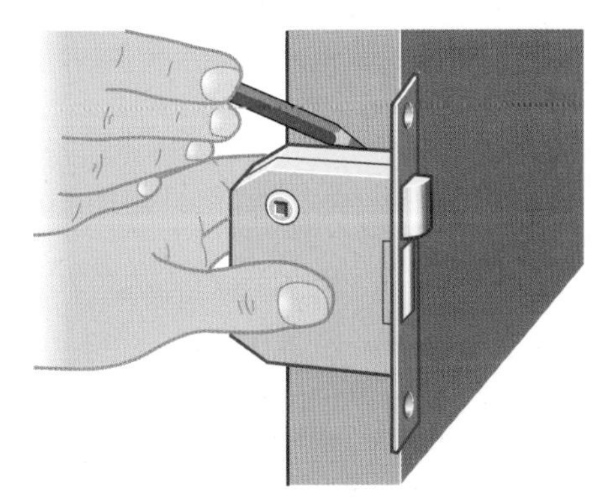

FIGURE 5.65 Stage 2 – Mark position on door edge

- Stage 3 – Use a marking gauge from either side of the door to determine the centre of the door, and mark a line along the edge of the door between the two points previously marked.
- Stage 4 – A series of holes will need to be drilled along the gauge line and between the lines drawn to establish the width of the mortise. This is achieved by measuring the thickness of the mortise lock body and using either an 'auger' or 'spade' drill bit a few millimetres bigger to remove the waste material. The holes should be drilled slightly deeper than the depth of the mortise lock. This can be controlled with the use of the depth stop on the power drill; alternatively, a piece of masking tape can be wrapped around the drill bit to mark the maximum depth.

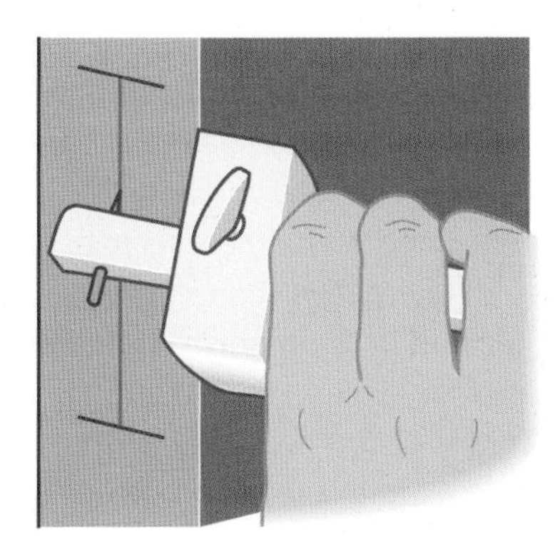

FIGURE 5.66 Stage 3 – Gauge centre line on door edge

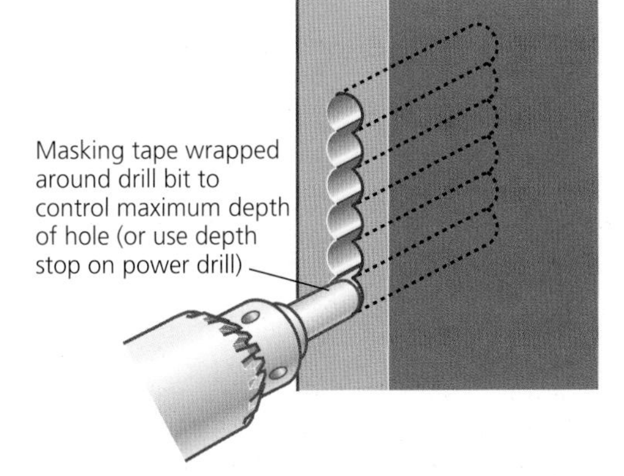

FIGURE 5.67 Stage 4 – Drill out to width and depth of lock

- Stage 5 – Use a sharp chisel to complete the mortise, by removing the waste material between the markings on the door edge. Offer the lock into the mortise, making sure that it slides easily into position without forcing or hitting it, as this will cause damage to the face plate.
- Stage 6 – Make sure the lock is centrally aligned in the mortise, and then use a sharp pencil to draw around the face plate. Remove the lock from the mortise and recess out the area to the thickness of the face plate with a chisel. Be careful not to damage the edges of the recess when removing the lock from the mortise after a trial fit.

TRADE SECRETS

'Cleaning out' the mortises required for locks is probably the most time-consuming part of their installation.

When drilling a series of holes along the gauge line on the edge of the door, make sure that they overlap slightly. This process is known as 'stitch drilling' and is particularly useful to allow the easy removal of the waste material from the mortise.

- Stage 7 – At this stage the hole for the spindle to operate the handles, and the keyhole, will have to be marked and drilled. This is achieved by holding the lock on the face of the door with the face plate flush with the edge of the door, and positioned at the same height as the recess. The position of the spindle is marked with a pencil through the top hole in the lock. It is difficult to mark the keyhole with a pencil because the hole in the lock is too small, so this is normally marked with a 'bradawl'.

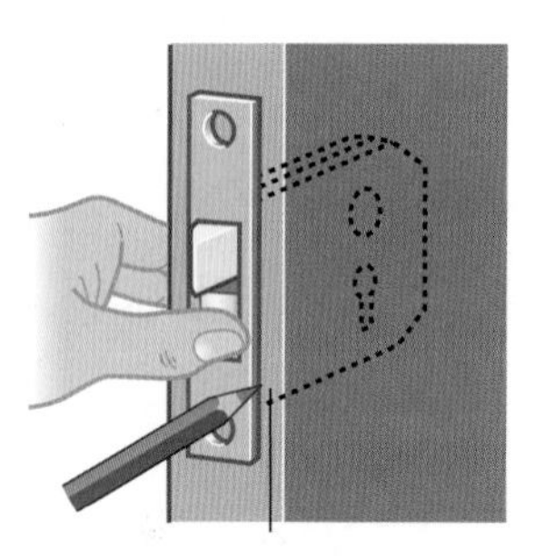

FIGURE 5.68 Stage 6 – Mark lock face plate

FIGURE 5.69 Stage 7

- Stage 8 – Use a 16 mm drill bit to drill the hole required for the spindle, working from both sides to avoid any damage to the face of the door. The keyhole can also be cut at this stage. This is achieved by drilling a 10 mm hole at the top of the keyhole position and a secondary 6 mm hole directly below. Use a chisel to remove the remaining waste material and complete the keyholes on each side of the door. (*Note*: an alternative sequence of drilling the spindle and keyholes prior to mortising prevents breakout in the mortise – splitting of the timber as the drill bit exits the opposite face to that being drilled.)
- Stage 9 – Place the mortise lock/latch back into position and secure it in place with the screws provided. Some mortise locks have a separate face plate to the main body of the lock; this prevents surface damage to the decorative face plate during the installation. These are secured to the body of the lock with short 'grub' screws after the lock has been fixed.

FREQUENTLY ASKED QUESTIONS

Why can't you drill one large keyhole instead of cutting a keyhole shape in the door?

Although the door handles may cover the holes cut for the key, an oversized hole will cause difficulty when trying to align the key in the lock. A neatly cut keyhole will assist the alignment of the key as it glides into the lock.

- Stage 10 – Slide the spindle through the mortise lock so that it projects equally each side of the door, and place the handles over the spindle before fixing in position. The metal spindle may have to be adjusted in length to suit the thickness of the door. Check the handles are correctly aligned, operating properly and the key freely opens and closes the deadlock.
- Stage 11 – The next stage involves fitting the 'striking plate' to the door frame/lining. Swing the door towards its closed position until the latch and open deadlock touch the edge of the door frame. Transfer the position of the lock and latch onto the edge of the frame with a pencil, and square these lines around to the face of the frame.
- Stage 12 – Measure the distance between the edge of the latch and the outer face of the door, and transfer this size onto the frame between the pencil lines previously marked.

FIGURE 5.70

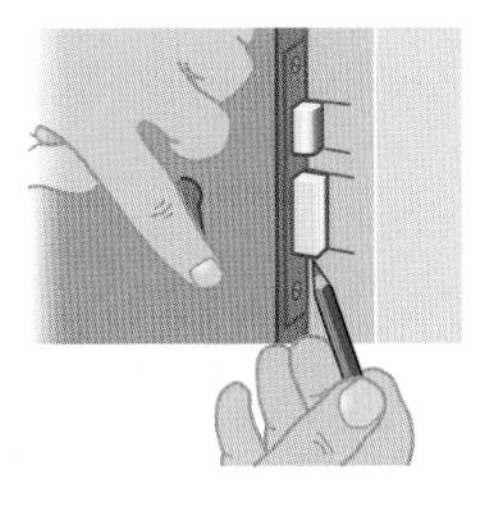
FIGURE 5.71 Stage 11 – Mark bolt position on frame/lining

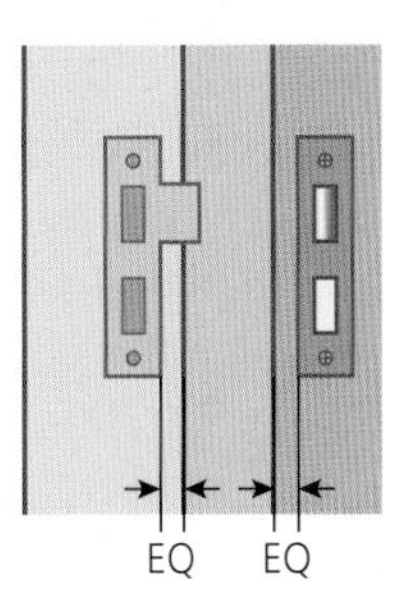

FIGURE 5.72 Stage 13 – Mark position of striking plate

- Stage 13 – Place the striking plate face down, over the marking out on the door frame. Align the cut front edge of the rectangular holes in the striking plate with the vertical line marked on the door frame. Draw a line around the inside and outside of the striking plate. Use a chisel to form the mortises required in the centre of the marking out on the face of the door frame.

- Stage 14 – Close the door into the frame, and double check that the deadlock engages the striking plate correctly. If it does not, the position of the mortise can be slightly adjusted at this stage, although the outline of the striking plate will also need to be realigned.
- Stage 15 – Recess the thickness of the striking plate into the face of the door frame. The leading edge of the recess may have to be removed to allow the striking plate to sit flush within the recess. Secure the plate to the frame using the screws provided. (*Note*: some striking plates have a box attached the rear side. This provides a neat finish to the mortise lock and prevents the need to leave a clean/smooth bottom to the mortise; it also improves security.)
- Stage 16 – Close the door into the frame, making sure that the latch and lock engage the striking plate correctly.

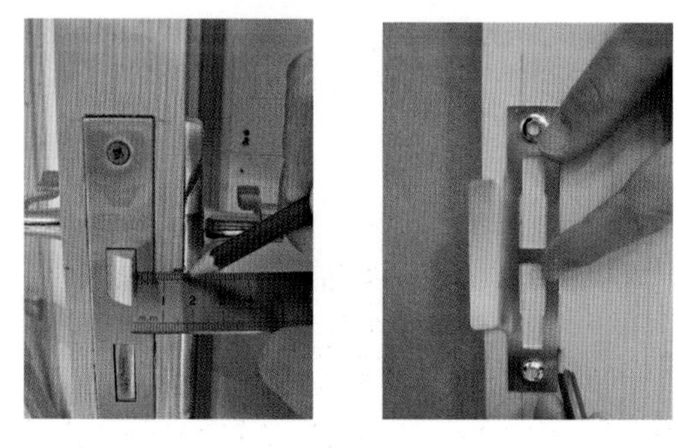

FIGURE 5.73 Mark out face plate

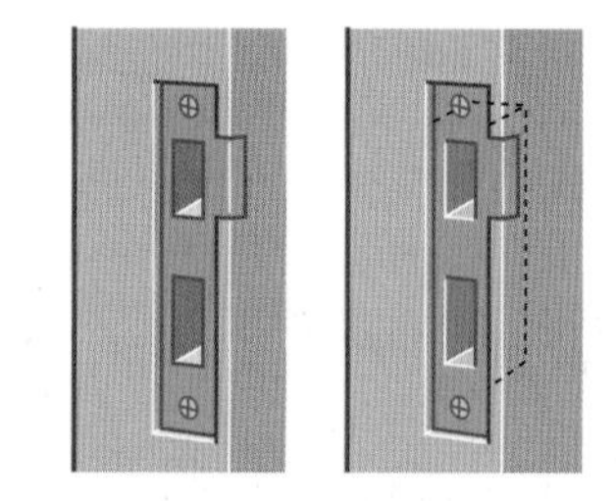

FIGURE 5.74 Let-in plate cut mortise for bolts

INSTALLING A CYLINDER NIGHT LATCH

The fixing height of cylinder night latches can vary between 1200 and 1500 mm to the centre of the cylinder. This position is determined by the height of the door and other locking devices fitted. The most common fixing height for a standard size 1981 mm door is 1500 mm.

- Stage 1 – Place folding wedges under the bottom edge of the door to hold it secure temporarily while fitting the lock.
- Stage 2 – Mark the height of the latch on one face of the door.
- Stage 3 – Determine the position of the centre of the cylinder to the edge of the latch and transfer this point onto the face of the door, to complete a cross. This point can be established with the use of the paper template supplied with the latch taped onto the face and folded over the edge of the door. Alternatively, if no template is available, a marking gauge can be adjusted to the same position and transferred onto the face of the door.
- Stage 4 – Use a 32 mm flat/centre drill bit to bore a hole through the centre point marked on the face of the door, working from both sides to prevent 'breakout' .

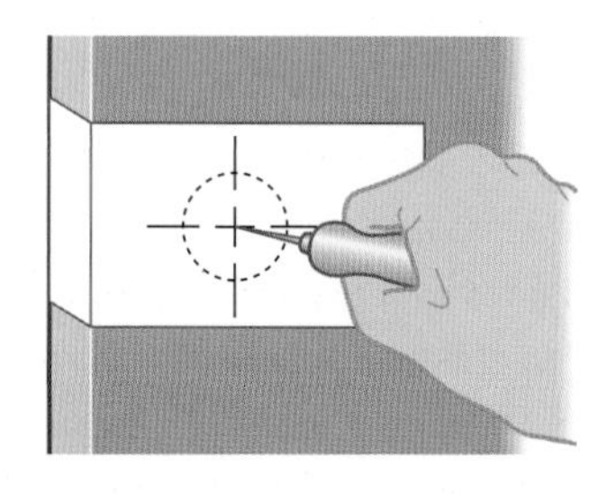

FIGURE 5.75 Stage 3 – Use template to mark centre of hole

FIGURE 5.76 Lock, backing plate and screws

- Stage 5 – Check the length of the flat bar coming through the mounting plate; this should generally be about 15 mm. If it is not, cut the bar to the correct length using a junior hacksaw.
- Stage 6 – Insert the cylinder through the 'rim' and into the hole previously drilled in the door. Secure the cylinder in position by placing the backing plate over the back of the hole in the door and fixing both parts together with the two machine screws provided. Check that the key slot on the front of the cylinder is vertical before finally tightening the machine screws.
- Stage 7 – Use the small countersunk screws provided with the lock to fix the backing plate in position. This will retain the keyhole in the cylinder in a vertical position and prevent it from moving during use of the lock.

FIGURE 5.77 Fixing the backing plate

FIGURE 5.78

- Stage 8 – The rear side of the lock body should have two arrow symbols towards the centre of the lock. One of the arrows is marked on the turntable; this will have to be rotated until it aligns with the other arrow marked on the back plate. Place the lock body over the flat bar, making sure that it locates with the hole in the centre of the turntable. Make sure that the body of the lock fits flat against the face of the door; if it does not, the flat bar may be too long and will need to be re-cut.
- Stage 9 – Drill several pilot holes through the fixing points on the body of the lock and secure it in position with the screws provided.
- Stage 10 – Place the key in the cylinder and check that it operates the lock correctly.
- Stage 11 – Swing the door to its closed position, and transfer the position of the top and bottom of the lock onto the edge of the door frame. Place the striking plate between the two marks indicated on the door frame. Use a sharp pencil to draw around the section of the striking plate resting in the rebate on the door frame.

- Stage 12 – Adjust a marking gauge to the thickness of the returned section on the striking plate and transfer this measurement onto the face of the door frame, between the two marked lines. Use a chisel to recess this area to the depths indicated. Place the striking plate into the recess on the door frame, drill several pilot holes through the fixing points and secure the striking plate in place with the screws provided.
- Stage 13 – Swing the door to its closed position. Check that the lock engages the striking plate correctly and that the key operates freely to lock and unlock.

FIGURE 5.79 Mark lock position on door frame

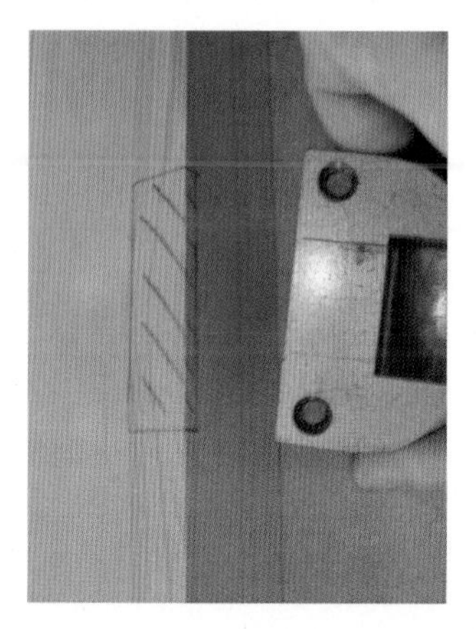

FIGURE 5.80 Mark striking plate on door frame

INSTALLING A FLUSH BOLT ON ONE LEAF OF A PAIR OF DOORS

Flush bolts are better fitted in a joiner's workshop than on site because they can be very labour intensive to install due to their depth and complexity.

- Stage 1 – Unscrew one leaf from the double door frame and lay it across a pair of trestles.
- Stage 2 – Measure the width of the flush bolt and transfer this size to the centre of the stile, at the top and the bottom of the door.
- Stage 3 – Mark the length of the flush bolt along the door stile. Set a mortise gauge to the width of the flush bolt indicated on the edge of the door and mark parallel gauge lines up to the length marks.
- Stage 4 – Measure the length of the returned end on the flush bolt and transfer this size to both ends of the door.
- Stage 5 – Insert a cutter equal to the width of the flush bolt into a portable powered router. Adjust the cutter to the depth of the flush bolt face plate. Use the router to recess along the face and end of the door, between the markings on both ends of the door. Use a sharp chisel to square out the rounded corners remaining after routing.
- Stage 6 – Lay the flush bolt face down on the door along the side of the recess. Transfer the length of the deeper section of the face plate, either side of the lever, onto the recessed area. Readjust the router depth to suit the deeper section of the flush bolts and recess between the points indicated on the door.
- Stage 7 – Lay the flush bolt on the face of the door, along the side of the recess. Move the lever on the flush bolt to the open and closed positions, and transfer the extreme positions of the moving parts onto the door.
- Stage 8 – Insert a cutter into the router equal to the width of the lever. Adjust the router several times to create a recess between the points indicated on the door. Insert the bolt

into the recess. Check it sits flush with the surface of the door, and opens and closes without fouling on the bottom of the recess. Use the screws supplied to secure it in position. Repeat this process on the other end of the door.

FIGURE 5.81 Alignment of the 'keep'

- Stage 9 – Refit the door back into the frame. Move the levers on the flush bolts to the open position and swing the door against the frame until the projecting bolts touch the face of the door frame. Use a pencil to mark each side of the bolts on the face of the door frame and square these positions around the frame, towards the rebate. Set a marking gauge from the face of the door to the centre of the flush bolts and transfer this position onto the inside face of the frame.
- Stage 10 – Hold the 'keep' centrally over the marking out on the door frame and draw around the outer and inner edge. Use a drill bit to bore a hole slightly bigger than the flush bolt, in the centre of the marking out. The depth of the hole will be determined by the amount that the bolt projects over the edge of the door.
- Stage 11 – Use a chisel to recess the shape of the 'keep' around the hole in the door frame. Use the screws provided to secure the 'keep' to the frame. Swing the door into the frame to its closed position and operate the flush bolts to check they function freely to secure the door.

INSTALLING A RACK BOLT

- Stage 1 – Place folding wedges under the bottom edge of the door to prevent it from moving during the installation of the rack bolt.
- Stage 2 – Measure down from the top of the door the intended position of the rack bolt. Adjust a marking gauge to the centre of the door edge and mark a short vertical line over the intended height position. Use a square to transfer the centre point from the edge of the door onto one face.

Gauge centre line on door edge

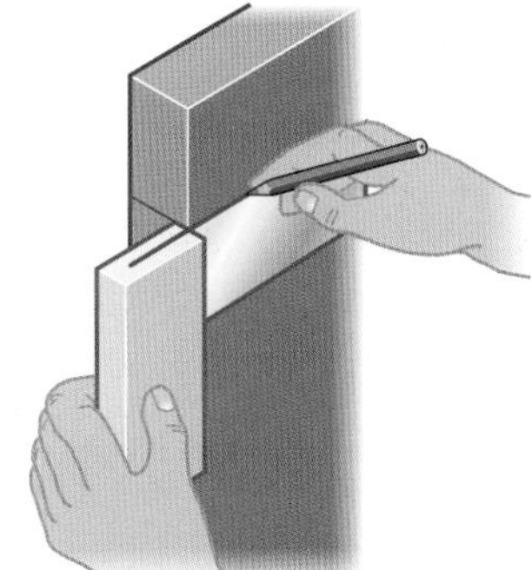

Mark centre line on edge and face of door

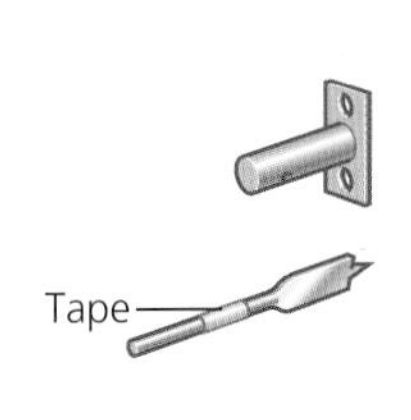

Select auger or spade bit to match bolt. Use tape to mark required depth

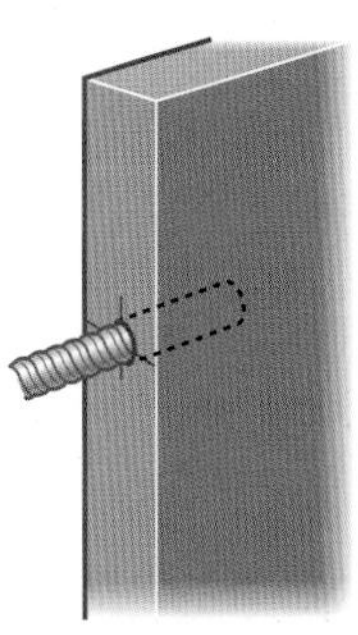

Drill hole in door edge

FIGURE 5.82 Installing a rack bolt

- Stage 3 – Use a flat/auger bit slightly bigger than the width of the bolt to bore a hole in the centre point indicated on the edge of the door. The depth of the hole should be approximately 6 mm deeper than the length of the rack bolt; this can be indicated on the drill bit with a piece of masking tape.
- Stage 4 – Place the rack bolt in the hole in the edge of the door. Use a sharp pencil to draw around the face plate. Remove the rack bolt from the door and recess the area marked on the edge of the door with a chisel.
- Stage 5 – Place the rack bolt on the face of the door and align it with the recess on the edge. Use a bradawl to mark the keyhole position through the rack bolt and onto the face of the door. Use a 10 mm drill bit to bore a hole through one side of the door, on the centre point marked.
- Stage 6 – Place the rack bolt back into the hole in the edge of the door and secure it in position with the screws provided. Insert the fluted key into the bolt and rotate it to check that it locates and operates correctly.
- Stage 7 – Remove the key from the bolt and place the keyhole plate centrally over the hole on the face of the door. Secure the keyhole plate with the screws provided.
- Stage 8 – Most rack bolts are designed with a small centre point, located on the end of the bolt. This can be used to transfer the centre point of the bolt onto the frame. This is simply done, by extending the rack bolt as far as possible once the door is in its closed position. This should then mark the centre point for the receiving plate. If the point is not clearly visible, then the end of the bolt should be marked with pencil before repeating the process.

TRADE SECRETS

To ensure a good fitting of any striking/keep plate, once the position of the plate has been determined, use short screws to secure it. Then use a craft knife to cut a line around the face plate on to the timber. Remove the plate and chop out the recess. (Note: a cut line allows a much more accurate position for the final chisel paring once the centre has been removed.)

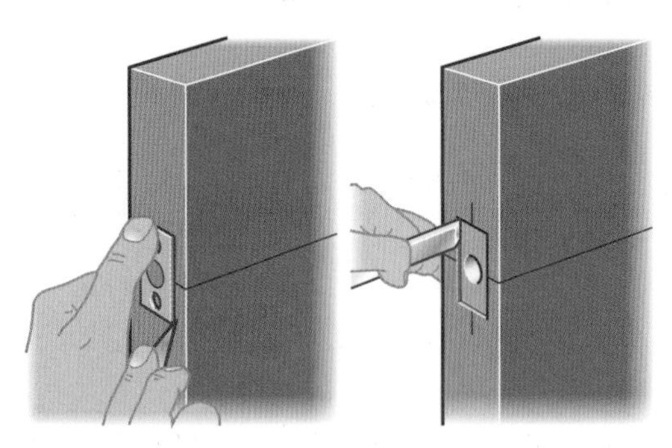

FIGURE 5.83 Stage 4

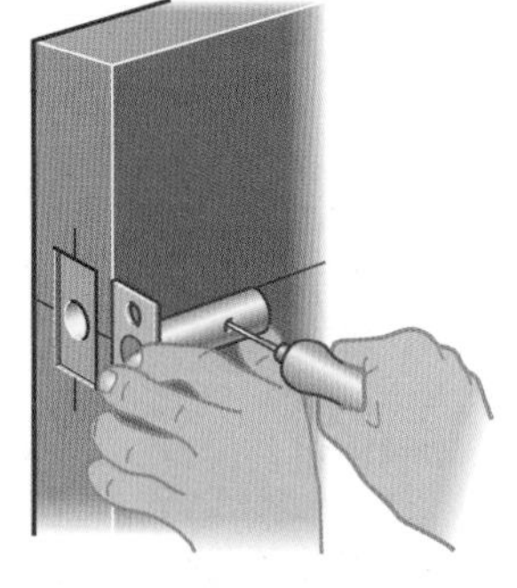

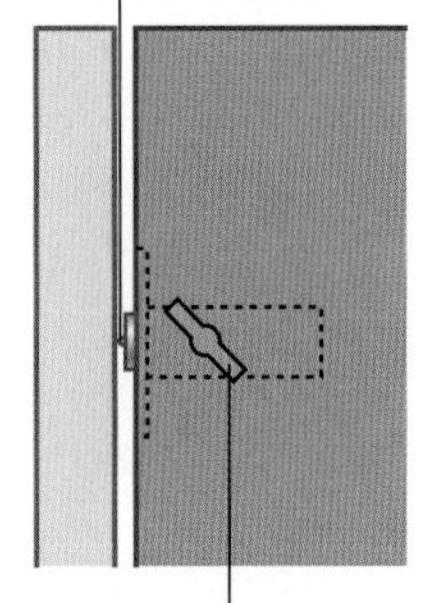

FIGURE 5.84 Stages 7 and 8

- Stage 9 – Bore a hole slightly bigger than the rack bolt frame through the centre point indicated on the frame.

- Stages 10 – Place the keep plate centrally over the hole drilled in the frame and draw a pencil line around the outside. Use a chisel to recess the area marked on the frame to the depth of the keep plate. Use the screws provided to secure the keep plate to the door frame.
- Stage 11 – Swing the door to its closed position and operate the rack bolt to ensure it correctly engages the keep in the door frame.

FREQUENTLY ASKED QUESTIONS

What is the purpose of the 'keep plate'?

The 'keep plate' is secured over the hole bored in the door frame to prevent damage to the hole through regular use of the rack bolt.

If the rack bolt is opened into the hole bored in the frame before the keep plate is fitted and fails to locate correctly, the keep plate can be used to realign the hole.

INSTALLING A LETTER PLATE

- Stage 1 – Establish the position of the letter plate on the door.
- Stage 2 – Lay the letter plate centrally on the face of the door and mark a pencil line around the outside.
- Stage 3 – Measure the length and width of the flap on the letter plate and add 5 mm to each size. These measurements should be transferred centrally onto the face of the door to indicate the cutout required in the door.
- Stage 4 – Use a 10 mm flat/auger bit to bore holes in all four corners of the marking out on the face of the door.

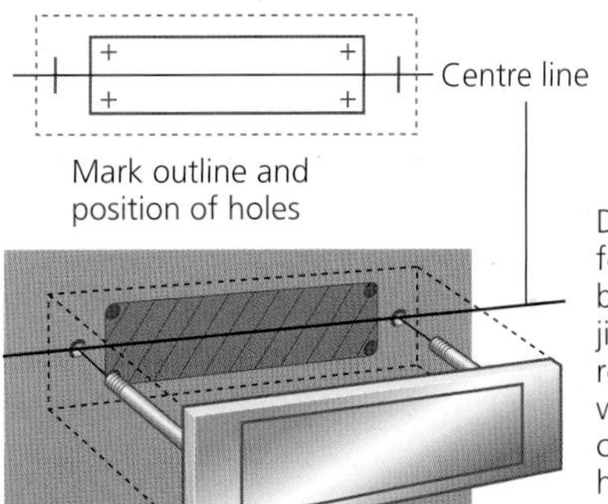

FIGURE 5.85 Installing a letter plate

- Stage 5 – Use a portable powered electric jigsaw to cut along the lines marked on the face of the door to remove the waste material. Working from both sides of the door to prevent breakout, use a sharp chisel to remove the remaining waste material to form square corners.
- Stage 6 – Clean up the inner face of the cutout and remove the sharp edges with some abrasive paper.
- Stage 7 – Measure the distance between the bolt holes in the back of the letter plate and transfer this position onto the face of the door. Use a powered drill to bore two holes, slightly bigger than the fixing bolts, through the door.
- Stage 8 – Screw the threaded bar provided into the back of the letter plate and locate the assembly through the holes in the door. The bar projecting through the back of the door may have to be cut to length, so that only 6 mm is visible. This will allow the washers and fixing nuts to be screwed over their ends to secure the letter plate in position.
- Stage 9 – Check that there is enough clearance between the edges of the flap and the cutout to allow it to function freely.

ACTIVITIES

Activity 20 – Installing side-hung doors and ironmongery

Read through the following questions and answer them as fully as you can to help you develop your underpinning knowledge of this subject area.

1. List the characteristics of an external door.
2. Explain the term 'binding'.
3. What is the purpose of a door schedule?
4. How can you identify the position of a 'lock block' in hollow core doors?
5. How can you identify a fire door?

INSTALLING SERVICE ENCASEMENTS AND CLADDING

Wherever possible, electrical cables and plumbing pipes should be buried behind walls and cupboard units to prevent an unsightly finish to the room. This will also prevent having to decorate around services mounted above the wall surface. Many older buildings and properties that have been modernised will have pipes and cables mounted on the surface of the walls or just above the finished floor level; this will prevent the need for 'chasing out' along the walls and minimise the damage and repair required. Encasing services is sometimes referred to as 'boxing in'; this is due to the arrangement of the panelling around the pipes and cables.

SERVICE ENCASEMENTS

ACCESS PANELS

Before totally encasing any services, it is important to establish whether or not access will be needed at a later date for servicing or maintenance. Access may be required to the full length of a pipe or cable, or just a small section. This can be achieved by any of the following methods.

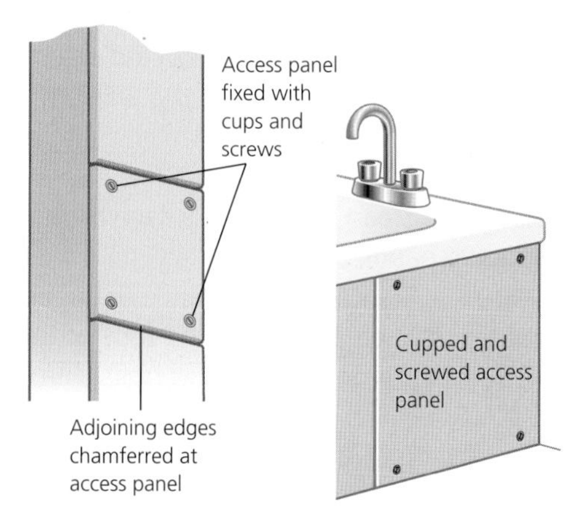

FIGURE 5.86 Access panels

Access panels may be hinged to permit viewing into an area, or they may require total removal from the encasement. To ensure that the panels relocate in exactly the same positions and do not show signs of wear, they should be secured with good-quality hardened screws and 'cups'. These will prevent the countersunk heads of the screws burying themselves deeper into the panelling each time the panelling is reattached to the encasement.

ENCASEMENT CONSTRUCTION

Service encasements should be built to hide pipes and cables while remaining as small as is practically possible. They are generally constructed on site with a skeleton frame built with prepared 32 × 32 mm softwood timber. Traditionally, simple halving joints would be used to construct the ladder frames required for larger encasements; smaller frames are usually butt jointed and screwed together. The framework is usually 'clad' (covered) with man-made timber-based sheet materials because:

- it is relatively cost-effective;
- can be used to cover wide areas without a need for jointing in its width;
- it is stable and easily worked.

In some situations solid timber is the best material for use to match existing panelling or timbers, and blend more discreetly into its position. Alternatively, uPVC (plastic) sections can be used in damp or wet areas such as bathrooms or kitchens. Plastic cladding can be cut, planed and fixed using carpentry hand and power tools without the risk of damage. Solid timber should never be used as a cladding for an encasement if tiles are going to be used as a covering material. The movement in the natural timber will cause cracking along the grouted joints and may result in some tiles working loose.

Clever design and consideration for the type of materials and encasement used may permit the encasement to be built in to a feature, such as a shelf.

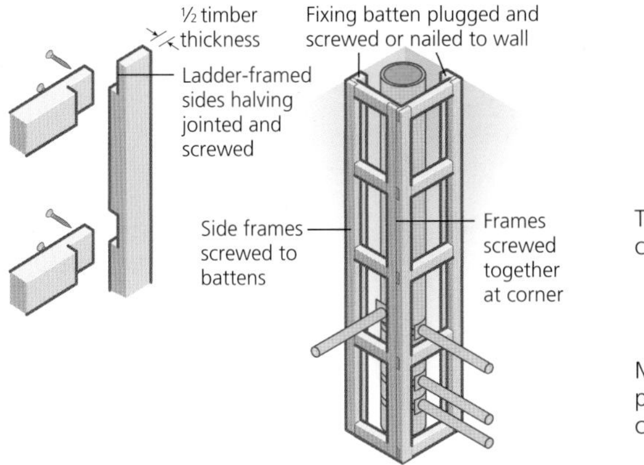

FIGURE 5.87 Marking out and fitting a vertical service encasement

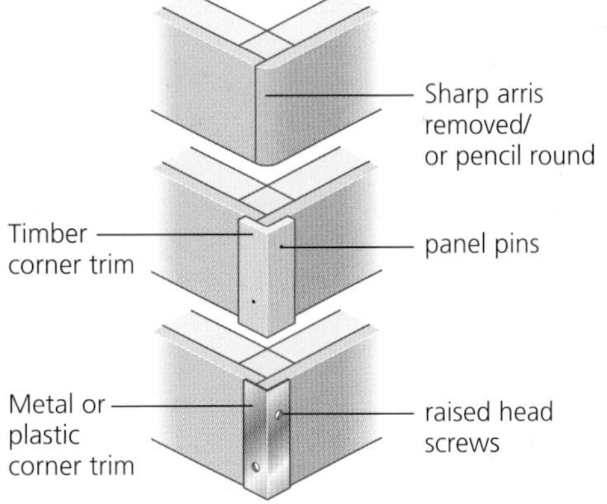

FIGURE 5.88 Finishing the external corner on service encasements

SEQUENCE OF CONSTRUCTING SERVICE ENCASEMENTS

EDGE PROTECTION

There are several methods commonly used to finish or protect the exposed external corners of service encasements.

SCRIBING CLADDING

In some cases the framework needed for an encasement may have to be built over pipes and cables, and may require the cladding to be scribed against or around a shaped wall.

There are several ways this can be achieved; these include:

- profile gauge – used to transfer the complex shapes of the pipes and cables onto the cladding before cutting;
- compass – used together with a pencil to transfer the profile of a wall directly onto the cladding, providing a neat line to follow when cutting the scribe;
- gauge slip – a small section of timber cut to the thickness of the scribe required to allow a panel or moulding to fit perfectly against a shaped surface;
- measurement – in some situations the pipes that require scribing around may be too big to allow a profile gauge to be used; in these cases the distance to the top, bottom and projection of the pipes should be measured and transferred onto the cladding; a compass should then be adjusted to the radius of the pipe, and held in the centre of the marking out to mark the circumference; alternatively, a short piece of the pipe could be used to draw around.

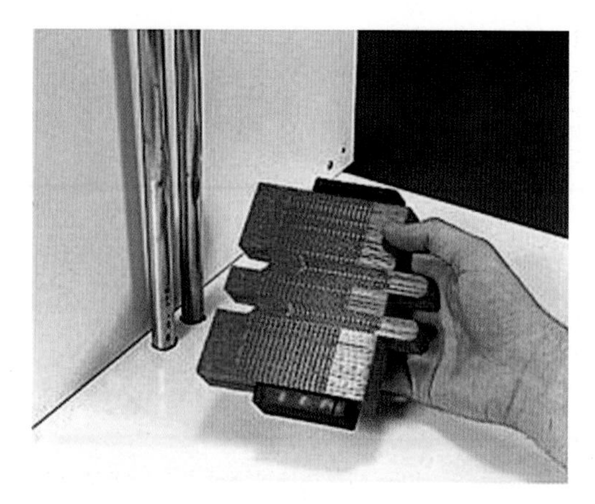

FIGURE 5.89 Profile gauge

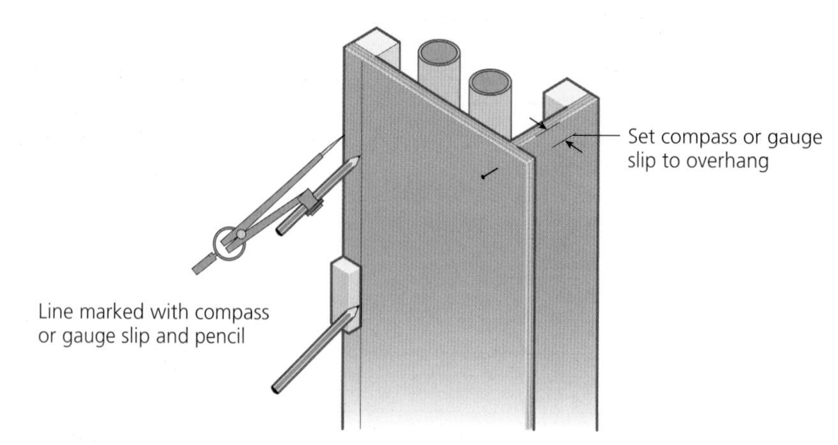

FIGURE 5.90 Using a compass

INSTALLING BATH PANELS

Many modern bathroom suites are supplied with an optional bath panel. Lower-cost panels are normally manufactured from acrylic (plastic) to complement the bathroom suite. The acrylic panels do not normally require any trimming to length or width, provided that the bath has been correctly installed at the right level. If they do have to be cut to size they can be trimmed with a portable powered jigsaw, fitted with a fine cutting blade, or by hand with a fine-point panel saw. Acrylic panels are installed by simply locating the top of the panel on the clips on the underside of the bath rim. The bottom of the panel should then be pushed towards a batten fixed previously on the floor. Once the panel is tight against the batten in a vertical position, it can be secured with the fixing screws provided. The fixing screws should be covered up with screw caps to hide any visible surface fixing; both items should be made from materials resistant to corrosion. Never use adhesive or sealant to secure a bath panel in position, as this will prevent easy access underneath the bath if maintenance is required at a later date.

PURPOSE-MADE BATH PANELS

This type of panelling can be constructed on site with a simple framework of 45 × 20 mm treated softwood, with a moisture-resistant timber-based sheet material covering applied. Halving joints should be used to connect the timbers in the ladder frames with screws and glue to strengthen the structure.

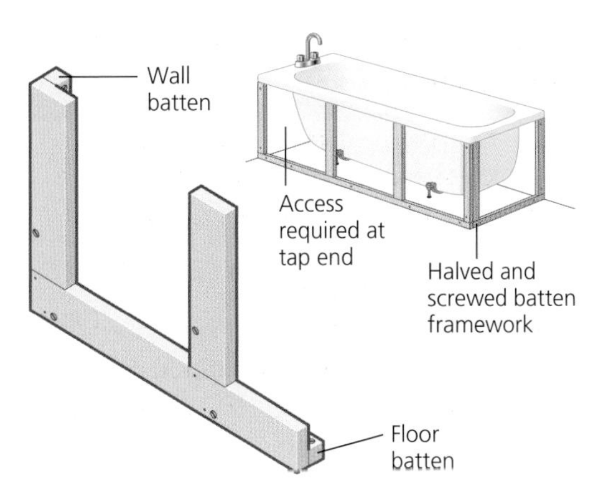

FIGURE 5.91 Building a purpose-made bath panel

FREQUENTLY ASKED QUESTIONS

How can I improve the appearance of the bath panels I have constructed on site?

There are many ways of improving the appearance of panels constructed on site. This can be achieved simply by applying small mouldings to the face of the sheet material to give the appearance of 'mock' panels. Another method regularly used to enclose the framework under the bath is to install 'V'-jointed tongue and groove boards.

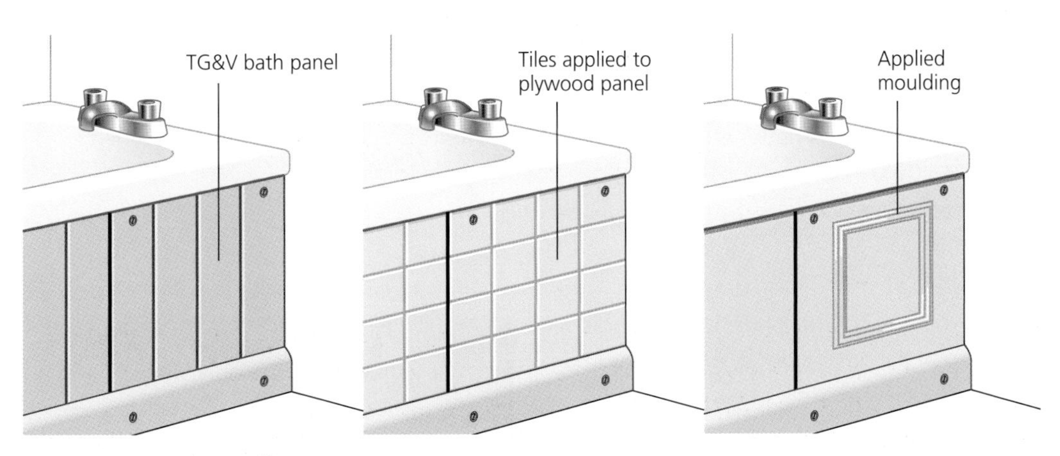

FIGURE 5.92 Bath panelling

Purpose-made bath panels can also be constructed by joiners off site in a workshop. These are normally manufactured with solid timber and joints that require specialist machinery to complete.

ACTIVITIES

Activity 21 – Installing service encasements and cladding

Read through the following questions and answer them as fully as you can to help you develop your underpinning knowledge of this subject area.

1. Why should access panels be built in to service encasements?
2. Identify a suitable material to construct an encasement in a bathroom suitable for tiling.
3. Sketch a section through a typical service encasement to demonstrate 'edge protection'.
4. Explain the purpose of a 'gauge slip'.
5. What section timber is commonly used to construct 'skeleton frames' for service encasements?

INSTALLING TIMBER FINISHINGS

INTRODUCTION TO TIMBER MOULDINGS

The term 'moulding' refers to the shape produced on a section of timber or man-made material. Some mouldings are produced to provide a decorative finish to a piece of work; others are functional shapes designed to do a job.

Timber mouldings and timber-based mouldings, also known as 'profiles', are commonly used for a number of different purposes; these include:

- protecting surfaces from damage;
- covering the joints between different materials;
- concealing gaps;
- to break up areas;
- to enhance the features;
- decorative purposes.

Timber mouldings are commonly available from suppliers in a wide range of standard shapes and sizes, for a variety of different applications.

FIGURE 5.93 A spindle moulder

PRODUCING TIMBER MOULDINGS

Traditionally, profiles were 'stuck' (produced) on timber by carpenters by hand with wooden moulding planes. Some of these planes are still used today by specialist joiners and cabinet makers, although the majority of commercial profiles are produced by multi-moulding machines.

Modern methods of producing timber mouldings include:

- portable powered router;
- spindle moulder;
- multi-head planer/moulder.

STANDARD PROFILES

Standard profiles are illustrated in Figure 5.94.

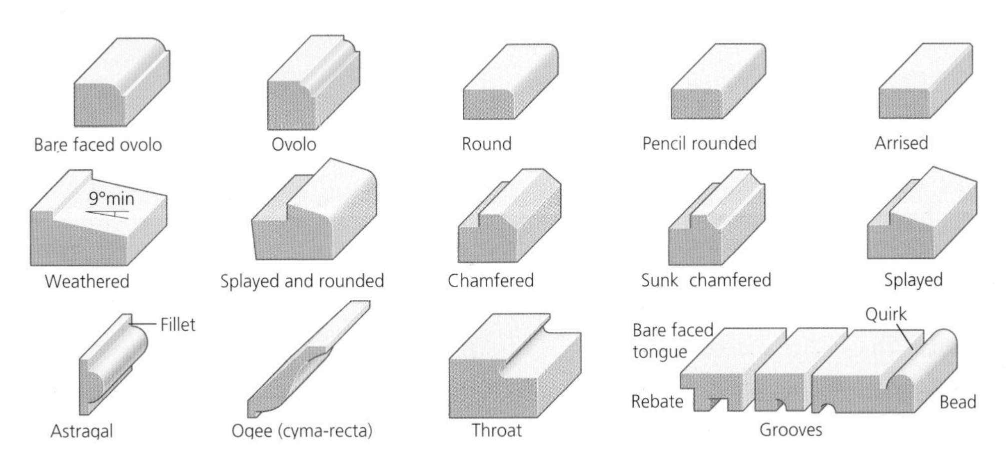

FIGURE 5.94 Standard profiles

MATERIALS

Commercial softwood and medium-density fibreboard (MDF) mouldings are commonly available from timber and builders' merchants. It would be uneconomical for suppliers to carry a range of hardwood mouldings because of the vast amount of species of timbers available. MDF mouldings are usually manufactured from moisture-resistant man-made board, with several coats of primer applied to all surfaces ready for finishing on site. There are several benefits to using man-made mouldings, including:

- no natural defects (e.g. knots, sap pockets);
- no twisting, warping or bowing;
- minimal shrinkage or expansion (although some in length if the moisture content is too high).

Materials commonly used for commercial mouldings include:

- softwood;
- hardwood (limited stock, manufactured to order);
- MDF;
- plastic.

Solid timber and man-made mouldings are usually stocked in sizes ranging from 1.8 metres up to 5.1 metres, with increments of 300 mm in between.

TRADE SECRETS

Machined timber is normally carried in stock by suppliers and is referred to by its 'ex' size. The term 'ex' simply means it is 'extracted' from the sawn size of the timber before it is planed – for example:

Skirting board listed as ex 25 × 125 mm (actual size 20 × 119 mm)

MOULDINGS AND TRIMS

ARCHITRAVES

Architraves are fixed to the faces of door frames/linings and loft hatches to mask the joint between the frame and the opening in the wall or ceiling. When fixing architraves, you should ensure that they are set back from the edge of the lining to form a 'margin'. The margin is usually between 6 and 9 mm, and should remain an equal distance all the way around the frame/lining. They are created to form a natural shadow over the edge of the frame/lining and to conceal any unsightly gaps that may appear between the back of the architraves and the frame that are impossible to hide if the joint is fixed flush.

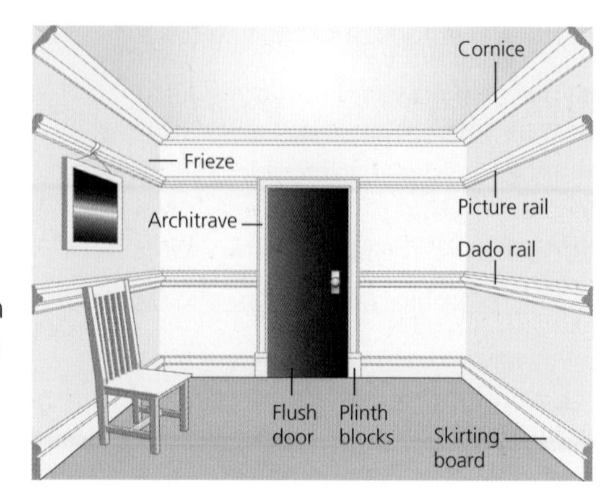

FIGURE 5.95 Mouldings and trims

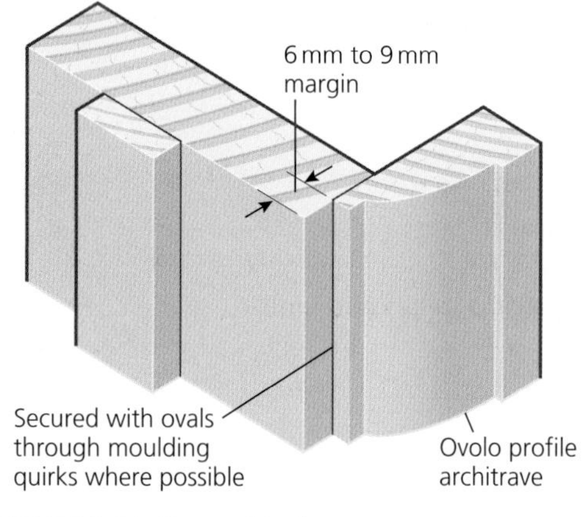

FIGURE 5.96 Corner blocks

CORNER BLOCKS

Corner blocks are a feature occasionally used in the corner of adjoining architraves. As well as creating an architectural feature, they also conceal the effects of movement between timbers sometimes encountered when mitring architraves together.

CORNICE

Cornices are manufactured in a variety of different shapes and materials, including plaster. Traditionally, they were used to hide the gaps and cracks revealed in the joints between the walls and ceilings as a result of materials moving in different ways. Nowadays, construction methods have developed with the use of flexible products to prevent these defects appearing; therefore cornices are now used to improve the 'aesthetics' (appearance) of a room.

DADO RAILS

Dado rails are traditionally fixed approximately 1 metre from the finished floor level to protect decorative wall surfaces from potential damage occurring from the backs of chairs. Dado rails are also known as 'chair rails' because of their alignment with the top rail of the tallest chair used in the room. Although they are rarely used for this purpose any more, they are still installed as a decorative moulding to add interest to rooms with high ceilings.

PICTURE/FRIEZE RAILS

As the name suggests, these traditional mouldings are used to hang pictures and paintings. Special clips and wire are fixed to the back of the picture frames, which allow them to hook over the top edge of the moulding and be displayed without the need to damage the finished wall. There are no suggested recommendations for the fixing height of picture rails, although they are generally located between 1.8 metres and 2.1 metres above the floor. The exact location of a rail should be determined by the height of pictures hung from it; ideally they should be at eye level.

PLINTH BLOCKS

Plinth blocks are traditionally used at the 'feet' of architraves to prevent damage to the lower section of the architraves. They are slightly bigger in section than the architraves that meet them and usually have a simple bevelled profile, which is easy to repair if damaged. Plinth blocks may also be used to conceal the joint between an architrave and a thicker section of skirting, as these can sometimes be difficult to finish neatly.

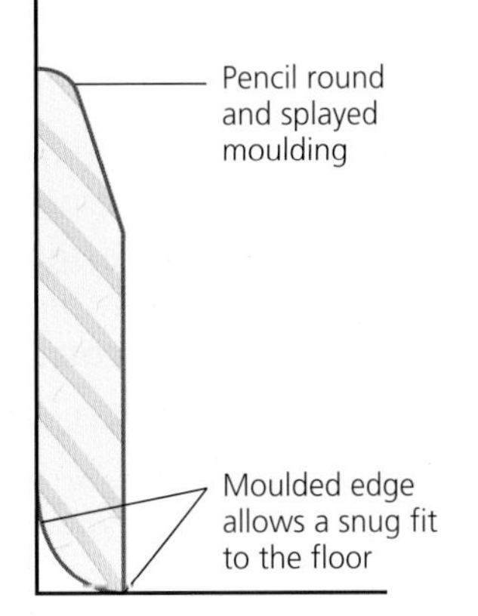

FIGURE 5.97

SKIRTING

Skirting is a section of moulding used to cover the joint between the bottom of a wall and the edge of the floor. It is also used to protect the lower part of plastered walls from damage caused by foot traffic and general wear and tear. Skirting is normally mass produced with different profiles on opposite sides of the board. This reduces the amount of stock that suppliers need to carry and allows the skirting to sit flat against the bottom of uneven walls.

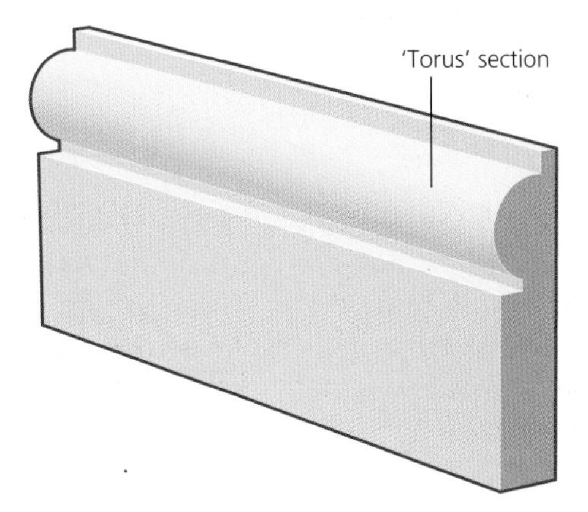

FIGURE 5.98 Skirting

INSTALLING MOULDINGS

ARCHITRAVES

Architraves are normally fitted to door linings after the doors have been installed. This allows any additional adjustments to be made to the packing and fixing of the frame, before finally covering over with the mouldings. They are usually positioned 6–9 mm back from the front edge of the lining unless the specifications state otherwise. There are two methods of returning architraves around the head of the door lining; these are:

1. mitre;
2. corner blocks.

Mitres are used to return architraves and various other mouldings because they allow the continuation of the profile around the frame. They are generally the preferred method of installation because they are quick to cut and fix, and do not require any complicated joints. Corner blocks are generally used nowadays only as an

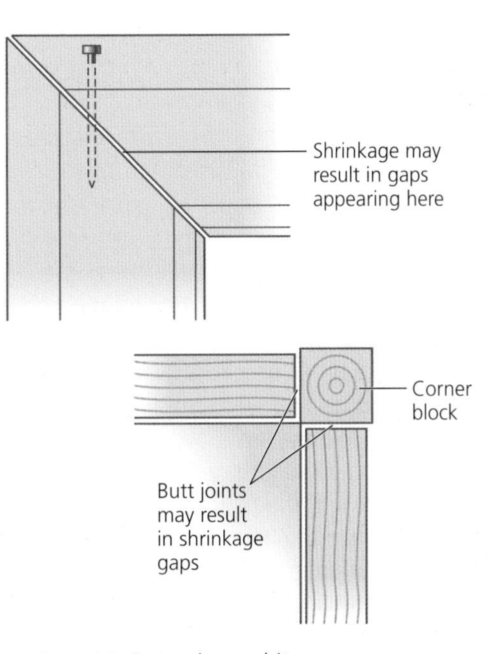

FIGURE 5.99 Returning architraves

architectural feature on work with a higher specification, or on restoration projects, where the existing arrangements have to be match or replaced.

FREQUENTLY ASKED QUESTIONS

How are corner blocks fixed?

It is important that the joints between the ends of the architraves and the corner blocks are designed to conceal the effects of shrinkage while providing a strong fixing. There are several ways that this can be achieved; these include:

- biscuit joints;
- dowel joints;
- housing joints.

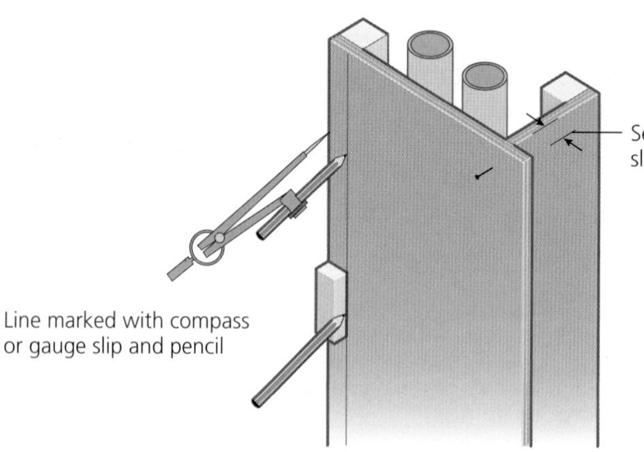

FIGURE 5.100 Joints between architrave and corner blocks

CUTTING AND FIXING

- Step 1 – Mark a 6–9 mm margin line all the way around the edge of the door lining using a pencil gauge or a combination square.
- Step 2 – Cut one mitre (45°) on the end of a length of architrave using a tenon saw and mitre block or 'chop saw' (powered mitre saw).
- Step 3 – Hold the architrave on the top of the door lining, making sure that the moulding is facing towards the door. Align the inner edge of the mitred saw cut with the margin on one end of the lining, and then mark a pencil line on the other to indicate the position of the next mitre to be cut.
- Step 4 – Cut a second mitre the opposite hand on the other end of the architrave.

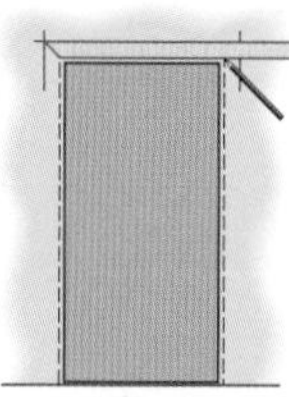

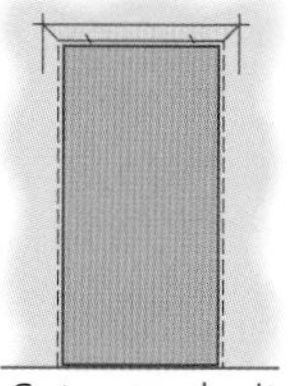

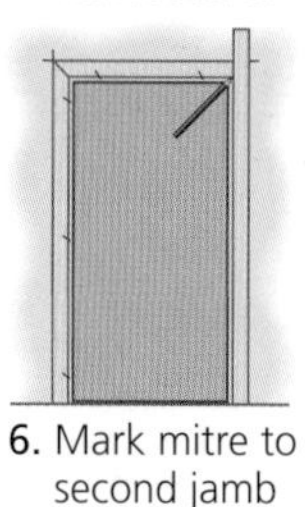

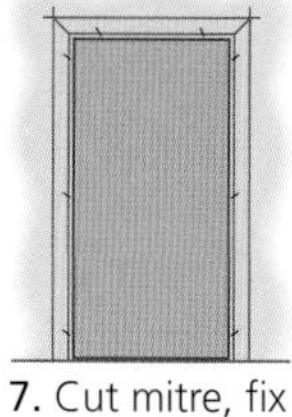

FIGURE 5.101 Marking out, cutting and fixing architraves

- Step 5 – Hold the cut piece of architrave on the top of the lining and align its bottom edge with the margin line marked along the length of the head. Check the mitres on either end of the architrave are correctly lined up with the margins marked on the jambs of the door lining.
- Step 6 – Fix the 'head' architrave to the door lining using oval nails, positioned at approximately 300 mm centres. Leave the heads of the nails slightly 'proud' of (above) the timber as a precaution, because adjustment may be required to the mitres before securing permanently.
- Step 7 – Hold another length of architrave against one vertical member of the door lining (the 'jamb') and mark the position of the mitre. Use a pencil to indicate the direction of the mitre required on the face of the architrave; this will prevent mistakes by cutting the wrong 'handed' mitre.
- Step 8 – Cut the mitre indicated on the 'jamb' architrave and place in position. If the mitres do not fit tightly, use a sharp block plane to close the gaps between the architraves. Fix the jamb in position, making sure that it is correctly aligned with the margin on the door lining.
- Step 9 – Repeat stages 7 and 8 for the other jamb to complete the set, making sure that the mitre is cut the opposite hand to the jamb already fixed.
- Step 10 – Fix through the head of the architrave into mitres to prevent movement between the mouldings. Try to avoid fixing the mitres through the jambs of the architraves, as this will remain visible after completion.
- Step 11 – Finally, carry out a visual check to ensure the mitres have remained tight, and drive the nails below the surface using a nail punch.

FREQUENTLY ASKED QUESTIONS

Why should you start installing the head of the architrave first?

Whenever possible, you should start installing the head of the architrave first. As well as being good practice, this will avoid having to check and possibly correct two mitres at the same time.

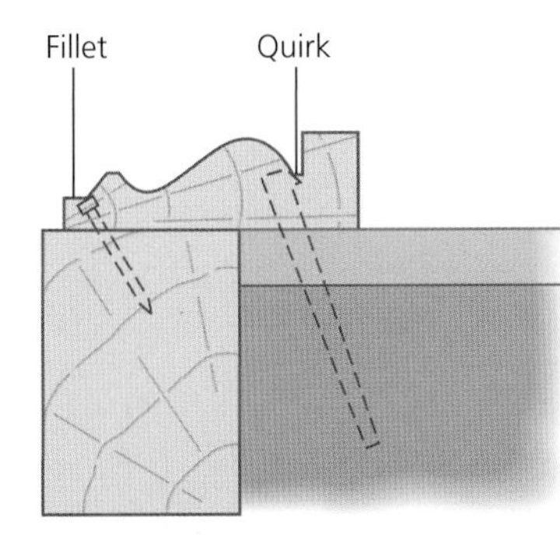

FIGURE 5.102 Concealing fixing

TRADE SECRETS

Wherever possible, care should be taken to conceal the method of fixing mouldings or trims to provide a professional and neat finish. There are a few ways that this can be achieved, including nailing through the quirks, rebates, grooves and fillets found on moulding.

CORNICE

Timber cornice is rarely used around the perimeter of domestic houses because it is not cost efficient. In most cases, plaster cornice (otherwise known as 'coving') is used between the ceiling and walls because it is easy to cut and fix, and relatively inexpensive. Plasterers, painters and decorators, and carpenters will at some point be expected to install plaster coving.

Coving is simply cut to length with an old handsaw and returned at the internal and external corners with a mitred joint. It is then fixed to the wall and ceiling with special cove adhesive, and any gaps filled. The shape of the cornice can sometimes cause problems when marking out and cutting mitred joints, so it is common practice to use one of the following methods:

- paper mitre template supplied with the coving;
- purpose-made mitre box;
- metal or plastic cornice mitre guide.

Timber cornice is commonly used in shopfitting and also for the trim across the tops of kitchen units and bedroom furniture. Deeper sections of timber cornice are often constructed in several sections to make it easier to machine the individual pieces. They also conceal the effects of shrinkage, sometimes resulting from bigger profiles.

DADO AND PICTURE/FRIEZE RAILS

Before installing either of these mouldings, it is important that they are accurately marked out on the walls at the correct height. This is simply done by referring to the architect's drawings and specification to establish the position of each moulding from the datum point. These measurements are then marked on to the wall and transferred around the room using one of the following methods:

- spirit level and straight edge;
- water level;
- laser level.

The smaller sections and complex profiles usually found on dado and picture rails make it difficult to scribe the internal joints, and as a short-cut they are often mitred.

Either masonry nails or screws with plastic plugs should be used to fix the mouldings on to brick or block walls. When fixing into timber stud walls, the vertical studs in the framework should be located and the moulding secured at these points with 'oval' or 'lost-head' nails. Alternatively, 'panel adhesive' can be applied to the back of the mouldings with a 'silicone gun' and glued onto the walls. The disadvantages of this method are that the moulding may have to be supported temporarily with nails until the adhesive dries. It may also be difficult to fix mouldings that are twisted or misshaped using this process.

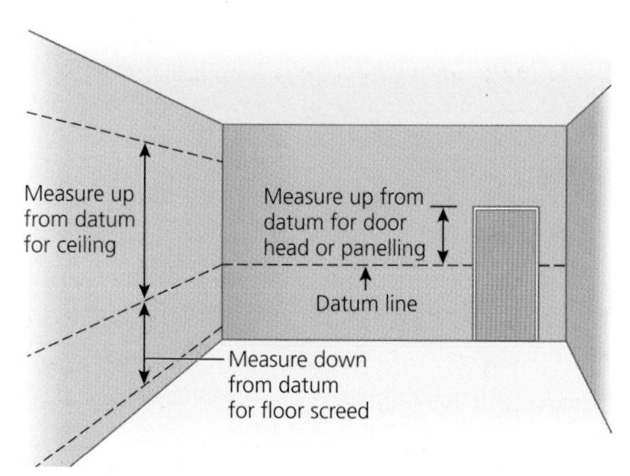

FIGURE 5.103

PLINTH BLOCKS

Plinth blocks are usually jointed into the feet of the architraves with the use of 'bare-faced tenons' and secured with several short countersunk screws through the back of the moulding. The profile of the skirting adjoining the plinth block is traditionally housed into its edge approximately 12 mm. This will prevent any unsightly gaps appearing between the two mouldings due to shrinkage in the timber.

FIGURE 5.104

SKIRTING

Timber skirting boards should be returned around all external corners with mitred joints. This allows the profile on the skirting to run continuously along the bottom of the wall and also prevents the end grain being displayed. It is bad practice to mitre the internal joints of skirting, because any shrinkage that occurs in the timber will result in unsightly gaps appearing in the corners. The best methods of minimising and concealing the effects of shrinkage are to 'condition' (second seasoning) the timber before fixing. This can be achieved by storing the mouldings in the room where they are going to be fixed until the 'equilibrium' moisture content is reached. Once this balance has been achieved between the timber and its environment, the amount of movement will be minimised. All the internal joints between skirting boards should be 'scribed' over the profile of the moulding. This will allow any shrinkage that takes place to occur under the joint, rather than on the face.

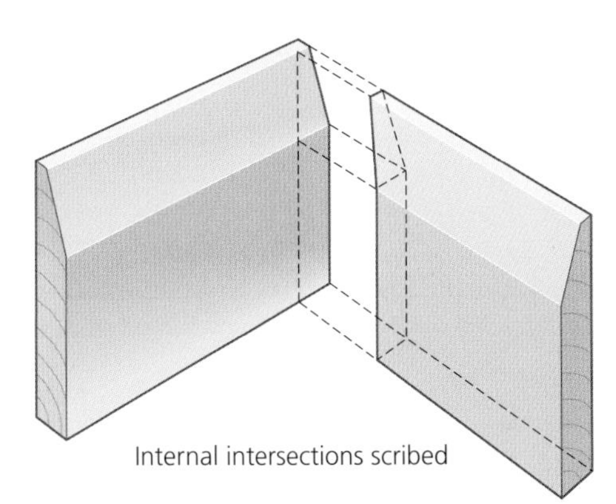

FIGURE 5.105

FIGURE 5.106 Effects of shrinkage (left) and a scribed joint (right)

JOINT POSITIONING

It is generally accepted that your eyes are always drawn towards any visible gaps between mouldings as soon as you enter a room. Carefully considering the position of the scribed sections of skirting board will help to hide the effects of shrinkage behind adjoining mouldings.

Long runs of skirting will require extending in length due to the limited size of the timber; this joint is known as a 'heading joint'. It is common practice to mitre the heading joints together rather than using 'butt' joints so that two sections blend together. The mitred heading joint also provides a larger gluing area and therefore a stronger joint, which is secured in position with oval or lost-head nails.

CUTTING A SCRIBED JOINT (TORUS PROFILE)

- Stage 1 – Determine which side of the skirting joint will butt against the wall and which one will be scribed over. Measure, mark out and cut the length of skirting required to butt into the corner of the wall.
- Stage 2 – Set up a powered chop saw to make a 45° cut. Place the second length of skirting board onto the 'bed' of the saw with the profiled face towards the operator, and complete the saw cut. Cutting with the moulded side up on the chop saw will avoid the timber breaking out on the face side.
- Stage 3 – Hold the skirting securely with the face side up, on a firm surface. Use a fine handsaw to make a saw cut across the timber, following the edge of the saw cut made on the chop saw. Continue the saw cut across the flat section of timber, making sure the saw cut is slightly 'undercut'. Make a second saw cut, parallel with the skirting, to remove a portion of the scribe.
- Stage 4 – Work from the top edge of the skirting profile to make a short saw cut square across the skirting, following the edge of the saw cut made previously on the chop saw.
- Stage 5 – The remaining portion of the scribe is curved and will therefore have to be removed with a coping saw. Care should be taken when cutting the scribe to ensure that the saw cuts are square to the face of the skirting while at the same time slightly undercut.

TRADE SECRETS

When ordering the mouldings required for a project, 10 per cent is normally added to the exact quantity to allow for saw cuts and waste. This prevents the need to use unsightly short pieces of moulding unnecessarily.

Always begin to install mouldings by establishing the longest lengths required and cutting them first; the shorter lengths will then be used up in smaller areas.

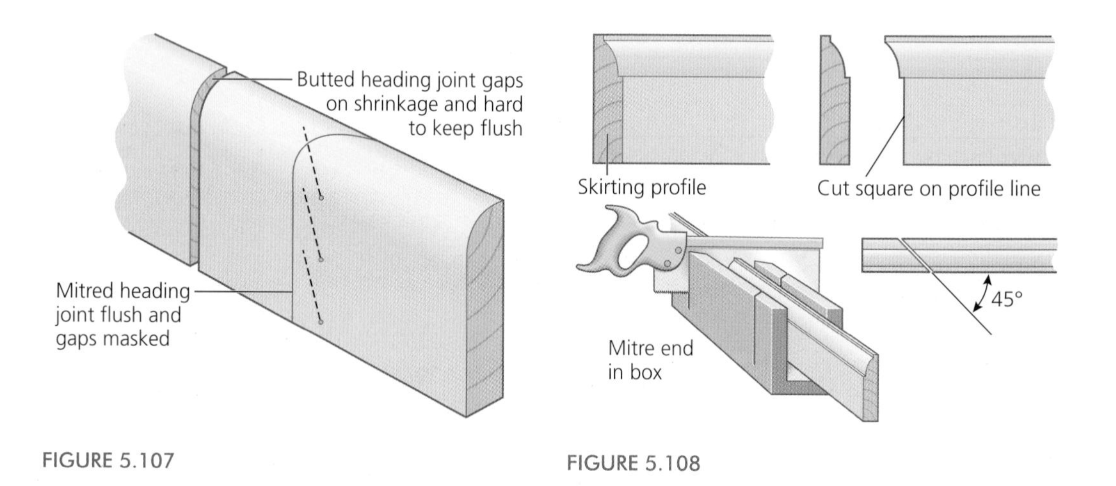

FIGURE 5.107

FIGURE 5.108

FREQUENTLY ASKED QUESTIONS

▶ What is the purpose of 'undercutting'?

The internal corners of solid and partition walls are rarely constructed perfectly square. Cutting scribed mouldings slightly out of square (under 90°) will ensure that the front edge of the joints remains tight when fitted.

CUTTING A SCRIBED JOINT (BULL-NOSED PROFILE)

Scribing the internal joints of some smaller profiles may result in a 'feathered' edge remaining. These types of scribe are difficult to cut and often result in a poor finish, due to the weakness of the diminishing timber. 'False mitres' should be used to return mouldings around internal corners to provide a neat joint.

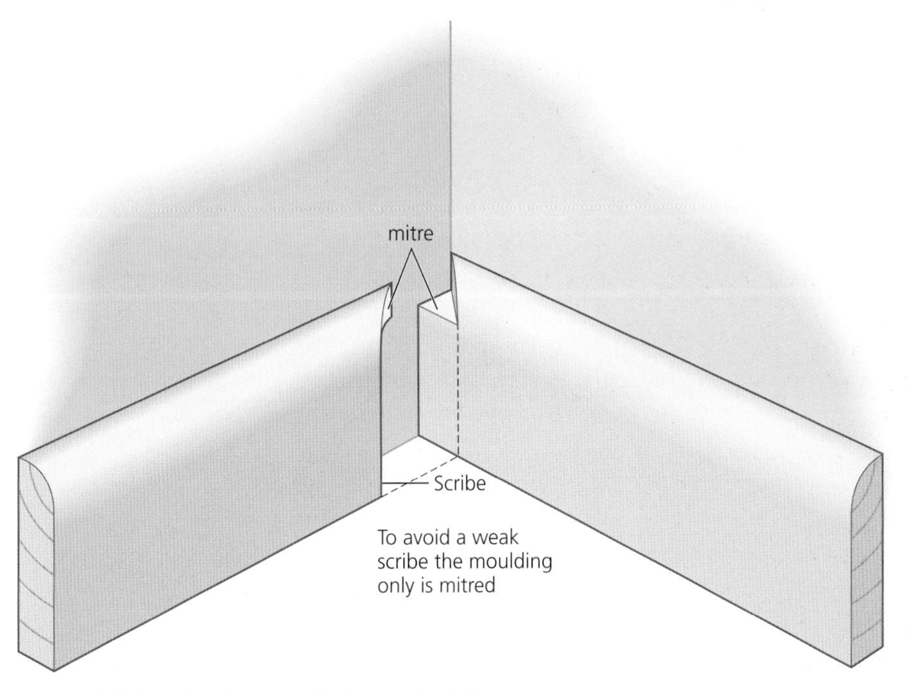

FIGURE 5.109 A false mitre between bull-nosed skirting

BISECTING ANGLES

It is a relatively straightforward task to mark out and cut scribed and mitred joints around square corners, but, as mentioned previously, not all walls are square. In these cases the exact angle of the walls must be determined and bisected to enable the joints to fit perfectly together. Marking out the intersection of the mouldings on the floor, around the 'acute' (less than 90°) or 'obtuse' (more than 90°) angles of the wall is the easiest method to determine the information needed. The bisected angles required to join skirting and other mouldings are usually established using the following method.

1. Sweep the floor area around the wall.
2. Placing a parallel section of skirting flat on the floor and up against one face of the wall, making sure that the skirting projects through the intended position of the joint.

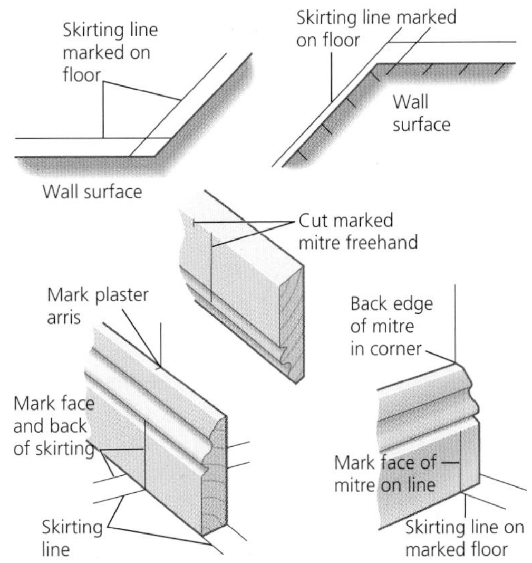

FIGURE 5.110 Bisecting an angle for skirting

3. Mark a pencil line on the floor along the outside edge of the skirting.
4. Repeat this process on the opposite wall, making sure that the pencil lines marked on the floor overlap each other.
5. Draw a straight line between the intersection of the two pencil lines towards the corner of the wall.
6. Place a sliding bevel against one face of the wall and adjust it to the bisected angle to transfer onto the moulding.

Stage 1

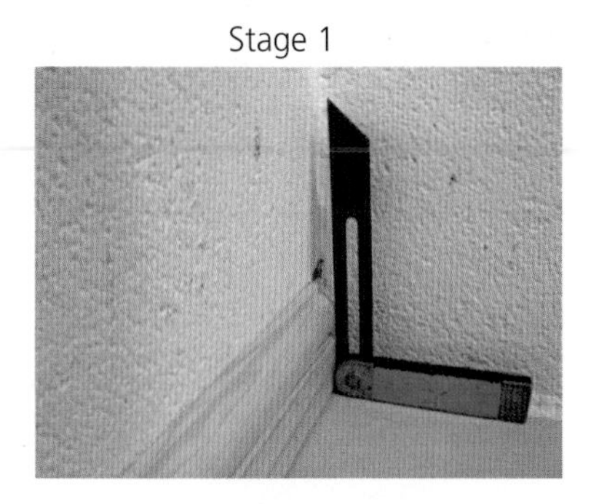

Stage 2

Stage 3

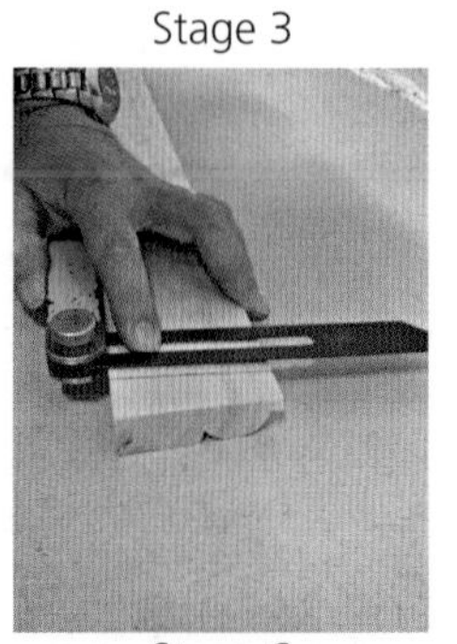

Stage 4

Stage 5

Stage 6

FIGURE 5.111 Scribing skirting joints against 'out of level' walls

SKIRTING FINISHES

The layout of some internal walls may prevent the continuation of skirting along their full length. Cutting the skirting square across the timber at these points may result in the unsightly end grain spoiling the finish of the moulding. Traditionally, the profile of the skirting can be returned over the end of the skirting in two directions to provide an aesthetically pleasing finish to the timber.

TRADE SECRETS

Trouble shooting – Occasionally timber skirting will have to be fitted to walls that are 'out of plumb'. In these situations the angle between the wall and floor will have to be established with a sliding bevel. This is then transferred onto the timber before marking out the mitre to create an accurate scribe joint.

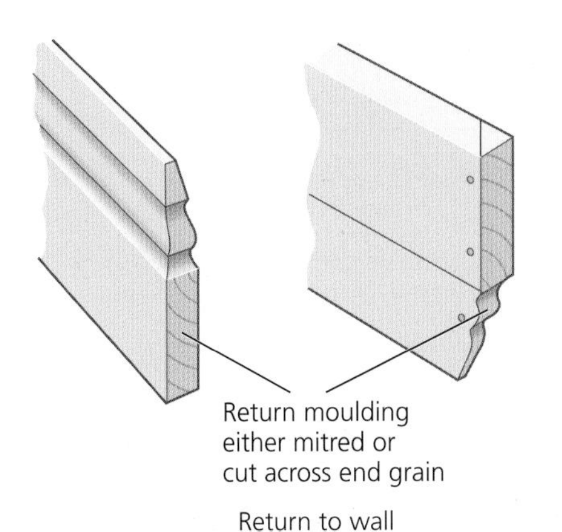

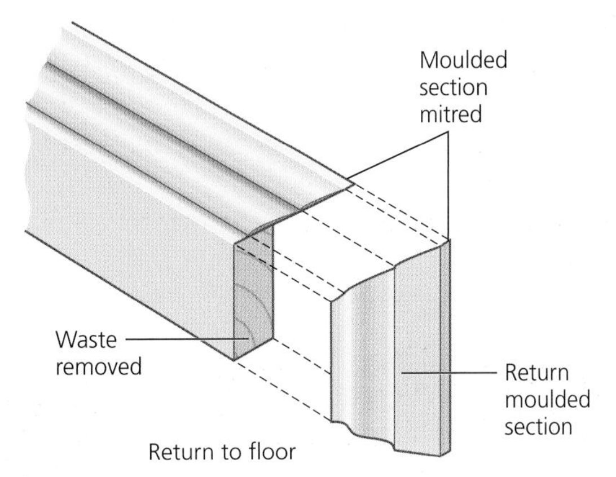

FIGURE 5.112 Skirting finishes

ACTIVITIES

Activity 22 – Installing mouldings

Read through the following questions and answer them as fully as you can to help you develop your underpinning knowledge of this subject area.

1. List four different materials used to create moulded sections.
2. What is the purpose of skirting boards?
3. At what height should dado rails be positioned on internal walls?
4. Sketch four different profiles and name them.
5. Explain the reasons for using 'plinth blocks'.

TIMBER GROUNDS

The term 'grounds' refers to the framework or battening used to support panelling and other types of wall cladding. Planed all round (PAR) low-grade treated softwood is normally used to construct the framework and battening for grounds, because it is readily available and relatively inexpensive. Treated timber should be used to prevent rotting caused by the trapped moist stale air between the back of the cladding/panelling and the wall. Using planed or 'regularised' (planed on a minimum of two opposite faces) timber ensures that the grounds are flat, which prevents potential difficulties encountered when packing behind the framework during their installation. Standard sections of timber such as 45 × 20 mm PAR are commonly used to batten walls and 45 × 45 mm to construct framed grounds.

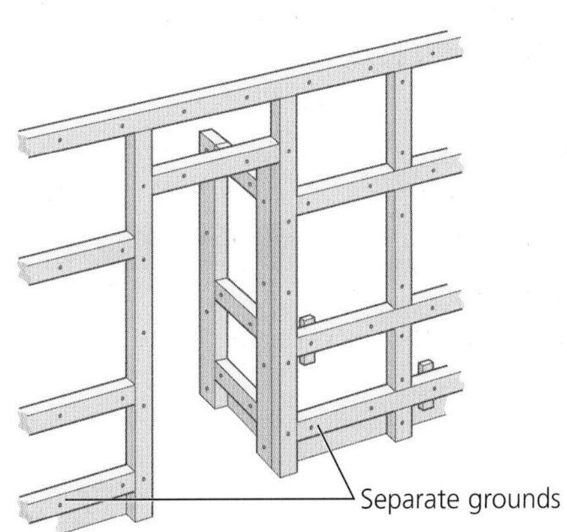

FIGURE 5.113 Timber grounds

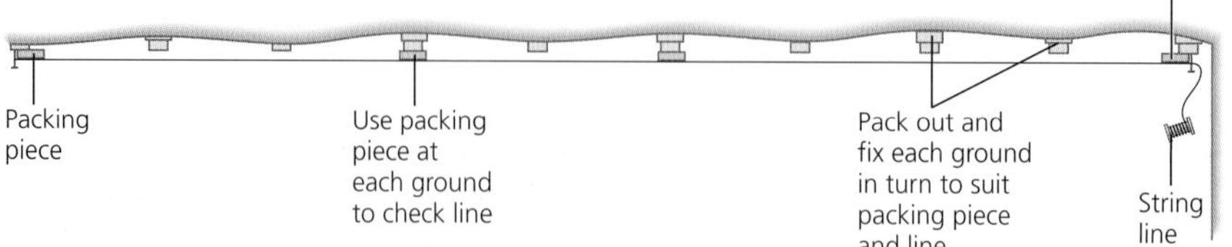

FIGURE 5.114 Use of packing from the high spot

Timber grounds may be used to straighten out uneven walls. This is done by establishing the most raised point on the wall, also known as the 'high spot', and regularising from this point. Working from the 'high spot' allows a fixed position on the wall to enable all the other lower areas to be packed to a straight line. This method of fixing the grounds uses only the minimum amount of packing.

Timber grounds allow wall coverings such as panelling, cladding and plasterboard to be easily secured with nails or wood screws.

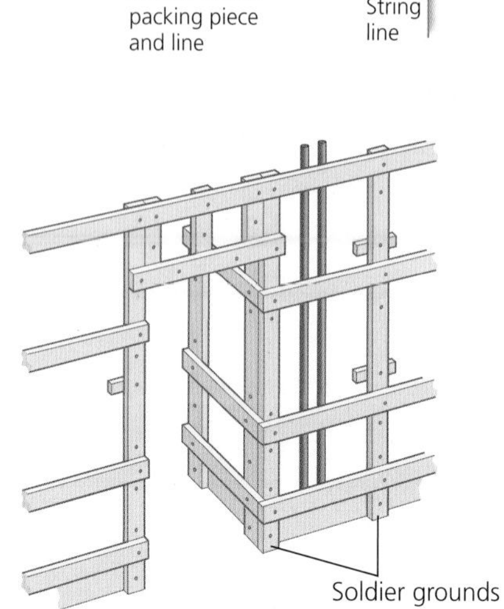

FIGURE 5.115 Counter-battening

The void created behind panelling and cladding with the use of timber battens can be used to hide service pipes and cables, although they can run only in the same direction as the battens. Fixing secondary battens across the initial rows allows the service pipes and cables to run vertically or horizontally behind the wall covering. This method of timber grounds is known as 'counter-battening'. Battens are normally spaced between 400 and 600 mm centres, depending on the type of cladding or wall panelling used.

FIXING TIMBER GROUNDS

The type of wall fixings needed to secure timber grounds will be determined by the wall construction. Failure to use suitable fixings may result in the weight of the wall panelling or cladding pulling the fixing out of the wall. Remember to check for buried services before drilling or screwing into any walls.

The types of fixing used in wall construction are as follows:

Wall construction	Type of fixing used
Brick	Cartridge tool fixing/screw and plastic plug
Block	Cartridge tool fixing/screw and plastic plug/masonry nails
Stone	Cartridge tool fixing/screw and plug
Concrete	Cartridge tool fixing/screw and plastic plug

WALL PANELLING AND CLADDING

TRADITIONAL WALL PANELLING

Timber panelling has traditionally been used in the building industry for hundreds of years. Its solid construction provides protection to plastered walls and also a decorative feature to enhance their appearance. Panelling is usually manufactured at three different heights; these are:

- dado-height panelling;
- three-quarter-height panelling;
- full-height panelling.

Three-quarter-height panelling is traditionally finished along its top edge with a 'plate shelf'. As the name suggests, this moulding was used to display plates and other ornamental items.

Solid timber wall panelling is normally constructed with top, bottom and intermediate rails, jointed together with stub mortise and tenon joints to form frames. The framework traditionally surrounds raised or sunk timber panels, which are grooved into the edges of the timber to hold it securely in position. The panels should never be glued into the frames because the movement between the components will cause the timber to split. Any movement that takes place with loose panels will occur in the grooves of the frame, rather than on the face of the panels. The proportions of the panels and sections of the mouldings, etc. are traditionally replicated on the window frames and doors to integrate them into the room.

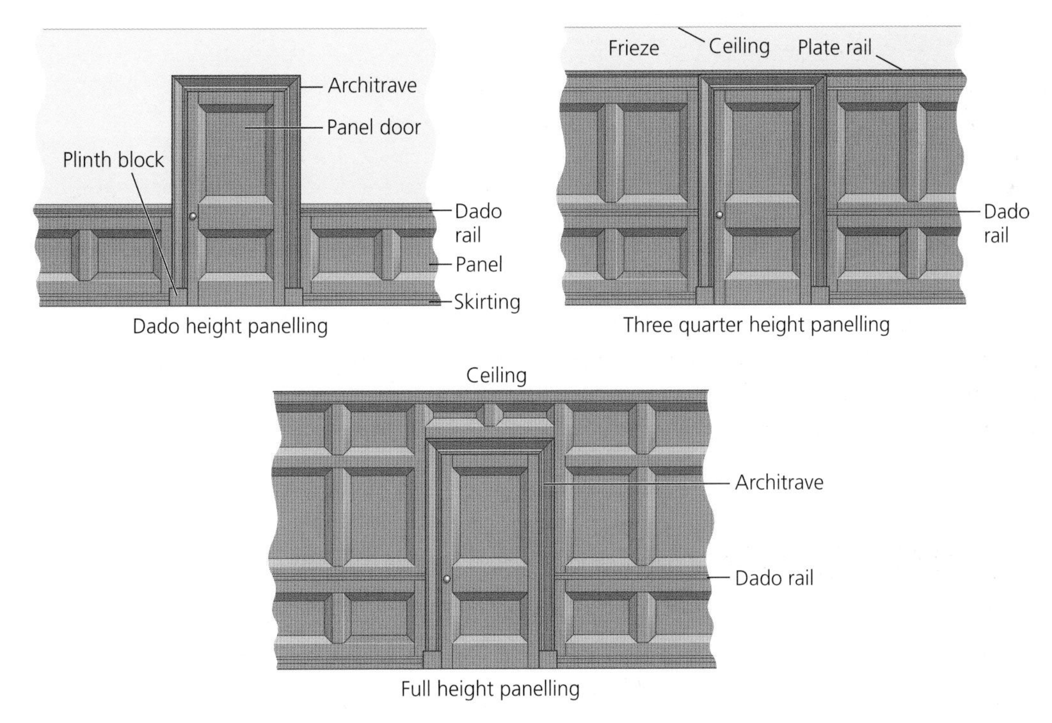

FIGURE 5.116

This style of panelling was traditionally used in banks, court houses, expensive houses and public buildings. Constructing panelling with solid timber is rarely used these days because of the material and labour costs.

MODERN METHODS OF CONSTRUCTING WALL PANELLING

Man-made timber-based sheet materials are commonly used these days to replicate the traditional styles of wall panelling at a fraction of the cost; they are also less likely to shrink or expand like solid timbers. The use of veneers and timber 'foils' allows materials such as MDF to be moulded and shaped to a variety of different designs and covered to replicate solid timber. Once these manufactured materials have been assembled it is very difficult to tell them apart from solid timber panels. Unfortunately, these materials are not as durable as solid timber, as knocks and scrapes usually damage the thin surfaces to reveal the core material. Similar damage occurring on solid timber panels can easily be sanded out and repaired.

Melamine-faced chipboard (MFC), laminated medium-density fibreboard (MDF) and veneered sheet materials are commonly used in the construction of modern panelling. These man-made boards are used because they are easy to work with and are available in a vast range of different colours and finishes.

MATCHBOARD PANELLING

Matchboarding is commonly used to clad the lower sections of walls up to dado rail height. It is secretly fixed to two rows of horizontal timber battens (grounds) through the tongues of each board before covering with the next adjoining board. Care should be taken to ensure that the oval nails used to fix the matchboarding are punched below the surface of the timber to ensure that the tongue and groove joint is not obstructed. The top edge of the matchboarding is covered with either a rebated dado rail or several sections of moulding. Skirting is normally used along the bottom of the panelling to cover the joint between the floor and the matchboarding. Deeper sections of skirting will require an additional row of battens at floor level; this will allow extra fixings across the width of the moulding and prevent the skirting from 'cupping'.

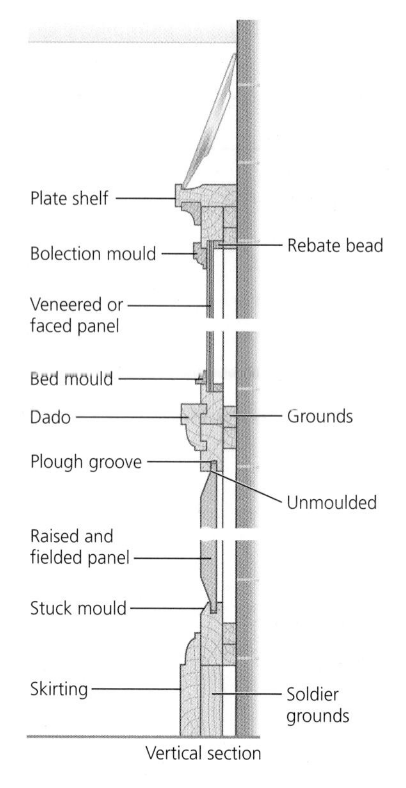

FIGURE 5.117

Apply the paint or stained finish to the panels before assembling and fixing the framework. This will prevent any shrinkage in the timber revealing an unfinished section of the panel.

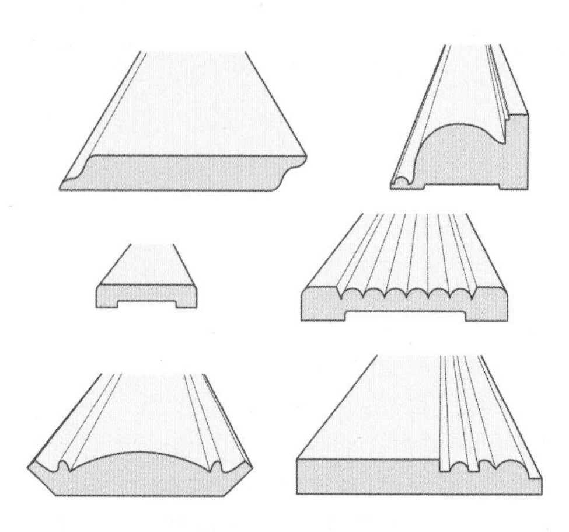

FIGURE 5.118 Foil-covered mouldings and veneered sheet materials

INSTALLATION OF WALL PANELLING

The back of wall panels and cladding should be sealed prior to installation to prevent moisture being absorbed into the timber. Failure to seal the timber correctly may result in the panelling distorting. Highly polished and decorative wall panels should have their methods of fixing concealed to maintain a flawless finish across their faces. Carefully planning the positions of the rails and trims at the design stage may permit some mouldings to cover the fixings used to secure the wall panelling. There are a number of alternative methods that could be used to secretly fix wall panels; these include:

- interlocking grounds, also known as 'split or hook battens';
- keyhole plates;
- screws and wooden plugs.

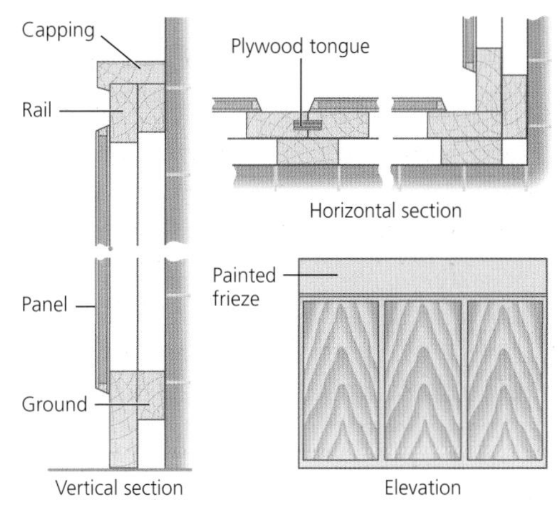

FIGURE 5.119

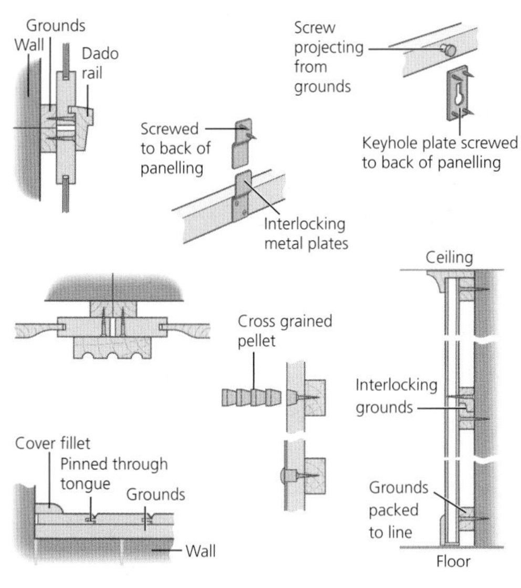

FIGURE 5.121 Spilt battens, keyhole plates and screws or wooden plugs

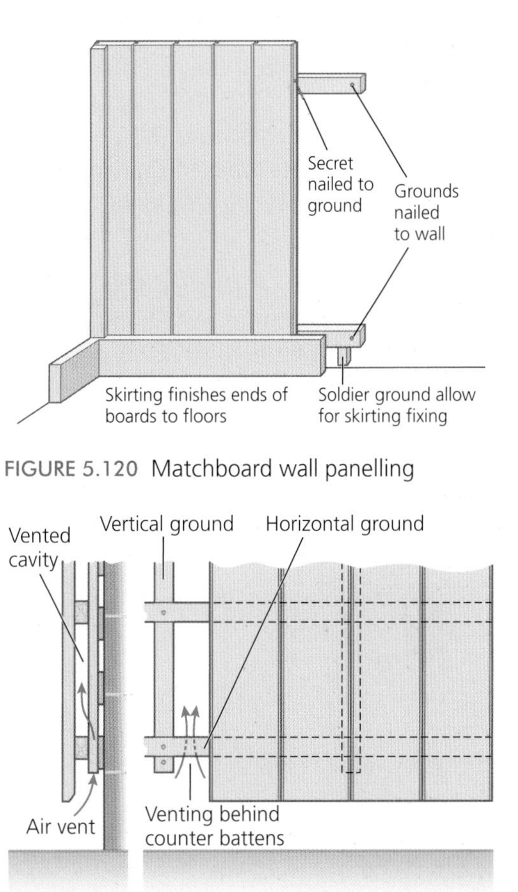

FIGURE 5.120 Matchboard wall panelling

FIGURE 5.122

EXTERNAL CLADDING

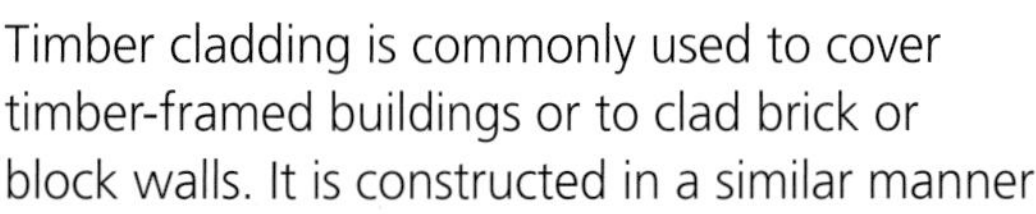

Timber cladding is commonly used to cover timber-framed buildings or to clad brick or block walls. It is constructed in a similar manner to internal cladding, except that the timber used in exposed positions must be treated or able to withstand the elements naturally without decaying. A vapour barrier is normally fixed between the back of the cladding and timber-framed walls to prevent water entering the building, while allowing moisture to escape. The void, or 'cavity', between the back of the cladding and the wall must be ventilated to prevent any moisture escaping the building being trapped and promoting the onset of timber decay.

INSTALLING WALL AND FLOOR UNITS AND FITMENTS

INTRODUCTION

Hand-crafted timber wall and floor units are rarely used in domestic houses because they are usually purpose built by cabinet makers or joiners, which makes them relatively expensive. Modern kitchens and bedroom furniture are mass produced on computer numerical control (CNC) machinery with timber-based sheet materials and simple mechanical joints. Mass-produced units can be supplied either 'flat packed' for assembly on site or 'rigid' (factory assembled).

- Flat packed units – this method of supplying units requires the installer to assemble all the components, following the manufacturer's assembly instructions. Budget ranges of furniture are normally supplied flat packed; this enables the cost of the units to be reduced due to them being only partially completed. Suppliers are able to increase the stock levels of units if they are flat packed because they require less space for storage and delivery. Depending on the quality, carcasses are usually manufactured from 16 mm melamine-faced chipboard (MFC).
- Rigid units – these types of unit are fully assembled and glued together during their manufacture. They are also higher quality, with 19 mm thick shelves and sides, and 9 mm back panels.

Standard kitchen base units are 600 mm deep and 900 mm high, and range in width from 300 mm up to 1200 mm. Wall-hung units range in width from 300 mm up to 1000 mm and 300 mm deep. This depth prevents people knocking their heads on the wall cupboards, while still allowing a dinner plate to be stored; it also allows the full width of the worktop to be viewed while standing close to the base units. Corner units are available in a range of different shapes and sizes to make the best use of the difficult space.

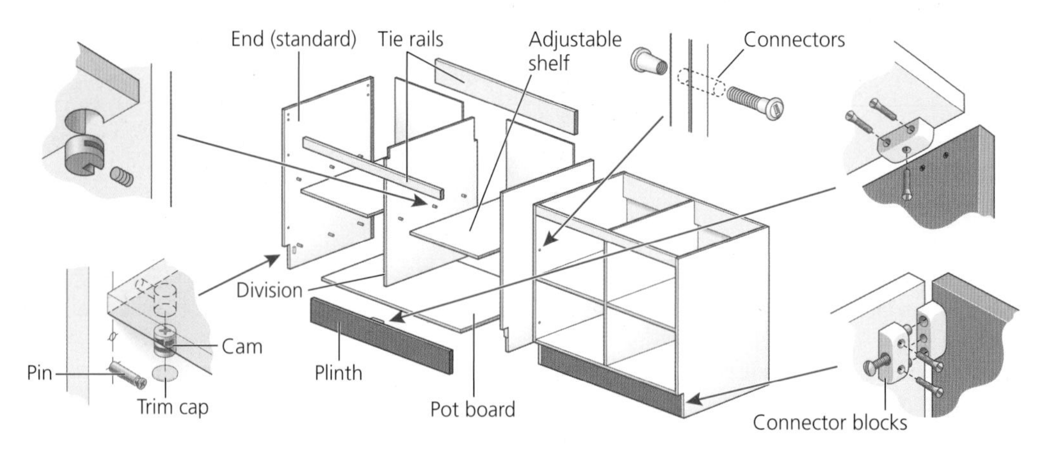

FIGURE 5.123 Kitchen base and wall unit

TRADITIONAL METHODS OF CONSTRUCTION (FRAMED)

Wall and floor units are traditionally constructed with a series of frames and rails fixed together with mortise and tenons, or dowelled joints. The end frames (or 'standards') usually have panels loosely fitted in plough grooves around their inside edges. High-quality units have grooves running down the back of the units to allow the back panels to be fixed. This will reduce the depth of usable space within the units, but allows service pipes such as water, gas and waste to be hidden behind the back panel.

There are two methods of fitting shelves into units: 'fixed' and 'adjustable'. Adjustable shelves are commonly used because they allow more flexibility within the unit. There are several methods used to allow shelves to be adjusted; these include:

- 'tonk' studs and strip;
- shelf studs.

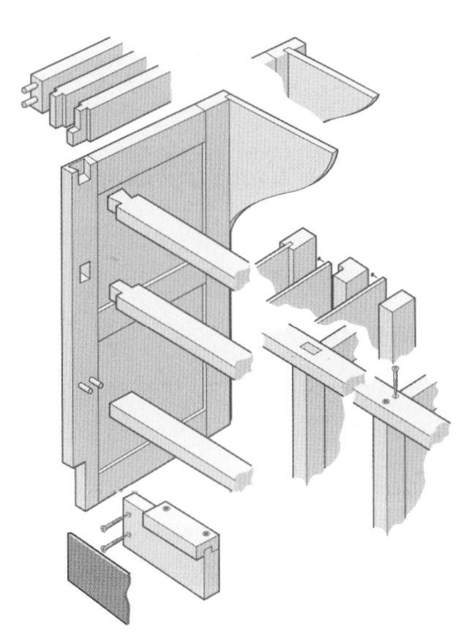

FIGURE 5.124 Construction of traditional framed unit

MODERN METHODS OF CONSTRUCTION (BOXED)

The majority of modern units are constructed with melamine-faced chipboard (MFC) in solid neutral colours or wood grains to match the doors and drawer fronts. Some more expensive units may have solid timber doors and drawers. The carcasses of these units do not require complex carpentry joints, because the man-made boards they are built with are not prone to warping or shrinkage. Each butt joint is located with wooden dowels and pulled tightly together to secure it in position with 'knock-down' (KD) fittings.

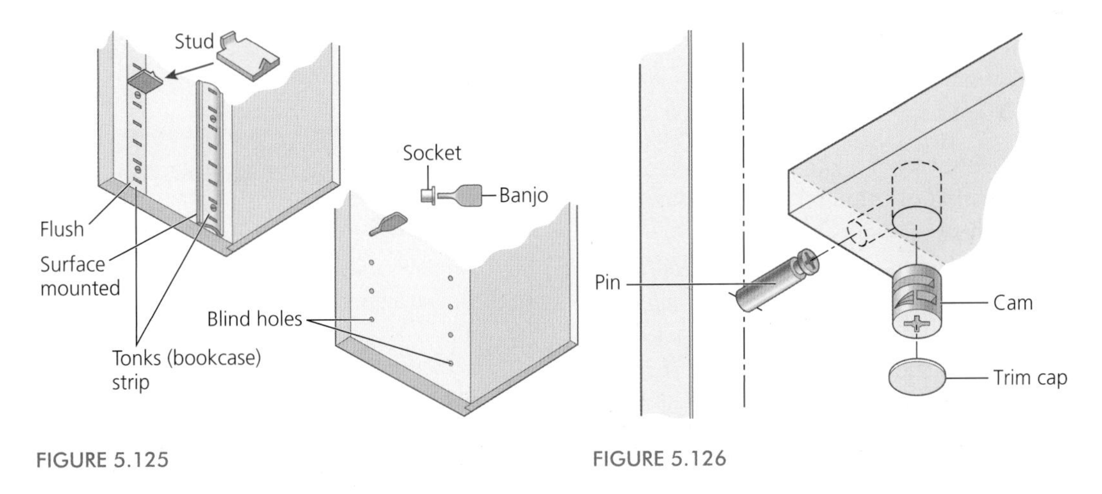

FIGURE 5.125

FIGURE 5.126

COMPONENT IDENTIFICATION

All floor-standing cupboards will require the 'potboard' to be built over the 'plinth', regardless of their method of construction. Raising the potboard off the ground prevents the unit from rocking or becoming unstable as a result of the floor being uneven. Some units may have a separate plinth, which may be scribed over an uneven floor surface before installing the unit. This method prevents having to adjust the base of heavy or large units during their installation.

FREQUENTLY ASKED QUESTIONS

Can cupboards be constructed out of any other materials?

Yes, cupboard units may also be formed using boxed construction with alternative timber-based materials such as:

- plywood;
- medium-density fibreboard (MDF);
- blockboard;
- solid timber panels.

These materials are more versatile than MFC because they can be painted, stained, polished, laminated and veneered.

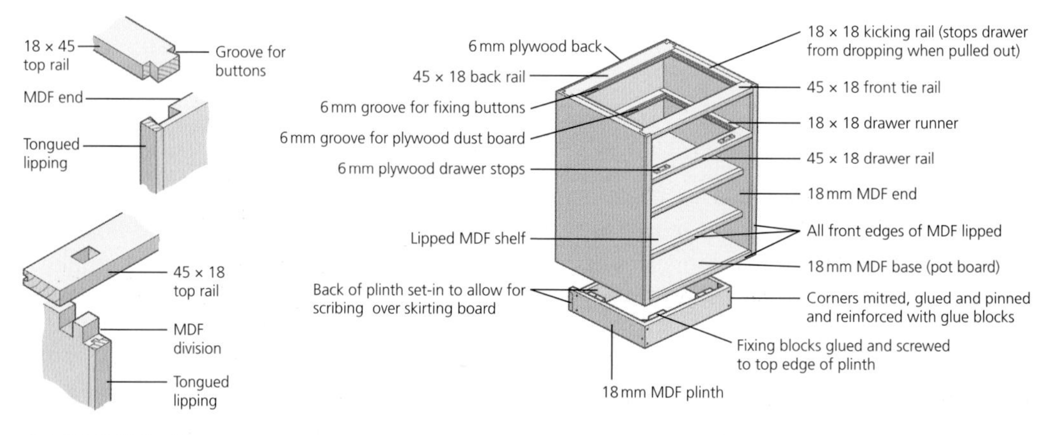

FIGURE 5.127 Components

INSTALLING KITCHEN UNITS

The layout out of a kitchen is planned around the position of the sink, fridge and oven; these three elements are collectively known as the 'working triangle'. Carefully planning and positioning these utilities will improve the efficiency and safety of the kitchen area. Before a kitchen can be installed correctly, you must ensure that you have a copy of the architect's or designer's plan and the specification. This information will provide the exact layout of the wall and floor units, as well as all the fixtures and fittings.

Good communication with the plumbers and electricians before installing the kitchen units may prevent mistakes at a later stage. It is important that the waste pipes, hot and cold water supplies, gas pipes, electrical cables, etc. are in the correct positions before encasing them with the units.

TRADE SECRETS

Most kitchen suppliers have a free, computer-based designing service. The consultant/designer will give advice and offer solutions to potential design problems. The kitchen plans are available and printed in two- (2D) and three-dimensional (3D) perspectives.

Some companies have similar software available free of charge, online. It is easy to use these facilities via the internet to accurately design and cost a kitchen, even with limited computer skills.

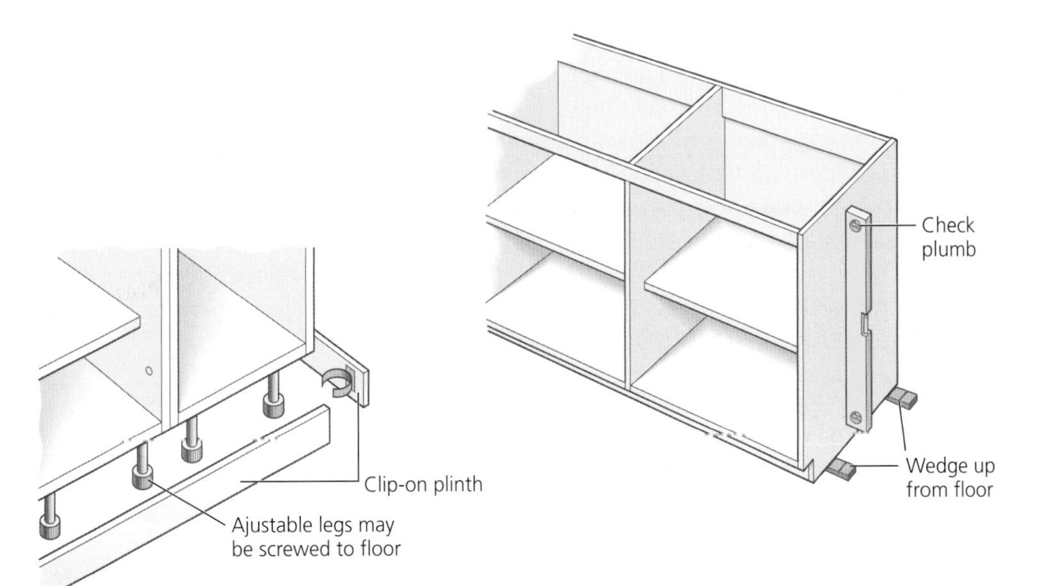

FIGURE 5.128 Adjustable legs

SEQUENCE OF INSTALLATION (RIGID BASE UNITS)

- Stage 1 – Mark a 1 metre high datum line around the perimeter of the room to provide a base level from which all other levels will be measured. This is usually transferred onto the walls with a rotary laser level, a water level or spirit level and straight edge.
- Stage 2 – Remove the doors, drawers and shelves from the corner base units. This will make them lighter during their installation and minimise the risk of damage to the decorative doors. Scribe the back of the units over any pipes etc. that may prevent the units going flat against the wall.
- Stage 3 – Level the corner units by turning the adjustable legs either clockwise to lower the units or anticlockwise to raise. (*Note*: the 'standards' on some units extend past the potboard to the ground level. These units are normally wedged under the bottom edge of the standards to a level position.)

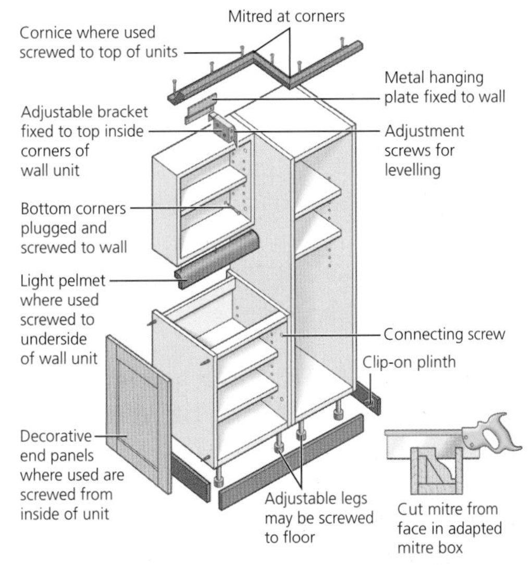

FIGURE 5.129 Installing base units

- Stage 4 – Position the remaining base units and adjust accordingly to level and align with the datum line. Use the 'connecting bolts' to fix the base units together through the standards. Secure the backs of the base units to the wall with metal fixing brackets, ensuring packing is used where the wall is hollow to ensure a straight line across the front of the units.

SEQUENCE OF INSTALLATION, CONTINUED (RIGID WALL UNITS)

- Stage 5 – Measure 450 mm plus the thickness of the worktop from the top of the base units. Transfer this height around the room, making sure it is parallel with the datum line.

- Stage 6 – Refer to the manufacturer's fixing instructions to establish the height of the 'cabinet hangers'. Transfer this height onto the wall, above the existing pencil line indicating the bottom of the walls units.
- Stage 7 – Mark the widths of the wall cabinets around the room. Measure in 25 mm from both sides of these positions and secure the metal hanging plates to the wall, using the fixings provided. Make sure that each wall unit has a pair of hanging plates located behind its intended position on the wall. (*Note*: it may not be possible to fix other wall units using this method. These cabinets will have to be secured by screwing directly through the rails running across the back of the unit into the wall, and covering the fixing with a 'screw cap'.)
- Stage 8 – Remove the doors and shelves from the wall units. Starting in the corner again, lift the cabinets on to the wall, making sure that the 'cabinet hangers' locate over the top of the hanging plates.
- Stage 9 – Use the adjustment screws located on the cabinet hangers to finely adjust the units to ensure they are level and correctly aligned. Clip the plastic 'cover caps' over the cabinet hangers to conceal the fixing.

FINISHING OFF THE UNITS

- Stage 10 – Cut the lengths of plinth to size and iron on the tape edging over the bare ends. Mark the centre positions of the adjustable legs on the back of the plinth, and secure the plinth clips with the screws provided. Attach the plinth to the adjustable legs by 'snapping' the clips into position
- Stage 11 – Insert the drawers back into the carcasses, making sure they are correctly aligned and functioning properly.

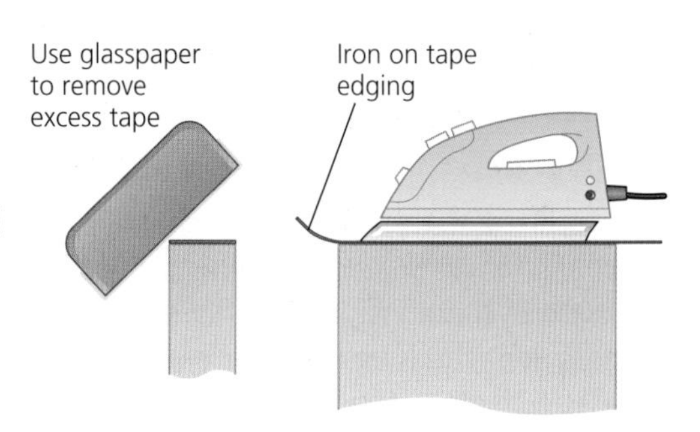

FIGURE 5.130 Ironing on tape edging

FREQUENTLY ASKED QUESTIONS

The floor under the kitchen units is uneven and the plinth will not fit. What can I do?

This is a common problem in kitchen fitting that is easily fixed by scribing the bottom edge of the plinth over the floor with the use of a 'gauge slip'.

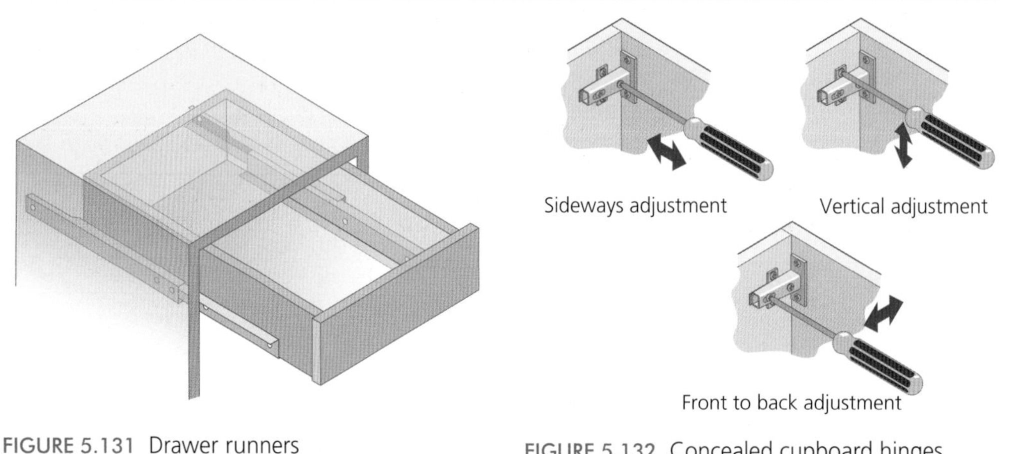

FIGURE 5.131 Drawer runners

FIGURE 5.132 Concealed cupboard hinges

- Stage 12 – Secure the decorative end panels to the outer standards of the units with fixing screws. The doors can be fitted to the units by clipping the concealed cupboard hinges to the mounting plates already fixed to the carcasses. (*Note*: other types of hinge may require a fixing screw to be tightened to hold the two parts of the hinge together.)
- Stage 13 – Finally, the lighting pelmet and cornice can be installed around the top and bottom edges of the wall units. These are simply cut to length or mitred and fixed together with fast-bonding mitre adhesive, and screwed into position with 'knock down' (KD) fixings. Mitre adhesive is commonly used to bond mouldings together because it sets in approximately 30 seconds once the activator has been applied to one face, providing an extremely strong joint.

WORKTOP SURFACES AND MATERIALS

Kitchen worktops are manufactured from a variety of different materials; these include those described below.

FIGURE 5.133

- Tiles – these are usually laid over a minimum thickness of 18 mm plywood. The grain in the plywood provides a good surface for the adhesive and tiles to bond. Tiled worktops are rarely used in this country because they are difficult to keep perfectly sterile.
- Laminates – post-formed laminated worktops are easily maintained and relatively inexpensive. They are commonly available in a vast range of different colours, surface finishes and materials.
- Granite, marble and stone – these materials have grown in popularity in recent years, although they are still relatively expensive. The choice of colours is limited, but the natural beauty and hardness of the material is difficult to replicate with other products. These products are usually cut to shape and size, and installed by specialists.
- Timber – worktops are normally made up in their width with narrow strips of wood, laminated together to minimise the amount of movement across the width of the board. The movement in the timber must be accounted for with elongated fixing points in the top rails of the base units. Timber worktops are usually supplied unfinished to allow them to reach their equilibrium moisture content 'in situ' before sealing with oil.
- Corian™ – corian is a solid surface material constructed with a blend of natural minerals, pigments and pure acrylic polymer. It is an extremely durable and non-porous material that can be shaped, moulded, cut and jointed just like timber, although it is usually fitted by specialists. When connected together, the joints become seamless and practically invisible. Corian is commonly used in areas where hygiene is of the highest importance, such as dental surgeries and hospitals.
- Stainless steel – stainless steel is normally used in commercial kitchens and canteen areas because it is extremely durable and hardwearing. The metal surface is not affected by hot items and is easily maintained.

WORKTOP EDGE DETAILS

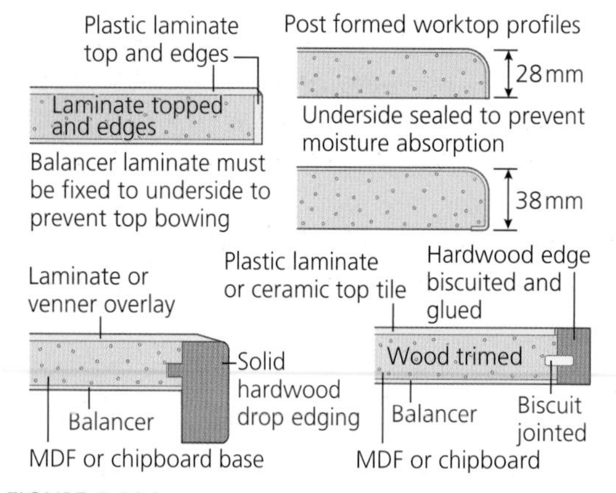

FIGURE 5.134

CUTTING AND JOINTING POST-FORMED WORKTOPS

The joints between adjoining sections of worktops can be connected with either 'metal jointing strips' or 'butt and mitre' joints. Metal jointing strips are simply screwed into position along the joint between two worktop sections.

Butt and mitre joints are formed with an electric router with a plunge facility and a worktop jig. The jig is used as a guide for the router to remove the moulded front edge of the worktop and square the end of the adjoining piece. Realignment of the jig will allow recesses to be machined across the joint to allow 'worktop connecting bolts' to be fitted. There are many different styles of worktop jigs, each one requiring a slightly different set-up, and capable of additional operations. It is advisable to read the manufacturer's instructions prior to cutting the joint to avoid costly mistakes.

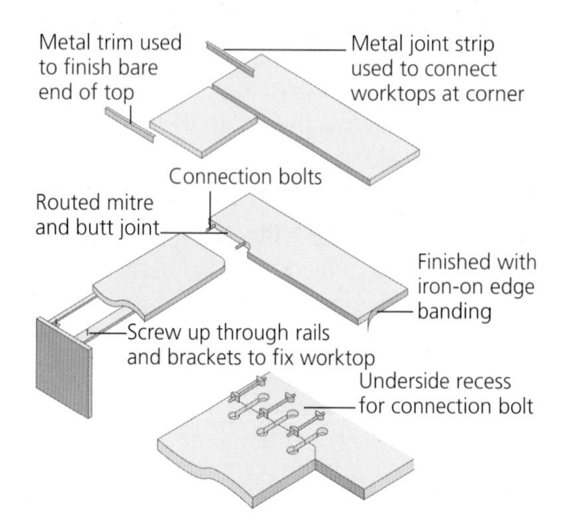

FIGURE 5.135 Joining post-formed worktops

FIGURE 5.136 Worktop jig

Care must be taken when cutting laminated kitchen worktops to ensure that no damage occurs to the faces and edges. There are several methods commonly used to overcome these difficulties they include those listed below:

- Use downward-cutting jigsaw blades to cut out the recesses needed for hobs and sinks. This operation should be done from the face of the worktop.
- Use upward-cutting jigsaw blades to cut from the back of the worktops. This method is sometimes preferred by installers, because cutout guidelines are clearly marked and visible on the chipboard surface. (*Note*: care should always be taken when cutting through thick material with a jigsaw, as they usually have a tendency to cut out of square, resulting in a wider cut on one face than the other.)
- Use a router to make several shallow cuts following a jig, until the full depth of the worktop is achieved. As a general rule, you should always rout into the edge of shaped or post-formed edges to prevent breakout.
- Portable powered circular saws should be used to cut worktops to length. Straight cuts are simply achieved by clamping a batten to the underside of the worktop and following this line with the base of the saw.
- Apply masking tape over the lines to be cut on the face of the worktop and use a fine upward-cutting jigsaw to remove the waste material.
- 'Plastic inserts' in the base of a jigsaw will minimise the amount of breakout on the upward strokes of the blade.

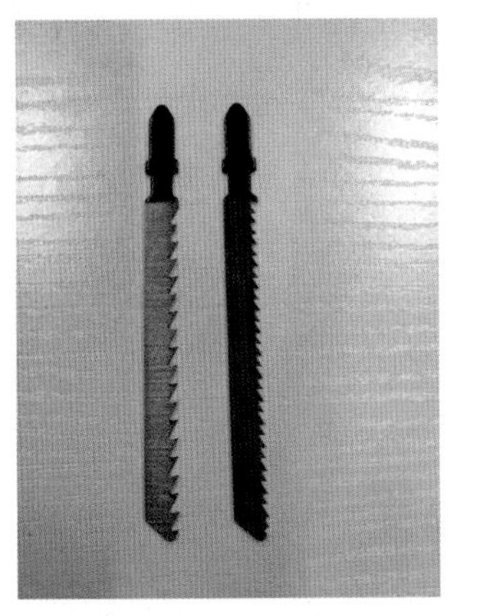

FIGURE 5.137 Downward and upward cutting jigsaw blades

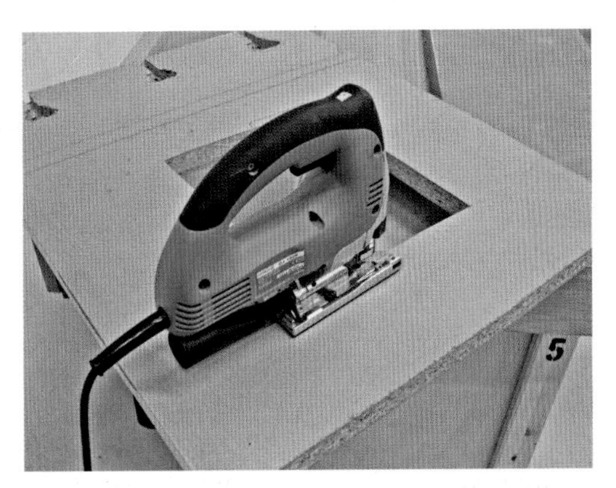

FIGURE 5.138 Cutting a worktop with a jigsaw

ACTIVITIES

Activity 23 – Installing wall and floor units and fitments

Read through the following questions and answer them as fully as you can to help you develop your underpinning knowledge of this subject area.

1. Name the two possible methods by which kitchen units could be delivered to site.
2. What materials can be used to construct 'boxed' units?
3. Explain the purpose of a datum line.
4. Identify the two possible methods of returning 'post-formed' worktops.
5. Explain one method of cutting openings in laminate worktops without damaging the visible surface.

MULTIPLE-CHOICE QUESTIONS

1 Which **one** of the following activities forms part of second fixing?
  a Roofing
  b Flooring
  c Door hanging
  d Stud partitioning

2 Which **one** of the following is a standard door size?
  a 1600 x 600
  b 1981 x 762
  c 1981 x 800
  d 2000 x 800

3 The thickness of an internal door is **most** commonly:
  a 25 mm
  b 30 mm
  c 35 mm
  d 40 mm

4 A 'Norfolk latch' is used to secure which **one** of the following door types?
  a Jib
  b Fire
  c Ledged
  d Revolving

5 The illustration below shows which type of lock?
  a Rim
  b Privacy
  c Secrecy
  d Mortise dead

6 Which **one** of the following rails is fixed horizontally to the wall at chair back height?
  a Dado
  b Frieze
  c Middle
  d Cornice

7 Plinth blocks are used between which **two** of the following components?
  a Dado and architrave
  b Frieze and architrave
  c Architrave and frieze
  d Architrave and skirting

8 Internal corners of skirting joints are best
  a butted
  b mitred
  c scribed
  d biscuited

9 Matchboarded panelling is best fixed by
  a screws with cover caps
  b using secret-headed nails
  c face fixing and filling the hole prior to painting
  d secret nailing through the shoulder of the tongue

10 Buried electrical wiring is best located using
  a drawings
  b a magnet
  c a services detector
  d residual current devices

CHAPTER 6

ERECT STRUCTURAL CARCASSING

LEARNING OUTCOMES

By the end of this chapter you should have developed a knowledge and understanding of:

- erecting truss rafter roofs;
- constructing gables, verges and eaves;
- installing floor joists.

INTRODUCTION

The aim of this chapter is for students to be able to distinguish between 'cut' and 'trussed' roofs, and to recognise their component parts. The chapter explains the methods used to install pitched roofs and timber floor joists at ground and upper floor levels. In addition, this chapter will identify and explain current Building Regulations applicable to this area of study, following all the relevant health and safety law and good working practices.

ROOF SHAPES

Roofs can be constructed in a variety of different shapes using several methods of construction. Each design and construction method is influenced by the following factors:

- cost;
- size (e.g. pitch – angle of the roof measured in degrees; span – width of the roof measured from the outside of the wall plates);
- type of building (e.g. listed, architectural design);
- position (e.g. in keeping with the surrounding area);
- materials used (e.g. the roof covering materials);
- living space;
- Building Regulations.

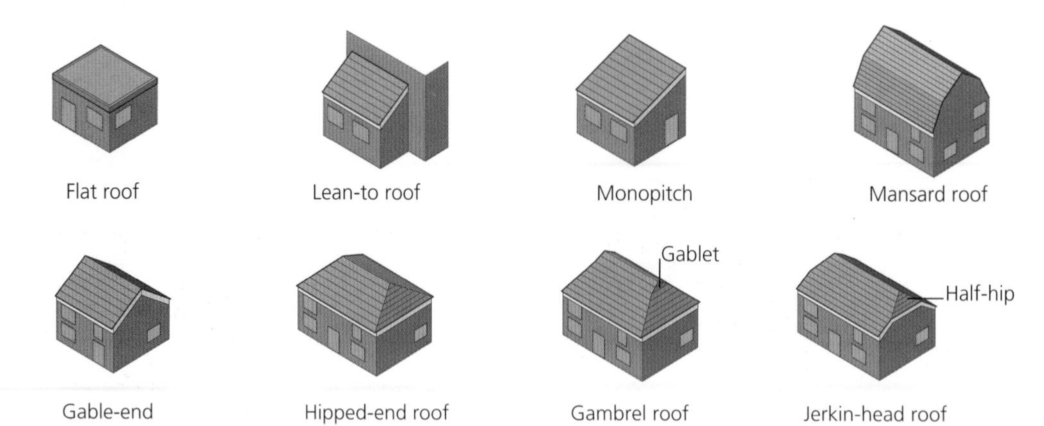

FIGURE 6.1 Roof shapes

TYPES OF ROOF

FLAT ROOFS

Flat roofs have a slight fall (between 1:80 and 10° depending on the covering material) or pitch to allow rainwater to run to the guttering. This will prevent rainwater from lying on the roof for long periods of time and, eventually, early deterioration. A roof with a pitch of 10° or less is considered to be a flat roof; any other roofs are known as pitched. The relatively low pitch of a flat roof means that rainwater runs off the roof a lot slower than pitched roofs. This is the disadvantage of flat roofs, because this slow distribution of water can often lead to high maintenance costs due to water penetration. The life expectancy for flat roofs is between 10 and 50 years depending on the covering material, although the development of roof coverings will improve this time. (There is more on flat roofs on page 233.)

LEAN-TO ROOF

Lean-to roofs have a single pitch, which is supported against a building that is higher than the ridge or apex of the roof (an abutting wall). They are relatively simple to construct and therefore a low-cost option; they are commonly used for:

- garages;
- small house extensions;
- porch or bay windows.

MONOPITCHED ROOF

Although this type of roof often gets confused with a lean-to it has several main differences. First 'mono' (meaning single) pitched roofs are independent; this means that they are unattached or free-standing at the ridge to another building. The detail at the ridge of a monopitch will be similar to the lower end of the roof, with fascia and soffit detailing, although guttering will not be required at the highest end of the roof.

GABLE ROOF

The brickwork on a gable-ended wall will be built up to the ridge (also known as the apex). A building may be end gabled, front gabled or a combination of the two; this is known as cross-gabled. Gable walls may endure high winds and are considered to be weak without the support of the timber-framed roof.

Gable roofs have a simple construction, with a series of common rafters and without any complex hip or valley timbers. There are two ways of finishing the gable end at roof height:

1. projecting verge;
2. flush verge.

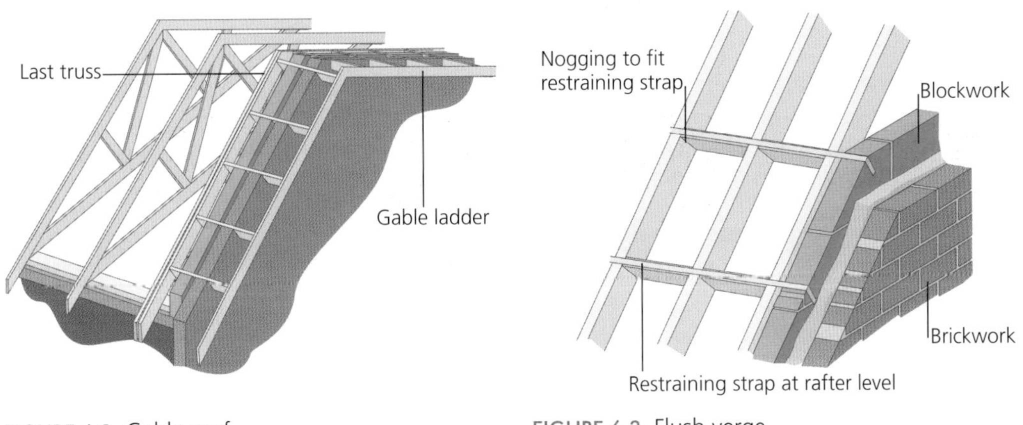

FIGURE 6.2 Gable roof

FIGURE 6.3 Flush verge

FREQUENTLY ASKED QUESTIONS

What is a 'verge'?

The verge is similar to the eaves on the lower part of a roof, but instead it returns up the gable end. The verge can be either projecting or flush and is covered with a barge board. Its purpose is to protect the gable end wall from the elements, and prevent water ingress (entering) between the wall and roof structure.

HIPPED ROOF

Hipped roofs require more complex methods of construction around the hipped rafters and the jack rafters. It is normal for the pitch to be returned around the end of the building on the hipped end.

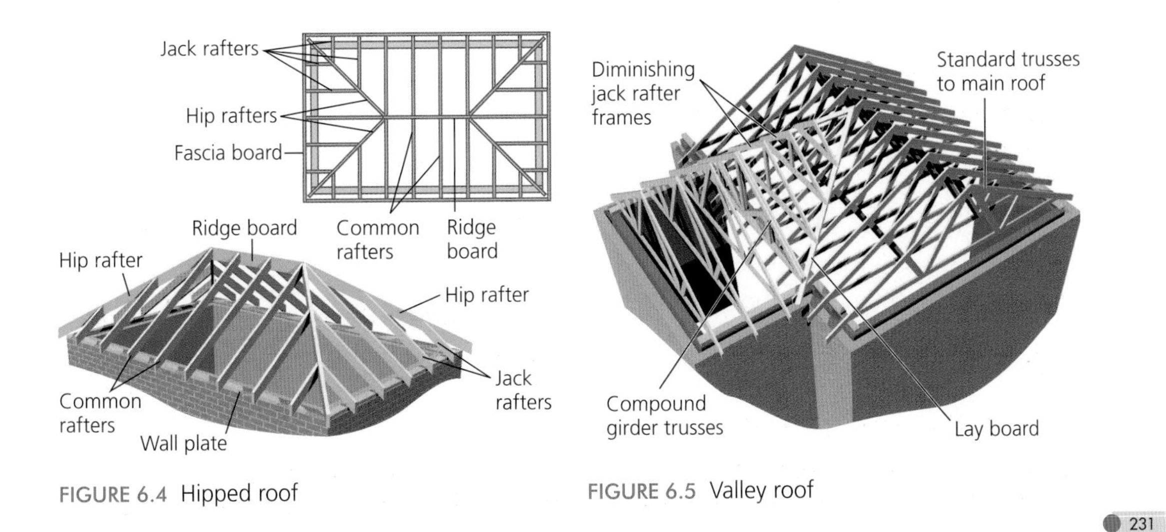

FIGURE 6.4 Hipped roof

FIGURE 6.5 Valley roof

VALLEY ROOF

Valleys are used to form the roof over an internal corner of the building. There are two ways that the valley can be formed, either with a 'valley rafter' or 'lay boards'.

VALLEY RAFTERS

A valley is very similar to a hip but in reverse; instead of forming an external corner on the roof, a valley forms an internal one. All the true lengths of the roof members are exactly the same on the valley as they are on the hip. The following table illustrates how this works:

Valley		Hip
The deductions that are made on the crippled rafters	=	The deductions made on the jack rafters
The true length of the valley rafter	=	The true length of the hip rafter

LAY BOARDS

Lay boards are the timbers used on top of the common rafters to form a valley. They are mostly used to extend off an existing roof structure and avoid the need to cut the existing roof timbers; for example, as used on most dormer roofs. Although lay boards prevent the need to cut the timbers in an existing roof, they restrict the usable room within a loft space.

TRADE SECRETS

After you have marked out the hip or valley rafter, use a tape measure to double check your measurements before cutting. Measure between the top of the ridge and the outer edge of the wall plate, and transfer this measurement on to the hip/valley rafter.

Remember – when using this method, you are measuring diagonally across the rafter length, not parallel to the rafter.

MANSARD ROOFS

Mansard roofs have either two sloping surfaces (the lower surface usually with a steeper pitch) or a flat roof on the top and a steep pitch on the lower part of the roof.

Mansard roofs provide additional space in the loft to accommodate extra rooms. Dormer windows are normally constructed to project from the lower pitch; this prevents them having to be installed the same pitch as the roof.

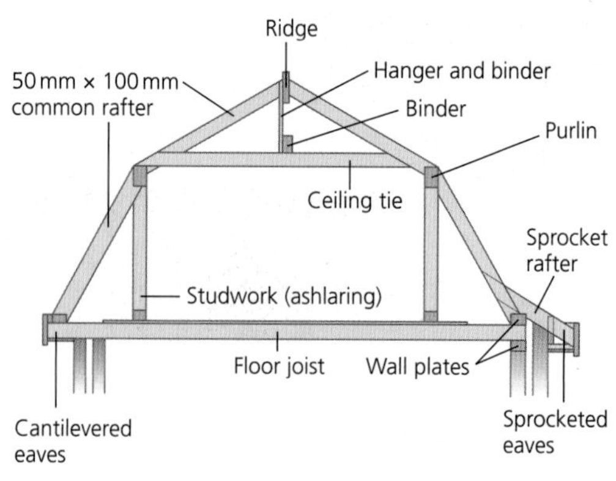

FIGURE 6.6 Mansard roof

DORMER ROOFS

Dormer roofs project from the surface of a roof to introduce natural light into the loft space/first floor. They are commonly used in loft conversions to improve the headroom in the roof space, although the amount of clear headroom will depend upon the size and type of dormer used. Dormers can also be used to break up large expanses of roof tiles to provide a decorative architectural feature. There are several ways to form dormers; these include:

- flat roof dormers;
- pitched roof dormers (gable end or hipped);
- eyebrow dormers.

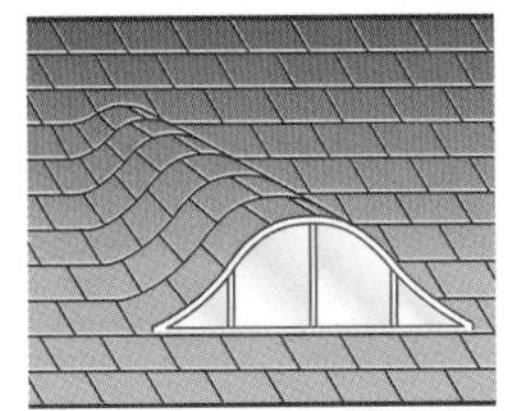

Segmental

FIGURE 6.7 Dormer roofs

FLAT ROOFS

As mentioned earlier, flat roofs are prone to leaks and rapid deterioration. It is essential therefore that they are constructed to allow the water to run off them effectively. This can be achieved by forming a single fall for roofs with a small span and double or stepped falls for larger spans. The increased span of a double roof will require additional support at the apex; this is normally provided by either a steel joist or an additional timber beam. Both of these methods of supporting the joists across the large span are known as 'principal beams'.

There are several methods that can be used to create a fall or pitch on a flat roof; these include the following:

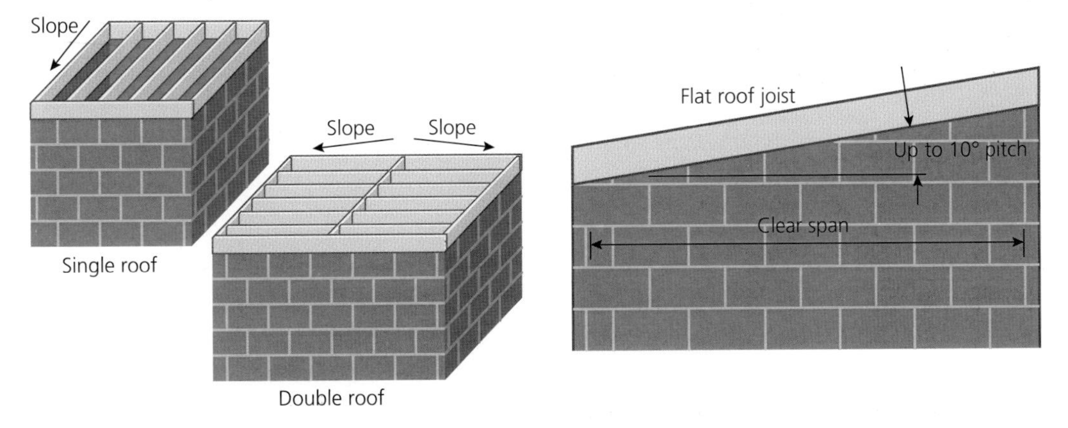

FIGURE 6.8 Double flat roof

FIGURE 6.9 Sloping joists

- Sloping joists – sloping joists are the simplest way of creating the fall on a roof. The disadvantage with this method is that the ceiling on the underside of the roof will have the same pitch.

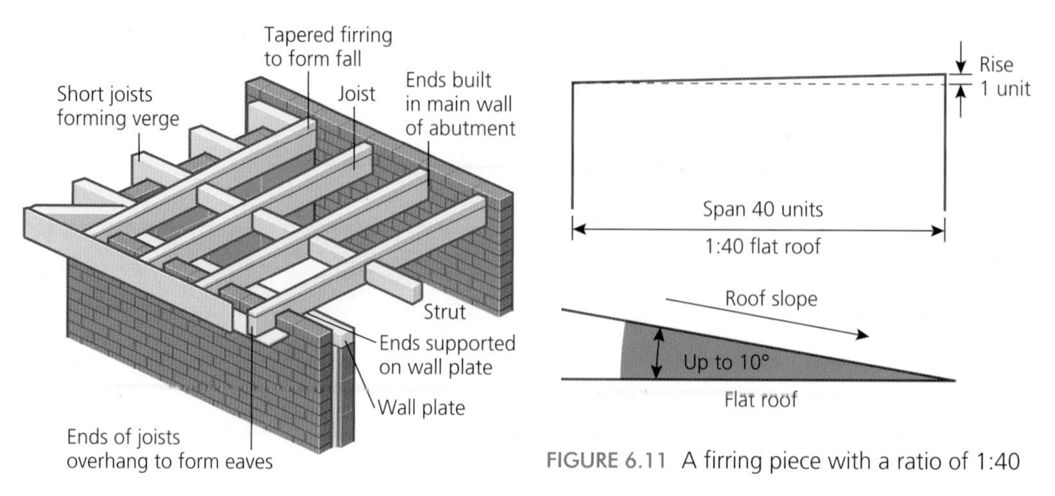

FIGURE 6.10 Firring pieces

FIGURE 6.11 A firring piece with a ratio of 1:40

- Firring pieces – firring pieces are the most common method of creating the pitch on a flat roof. Firring pieces are available from most builders' merchants; they can also be made on site. The thinnest end of the firring must be not less than 25 mm and should have a ratio of 1:40.

FREQUENTLY ASKED QUESTIONS

What do you mean by a ratio of 1:40?

Ratio means a comparison of numbers or measurement expressed as a fraction. In this case it is expressed as 1 in 40.

Example: Firring pieces – for every 1 unit that a firring piece rises, 40 units run along its length.

- Diminishing pieces (also known as 'counter-firrings') – the use of diminishing pieces eliminates the need for tapered firrings along the top of the joists. This method is used when the joists span across the width of the roof, but is prone to sagging between the joists. Sagging can lead to pools of water holding on the roof for long periods of time.

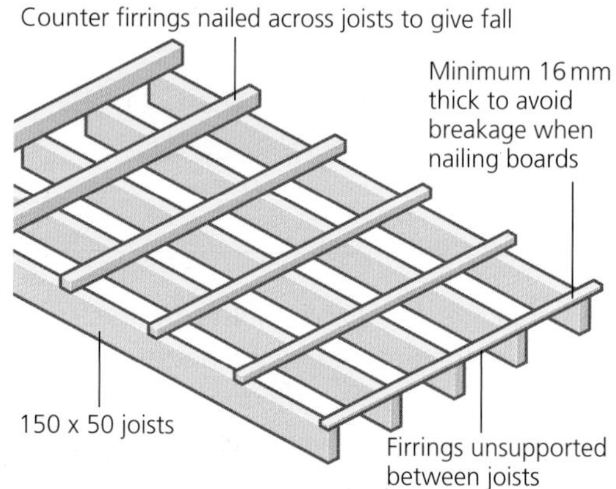

FIGURE 6.12 Diminishing pieces on flat roof joist

FIXING

The layout of flat roof joists is similar to the layout of floor joists. Joists are normally positioned at 400, 450 or 600 mm centres, depending on the thickness and size of the decking material used and the sectional size of the joists. These centres will ensure that the edges of the sheet material used for the decking will span from the centre of one joist to the centre of another to provide a secure fixing point. There are a number of ways to secure the joist in position as shown in Figure 6.13.

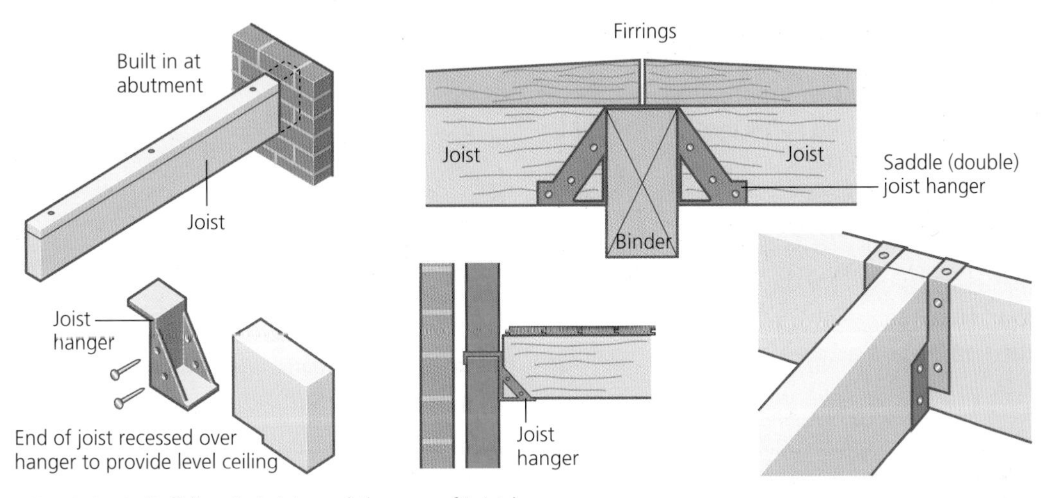

FIGURE 6.13 Building-in joists and the use of joist hangers

Size of roof joist	Spacing between joists 400 mm	450 mm	600 mm
50 × 97	1.89	1.86	1.78
50 × 122	2.53	2.49	2.37
50 × 147	3.19	3.13	2.97
50 × 170	3.81	3.73	3.47
50 × 195	4.48	4.36	3.97
50 × 220	5.09	4.90	4.47
75 × 170	4.44	4.33	3.96
75 × 195	5.13	4.95	4.53
75 × 220	5.76	5.56	5.09

Approved document A of the Building Regulations (Structure) considers the following factors when stating the minimum sectional size of flat roof joists:

- span;
- joists covering (decking, roofing felt, etc.);
- possible snow loads.

FREQUENTLY ASKED QUESTIONS

Can any timber be used to construct a flat roof?

No, only stress-graded timber is permitted for flat roofs. Joists are available in a range of different grades depending on their structural properties. Although grading can be carried out visually, a more effective method is machined stress grading.

STRUTTING

Noggings are introduced in rows at 1.8 metre centres; this prevents joists from twisting, and provides strength through interconnection. This type of movement in the joists will damage the finished surface of the ceiling by revealing hairline cracks.

Strutting is normally spaced at 1200 mm centres. This will ensure that the edges of the plasterboard used for the underside of the flat roof joists (the ceiling) will land in the middle of the timbers and achieve a good fixing point.

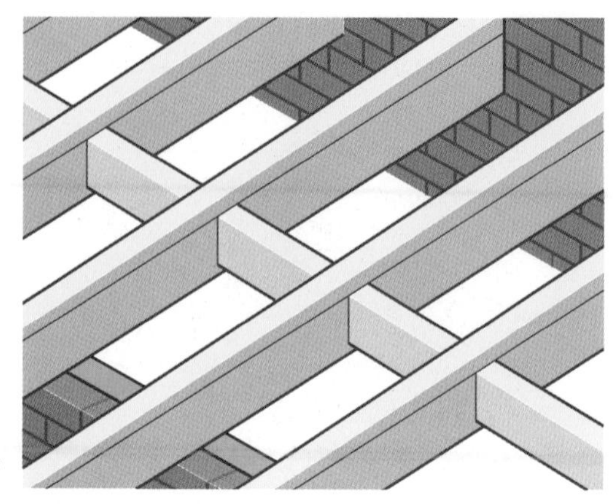

FIGURE 6.14 Strutting

DECKING AND ROOF COVERING

Decking is the term used for the material that covers the joists before the roof covering is applied. In general, the materials used are exterior-quality timber-based sheet materials, although some roofs will still contain softwood tongue and groove (T&G) boards. Tongue and groove boards are an expensive option, and can 'cup' to hold pockets of water if they are positioned across the fall of the roof.

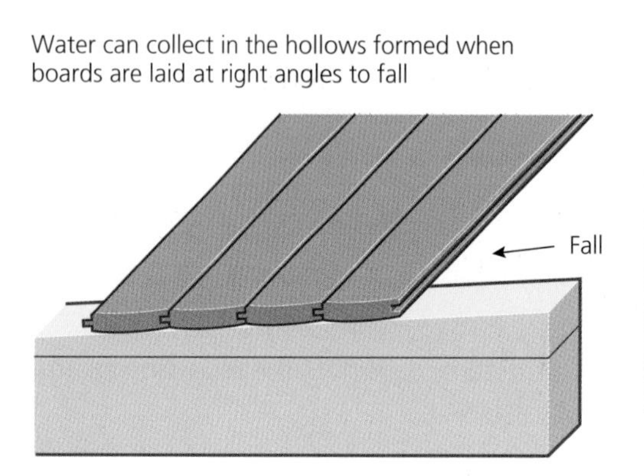

FIGURE 6.15 Cupped T&G decking

TRADE SECRETS

When fixing T&G boarding ensure that the annual rings running through the timber are facing downwards. If the boards are fixed through their centres into the joists, this will minimise the amount of cupping in the boards.

The overall size and thickness of the decking will vary depending on the spacing of the joists and the materials used. The following timber-based sheet materials are commonly used:

- exterior-grade chipboard 18–22 mm;
- orientated strand board (OSB) 15–18 mm;
- weather- and boil-proof plywood (WBP) 16–19 mm;
- wood wool slabs 50 mm.

FIGURE 6.16 Exterior-grade chipboard

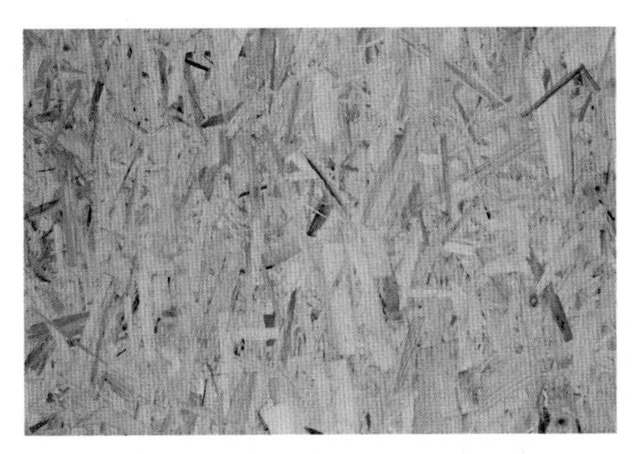

FIGURE 6.17 OSB

FIGURE 6.18 WBP plywood

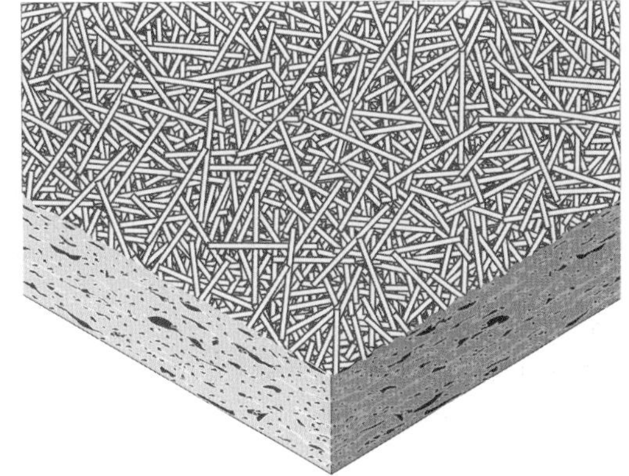

FIGURE 6.19 Wood wool slabs

DECK COVERING

Traditionally, flat roofs were covered with roofing felt, bitumen and gravel, and had a life expectancy of around 15–20 years. In recent years roofing materials have been developed to last in excess of 60 years; these materials include:

- thermoplastic polyolefin (TPO) – life expectancy 25 years;
- elastomeric membranes (EPDM) – life expectancy 20 years;
- PVC membranes – life expectancy 20 years;
- mastic asphalt – life expectancy 60 years;
- built-up felt – life expectancy 15–20 years.

FINISHING THE EAVES

The eaves are the portion of a pitched or flat roof at wall plate level. Overhanging eaves provide the means of closing in the joisting and the provision of a fixing for the guttering. The eaves type can dramatically change the appearance of a building.

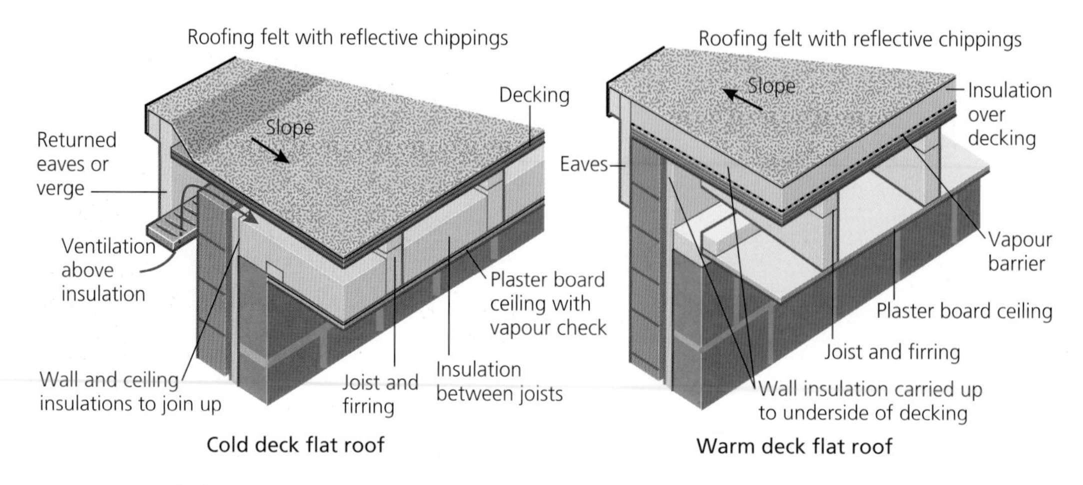

FIGURE 6.20 Flush and overhanging eaves

The fall on a flat roof will direct rainwater into the guttering at one end on a single roof and two ends on a double roof. Angled fillets fixed along each end create a curb, preventing the flowing rainwater from running over the edge of the fascia without a gutter. In addition to this, a 'drip' batten is fixed along the top edge of the fascia board at the lowest point on the flat roof. The drip batten will have the roof covering wrapped around it to ensure that any water that runs down the roof falls directly into the guttering.

FASCIAS AND SOFFITS

Fascias are the sections of either timber or uPVC (plastic) that are fixed directly onto the front of the eaves. They provide weather protection to the roof and a fixing point for the brackets to hold the guttering in position. The soffit is the section of material fixed to the underside of overhanging eaves. Soffits are used to prevent wildlife (e.g. birds and bats) entering the roof space; they also control the amount of warm air in the summer and cold air in the winter.

FREQUENTLY ASKED QUESTIONS

How are uPVC fascias and soffits fixed to the roof?

Plastic fascias and soffits are usually fixed with plastic-headed ring shank nails. The heads of the nails should match the colour of the materials being fixed. The ring shank along the nails provides a strong fixing that is resistant to pulling.

VENTILATION

Building Regulations (Document F) state that flat roofs should have permanent ventilation through the roof space equal to a continuous gap along the eaves of 25 mm. There are many different products available from builders' merchants to achieve this ventilation gap. The majority of the products available are manufactured from plastic in a wide range of colours and sizes.

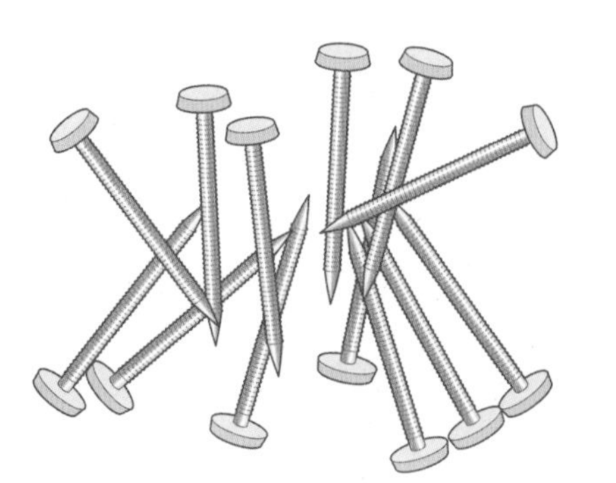
FIGURE 6.21 Poly pins

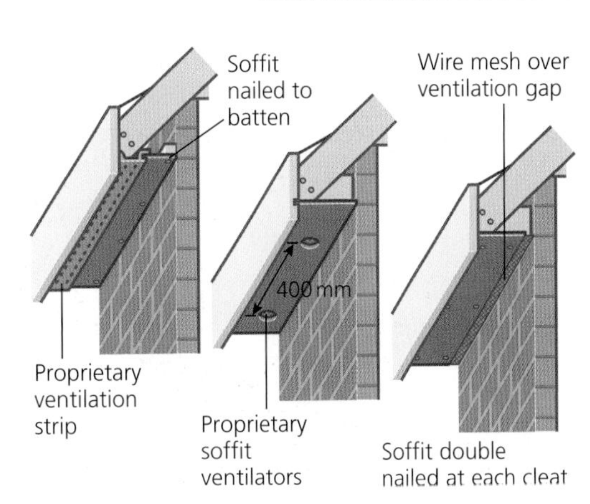

FIGURE 6.22 Soffit ventilation

ACTIVITIES

Activity 24 – Constructing gables, verges and eaves

Read through the following question and answer it as fully as you can to help you develop your underpinning knowledge of this subject area.

1. Draw a suitable finish for the eaves on a trussed roof and identify the following components: fascia, soffit, guttering and cradling.

CUT ROOFS

The term 'cut roof' generally means that all the roof timbers are marked out and cut on site. The methods used to construct roofs have been developed over the past 50 years to:

- reduce the time taken on site to mark out and cut complex joints;
- reduce the time taken to fix the roof timbers on site;
- reduce the amount of heavy timbers used in roof construction;
- increase productivity and lower costs.

FIGURE 6.23 A typical cut roof

Trussed roofs have been developed in recent years to replace cut roofs by providing a labour-saving and cost-effective alternative (see 'Fixing trussed roofs', page 253).

CUT ROOF TERMINOLOGY

Wall plate

The wall plate runs around the top of the inner cavity wall. It is used as a fixing point for the roof timbers and to spread the weight of the roof evenly. The wall plate is usually positioned on a bed of mortar and fixed into position with galvanised wall plate straps by the bricklayers. Wall plates are usually either 75 × 50 mm or 100 × 50 mm; this will depend on the size and weight of the roof. Half lap joints should be used to return the wall plate at the corners and to extend their length as required.

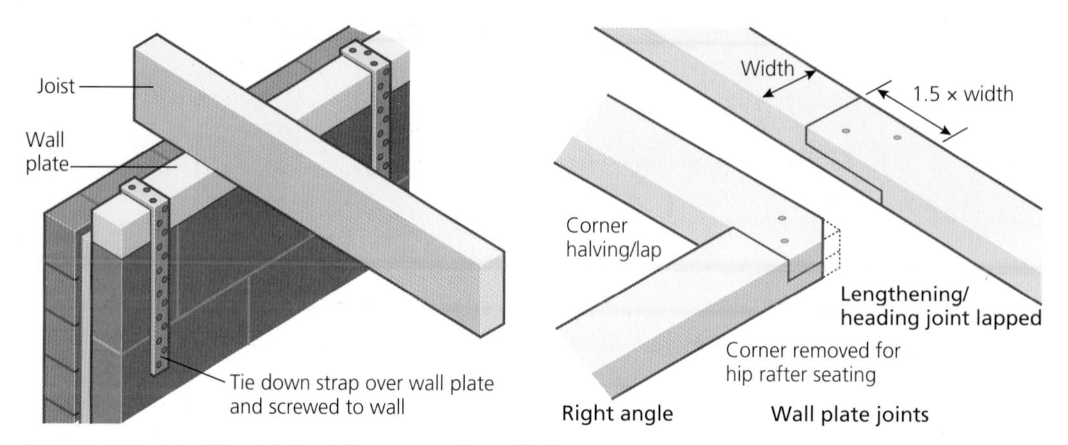

FIGURE 6.24 A half lap joint used to return the wall plate

Common rafter

A common rafter is the shortest timber fixed between the ridge and the wall plate.

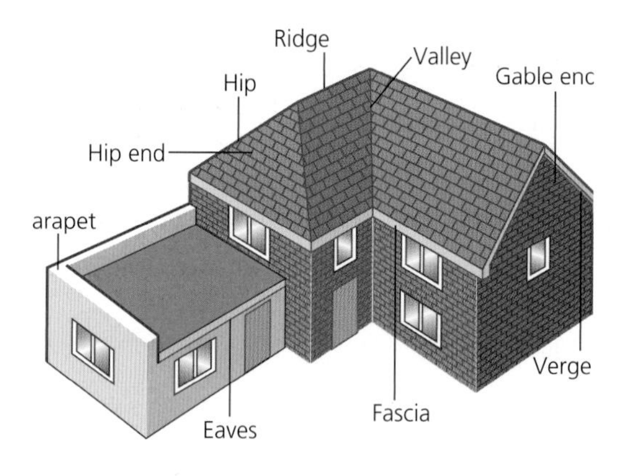

FIGURE 6.25 A roof

Ridge board

The ridge board provides the spine of a roof, with all the main timbers (e.g. common rafters, hip rafters, valley rafters, cripple rafters) branching off.

Crown rafter

The crown rafter is fixed in the centre of the hipped end and provides strength by bracing the exposed end. The crown rafter is marked out in exactly the same way as the common rafter. The reductions made to its length may vary depending on the thickness of the materials used for the ridge and saddle board.

Hip rafter

The hip rafter is main structural timber used to form a hipped roof. The hip rafter is usually deeper than the common rafters to accommodate the plumb cut on the upper ends of the jack rafters.

FREQUENTLY ASKED QUESTIONS

What is a compound cut?

A compound cut is an angled cut through both the face and edge of a piece of timber. There are many compound cuts on a cut roof; they can be found on jack rafters, crippled rafters, hip rafters and valley rafters. They can be cut with either a hand saw or a good-quality portable chop saw.

Jack rafters

Jack rafters are simply common rafters reduced in length to diminish into the hip rafter; they provide structural support on either side of the hip rafter.

Valley rafters

Valley rafters are the main central timbers located between an internal corner and the ridge.

Cripple rafters

Cripple rafters are the shortened common rafters that diminish into the valley with no 'foot' (bird's mouth joint).

Purlins

Purlins are normally found in roofs with large spans. They provide intermediate support to the rafters to prevent sagging along the length of the roof. They can be made from either large sections of heavy timber or steel joists with a timber wall plate fixed to the top. A cut roof that contains purlins is known as a 'double roof'. Roofs with smaller spans may not need purlins because the lengths of the rafters are usually smaller and are less likely to distort.

TRADE SECRETS

Compound cuts can be difficult to cut accurately on site, especially for trainees. Using the following techniques will improve your results:

1. *Position the timber with your marking out in a vertical position.*
2. *Position your saw over the cut to be made.*
3. *Look along the top edge of the saw and align it with your marking out.*
4. *As you start your first cut, continue to sight the top edge of the saw with your marking out.*

Remember – always cut on the waste side of your marking out.

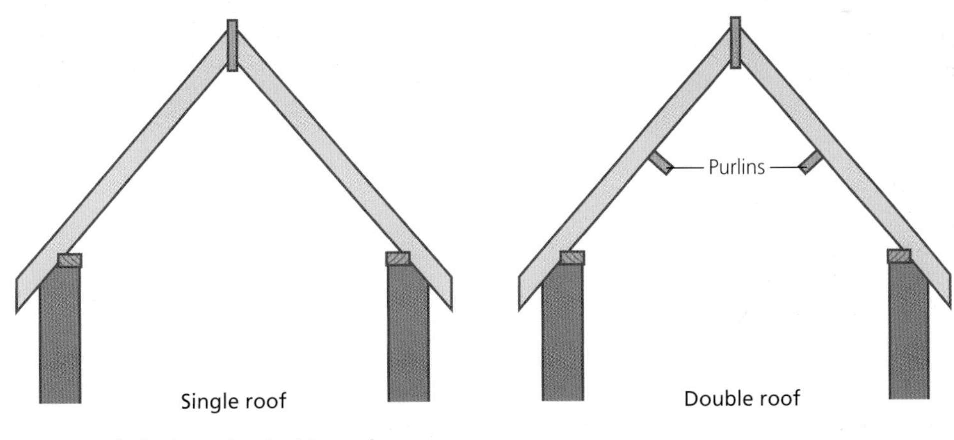

FIGURE 6.26 A single and a double roof

There are two methods of fixing the purlins to the underside of the rafters:

1. 'square' to the underside of the rafters (this is the simplest and most cost-effective method of supporting the rafters with purlins);
2. 'plumb' with the rafters and bird's mouthed over (a strong fixing through the bird's mouth into the wall plate will prevent any movement between the purlin and the rafters).

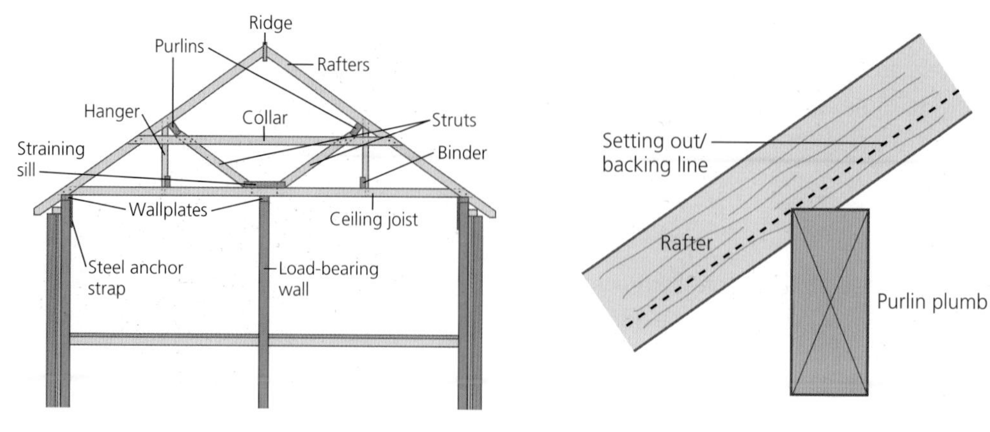

FIGURE 6.27 'Square' purlins

FIGURE 6.28 'Plumb purlins'

Sprocket piece

A sprocket piece is a section of roof timber fixed to the lower part of the rafters. Sprockets are normally fitted to roofs with a steep pitch; this will reduce the speed of rainwater running off the roof into the guttering.

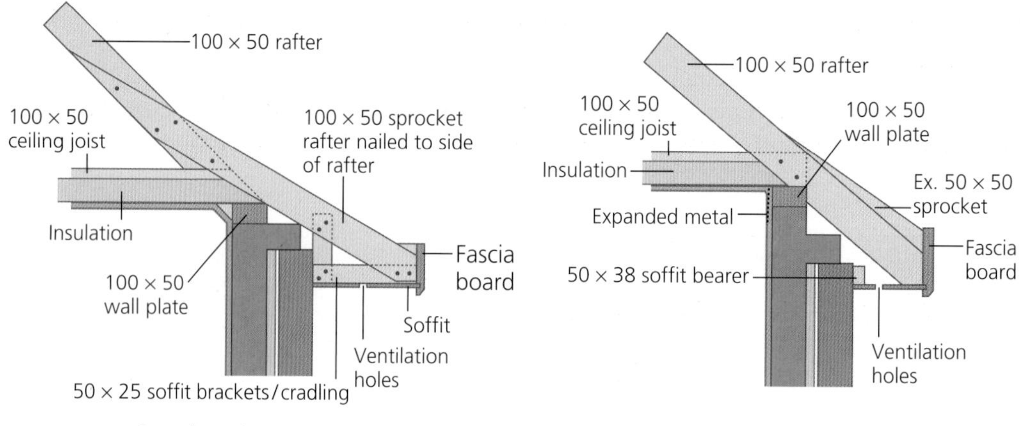

FIGURE 6.29 Sprocket piece

Ladder frame

Ladder frames are normally constructed with two common rafters and a series of timbers fixed in between, known as 'noggings'. Ladder frames provide an overhanging section of roof at the gable end, known as the 'verge', and provide a fixing for the barge board and soffit.

Strutting

Strutting is sections of timber that provide structural support to the purlins by bracing off load-bearing walls or straining sills.

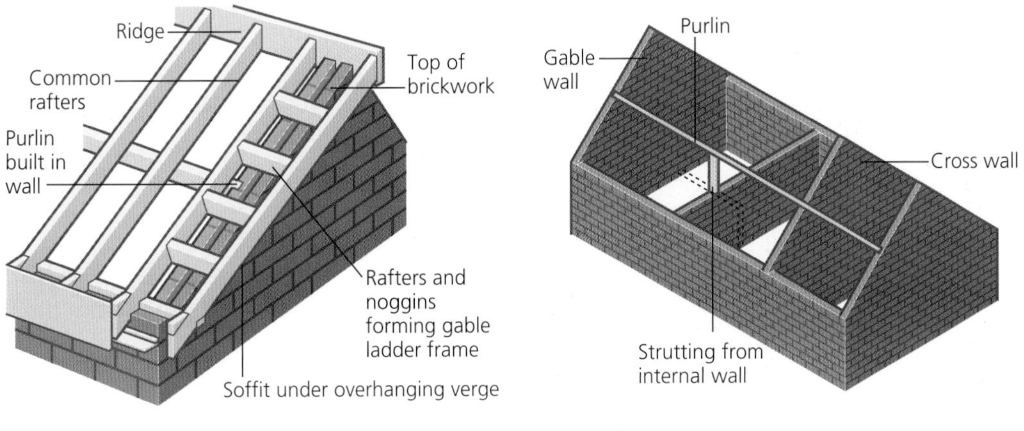

FIGURE 6.30

FIGURE 6.31

Ceiling joists

Ceiling joists span across the underside of a roof; they sit directly on the top and are skew nailed to the wall plates. Ceiling joists provide the level surface onto which the plasterboard is fixed. They should be fixed at centres that accommodate a fixing for the edge of the boards. A roof with ceiling joists at wall plate level is known as a close couple roof. Ceiling joists fixed to the wall plate will help to prevent the outer walls from spreading as they tie the feet of the rafters together.

Some roofs with small spans do not need to have ceiling joists because the risk of the outer walls spreading is reduced; these types of roof are known as 'couple roofs'.

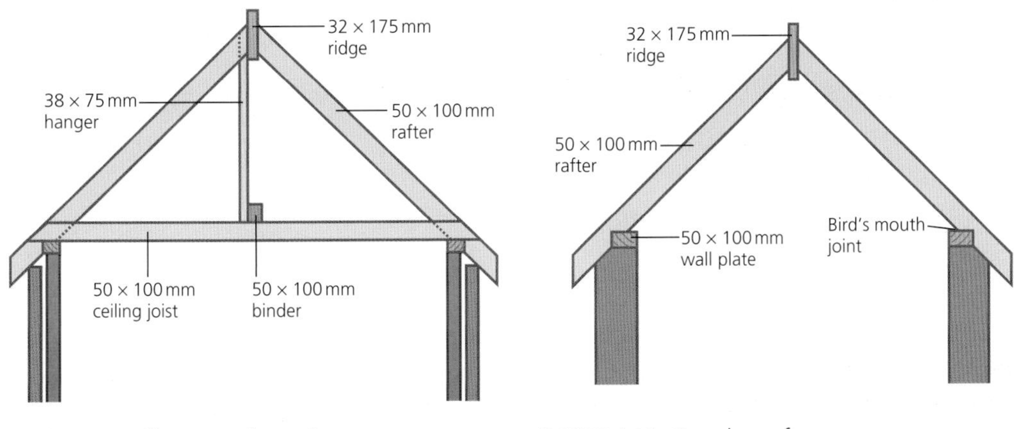

FIGURE 6.32 Close couple roof

FIGURE 6.33 Couple roof

Dragon or angle ties

These types of tie are fixed across the corners of a wall plate with a hipped rafter notched over the top. They strengthen the half lap joints on the corner of the wall plate, and prevent the weight of the hip rafter slipping from its intended position and pushing the corner of the building out.

SETTING OUT (THE PROCESS OF FINDING THE LENGTHS AND BEVELS FOR EACH ROOF MEMBER)

Before the lengths and angles can be established for a cut roof, you will need to know the following information:

- the run of the rafter;
- the rise of the roof or pitch.

To work out the run of the common rafter you must measure the distance across the width of the roof; this distance is known as the 'span' of the roof. The span is measured from the outer edge of the wall plate to the outer edge of the opposite wall plate.

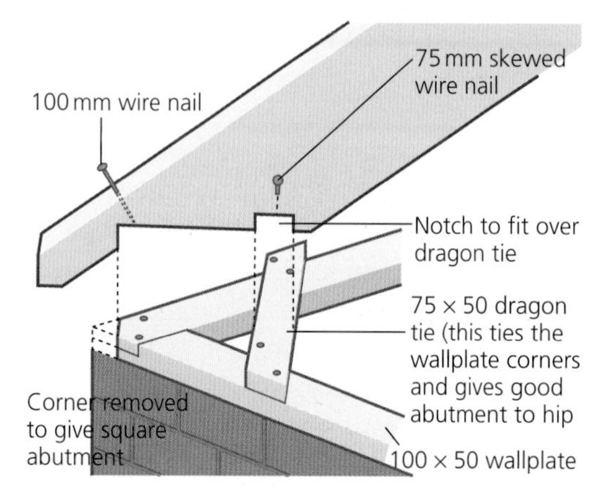

FIGURE 6.34

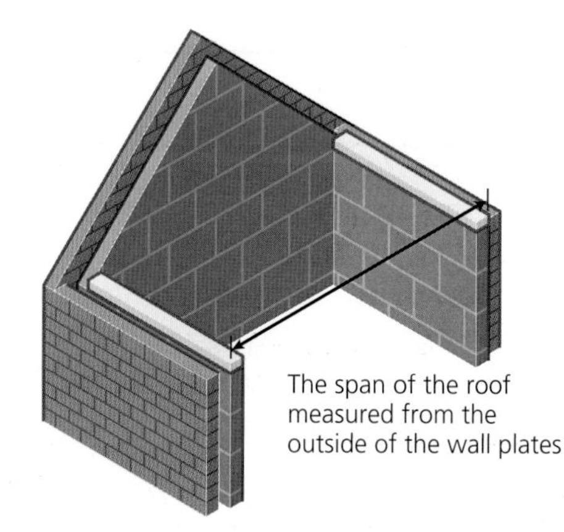

FIGURE 6.35 Measuring the span of a pitched roof

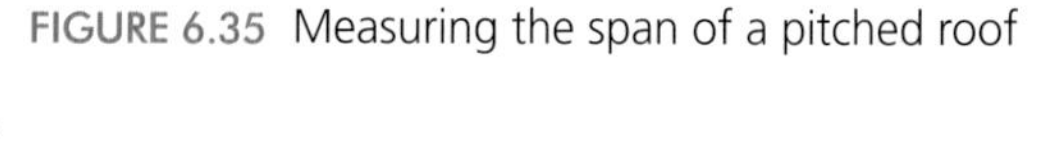

The run of a common rafter is half the span of the roof – for example, a roof with a span of 4 metres would have a rafter run of 2 metres.

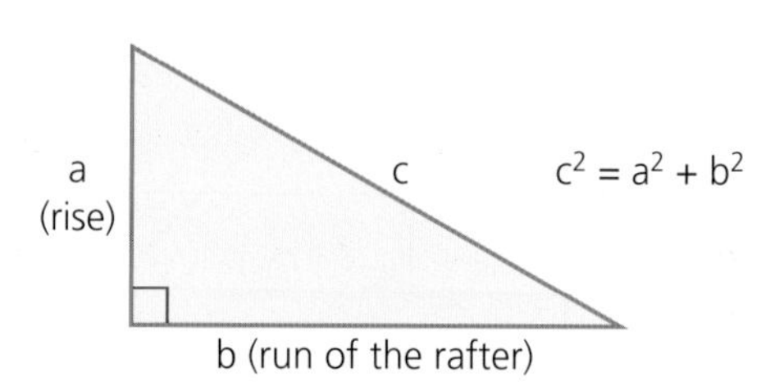

FIGURE 6.36

As the rise or pitch of a roof is normally specified on the architect's drawings, together with the run of the rafter, you will have all the essential information needed to work out the roof components.

While there are several methods of establishing the lengths and angles needed to construct a cut roof, it is important to understand the basic principles. The majority of the timbers within a pitch roof can be worked out using Pythagoras' theorem.

THE BACKING LINE (SETTING-OUT LINE)

The majority of the rafters contained in a cut roof are bird's mouthed over the wall plate; this provides a definitive fixing position for the rafters and prevents the timbers slipping over the wall plate. The depth of the bird's mouth is measured from the top edge of the rafter to the backing line.

Failure to mark the backing line in the correct position will result in the situations shown in Figures 6.37–6.40:

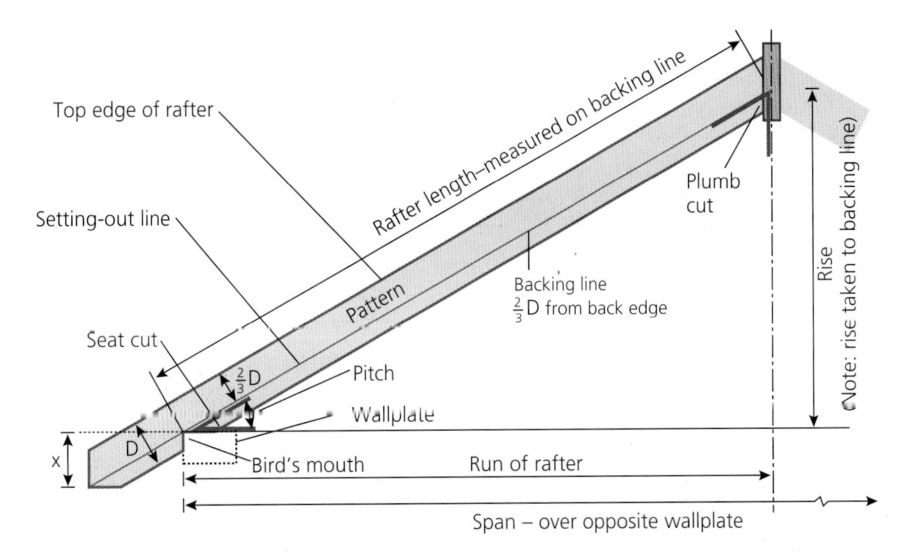

FIGURE 6.37 Use of the backing line along the length of a common rafter and a hipped roof

Figure 6.37 demonstrates how the rise of the roof and the run of the rafter will provide all the detail needed to set out a common rafter. Using a right-angled triangle, with the rise marked on one side and the run of the rafter on the other, the following information can be found:

- the true length of the common rafter;
- the angle of the vertical cut, known as the plumb cut;
- the angle of horizontal cut, known as the seat cut.

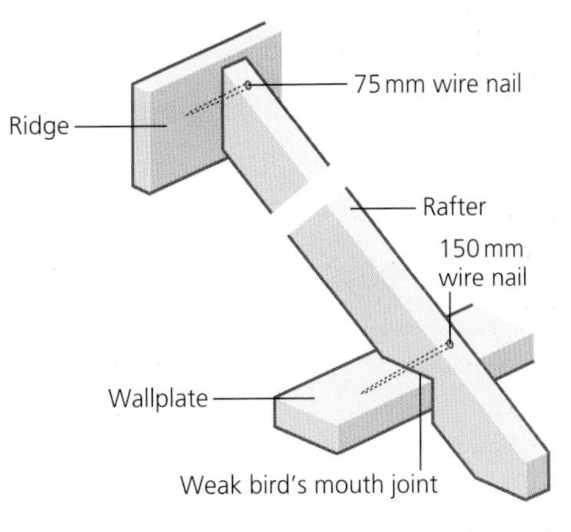

FIGURE 6.38 A poor example of a bird's mouth

FREQUENTLY ASKED QUESTIONS

What is meant by the true length of the rafter?

The true length of a rafter is the length from the outer edge of the wall plate to the centre of the roof. A deduction will have to be made to the top of the rafters to accommodate half the thickness of the ridge board. Deductions will also have to be made to the true lengths of the hip and valley rafters. The most accurate way to determine the deductions to be made is to draw a plan of the intersection between the timbers.

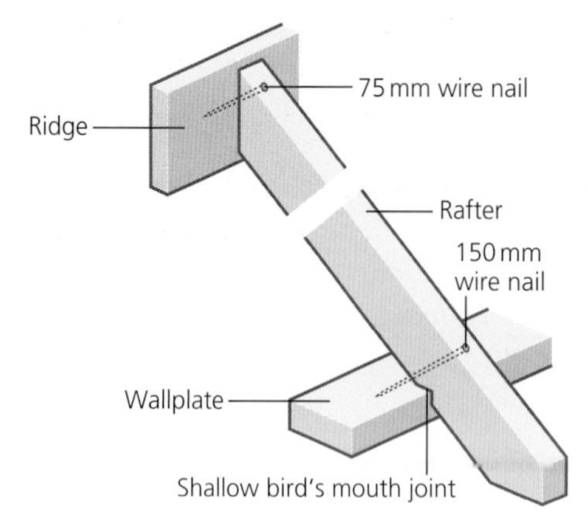

FIGURE 6.39 A poor example of a bird's mouth

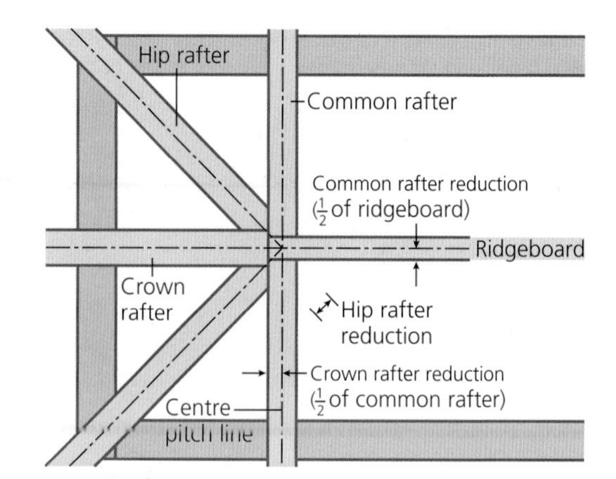

FIGURE 6.40 Deductions to the common and hipped rafters

DEDUCTIONS TO THE TRUE LENGTH

In addition to the deductions mentioned, the abutment of the bird's mouth over the wall plate on a hip rafter will also have to be adjusted. To achieve a good joint between the bird's mouth and the corner of the wall plate, the corner of the wall plate will have to be removed. The reduction made at the corner of the wall plate will have to be added on to the bird's mouth joint.

As mentioned previously, there are several methods that can be used to find the details needed to mark out the lengths and bevels for the roof members; these are:

- geometry (either manually or with computer-aided design, CAD);
- a roofing square;
- a roofing ready reckoner.

Geometrically

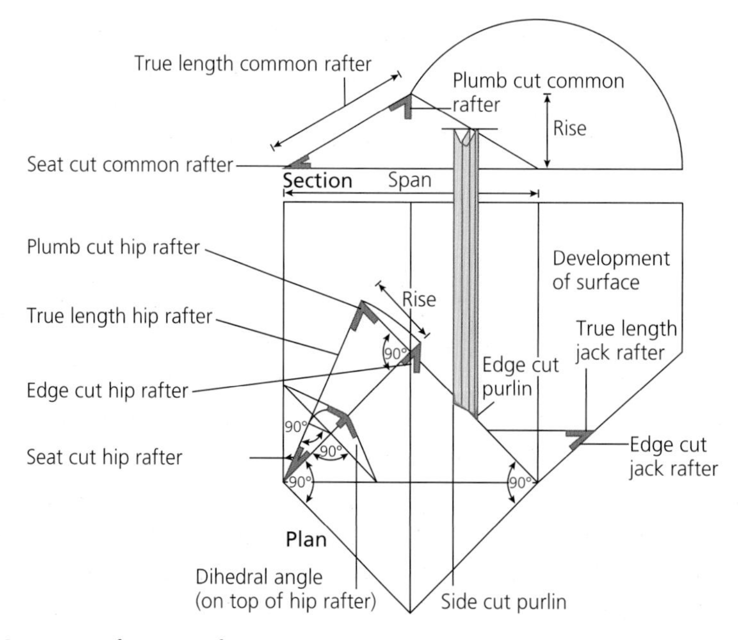

FIGURE 6.41 Geometry of a cut roof

This method involves setting out the roof to scale on a series of working drawings. Each drawing will have to be scaled up to full size and have each angle transferred from the paper with a sliding bevel. It is possible to draw the development of all the roof surfaces on one drawing, but this can lead to mistakes due to the number of lines that have to be drawn and the fact that many of them have to be superimposed (drawn over one another). To improve the clarity, several drawings will have to be produced, although this will be subject to the design of the roof and abbreviations will be used.

GEOMETRICALLY SETTING OUT CUT ROOFS

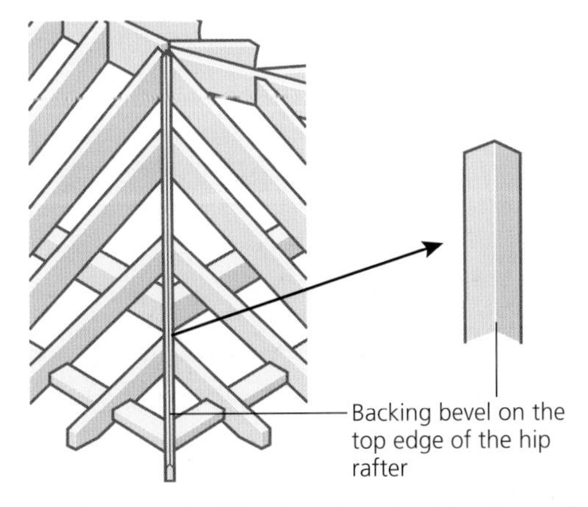

FIGURE 6.42 Backing bevel on the hip of a roof

FREQUENTLY ASKED QUESTIONS

What is the backing bevel used for?

The backing bevel is formed on the top edge along the length of the hip rafters. It allows the hip rafters to sit flush with the rest of the roof surface. The backing bevel or 'dihedral angle' will allow the roof battens to run over the hip rafter. The backing bevel is rarely used these days because the hip tiles cover this area of the roof.

Roofing square

This is another method of determining the true lengths and angles needed to build a cut roof. All the angles and lengths are established with the use of a steel roofing square (Figure 6.44). The roofing square is a steel, right-angled triangle, constructed with tables of imperial and metric measurements along the two edges on each side of the square. The square is used in conjunction with an adjustable fence or a pair of steel square buttons; these allow the square to be adjusted to the various settings required.

The roofing square is used in a similar way to a scaled geometrical drawing of the roof. The roof must first be scaled down to enable the measurements from the roof to be transferred on to the roofing square; this is achieved by dividing the roof measurements by 10. The rise of the roof is normally transferred on to the tongue and the run of the rafter on the longer section on to the blade.

FIGURE 6.43 A roofing square

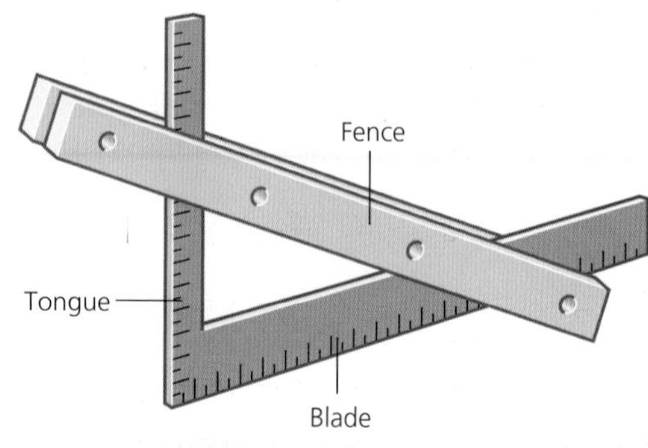

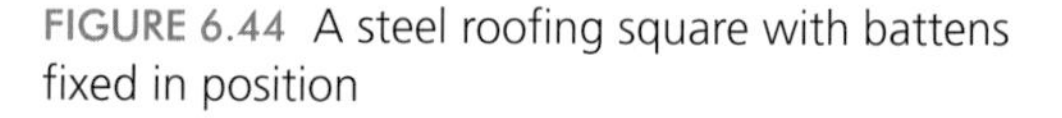

FIGURE 6.44 A steel roofing square with battens fixed in position

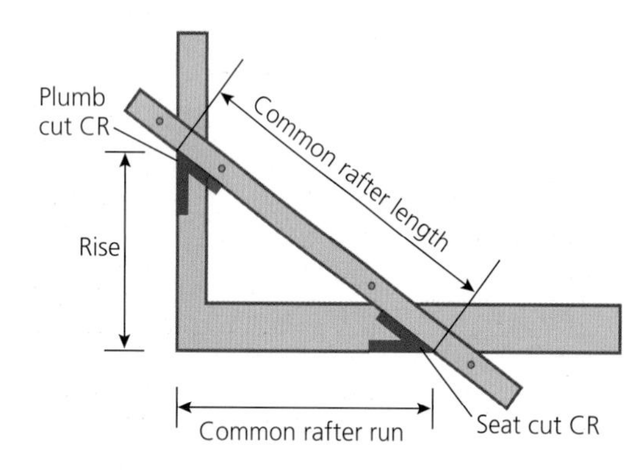

FIGURE 6.45

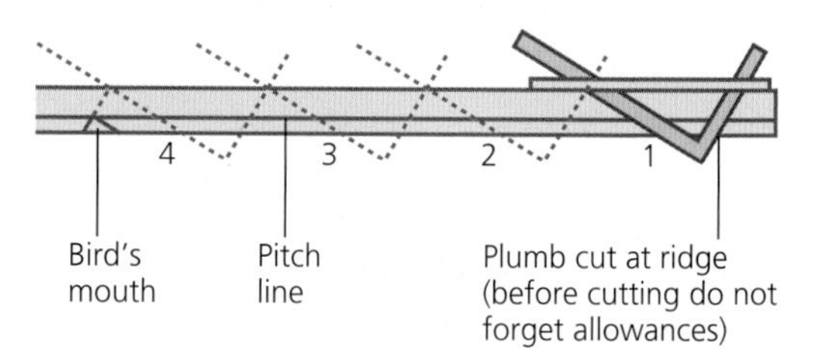

FIGURE 6.46

Figure 6.46 demonstrates how the roofing square is used to find the common rafter true length, plumb cut and seat cut.

The remaining components for the roof can be worked out using similar principles to the common rafter. As each plumb cut, seat cut and true length is worked out, it is transferred on to a piece of timber so that it can be referred back to without having to set the roofing square again.

While the jack and cripple rafter lengths can be worked out using the steel square, they are often measured directly off the roof as it is constructed.

The reductions in the lengths of the jack and cripple rafters can also be calculated using the following formula.

1. Work out the number of jack or cripple rafters needed to complete one side of the hip or valley rafter, and then add one.

TRADE SECRETS

Once the plumb cut, seat cut and the true length have been established and transferred on to the common rafter, and the deductions have been made for half the thickness of the ridge, the rafter can be cut. This rafter should be used as a pattern and labelled the 'pattern rafter'. The pattern rafter is used to mark the remaining rafters ready for cutting. This will avoid having to mark the backing line on each common rafter and the repetitive use of a sliding bevel. Drawing around the pattern rafter will also ensure consistency in the size of the rafters and avoid possible mistakes made by setting out.

2. Divide this number by the true length of the common rafter.
3. This figure is the length of the shortest rafter and the size that should be added to each successive jack or cripple rafter.

Roofing Ready Reckoner

A roofing ready reckoner is a reference book that contains tables in both metric and imperial measurements. The tables contain all the plumb, seat and edge cuts required to construct a pitched cut roof, as well as the true lengths of the common, hip and valley rafters. Each page is referenced by the angle of the roof or pitch. The detail contained within the tables can be converted to provide all the information needed for any span of roof, and all the most common pitches of roofs. The ready reckoner is the preferred method of establishing the roof parts because it is pocket sized, and easily referenced for new work and additions to existing roof structures. Figure 6.48 shows a sample page from a roofing ready reckoner.

The following information should be established before using the ready reckoner:

- pitch of the roof;
- run of the rafter (half the span);
- the rise of the roof.

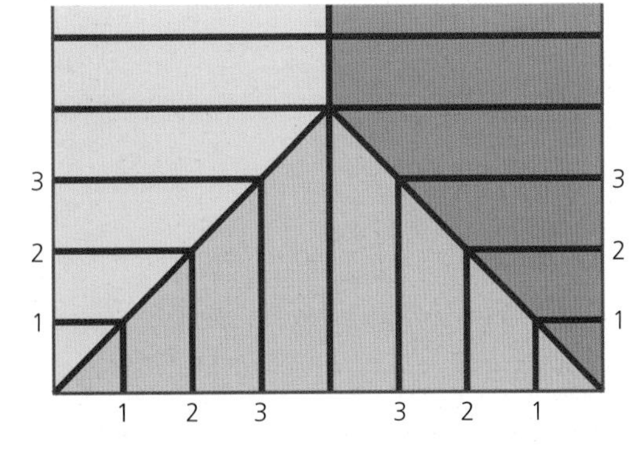

FIGURE 6.47

Rise of common rafter

0.727 m per metre of run

Pitch 36

Run of rafter	Length of rafter	Length of hip
0.1	0.124	0.159
0.2	0.247	0.318
0.3	0.371	0.477
0.4	0.494	0.636
0.5	0.618	0.795
0.6	0.742	0.954
0.7	0.865	1.113
0.8	0.989	1.272
0.9	1.112	1.431
1.0	1.236	1.59

Bevels:		
Common rafter	Seat	36
	Ridge	54
Hip or valley	Seat	27
	Ridge	63
Jack rafter	Edge	39
Purlin	Edge	51
	Side	59.5

Jack rafter	Centres decrease
333 mm	412 mm
400 mm	494 mm
500 mm	618 mm
600 mm	742 mm
(in mm to 999 and thereafter in m)	

FIGURE 6.48 A sample page from a roofing ready reckoner (Pitch 36°)

USING A READY RECKONER

The best way to explain how to use a roofing ready reckoner is to use an example pitch; in this case we will use the 36° pitch shown in Figure 6.49. The plumb, seat and edge cuts can be clearly seen in the example but the true lengths for each roof member will have to be worked out.

The true length of the common rafter can be worked out by referring to the table. The run of the rafter is broken down between 0.1 (100 mm) and 1.0 (1 metre). If the largest run of the rafter is 1 metre, for a roof with a span of 4 metres the length of the rafter must be multiplied by 4:

1.0 metre run of the rafter equals 1.236 metres for the true length of the common rafter

$1.0 \times 4 =$ metres run of the rafter $1.236 \times 4 = 4.944$

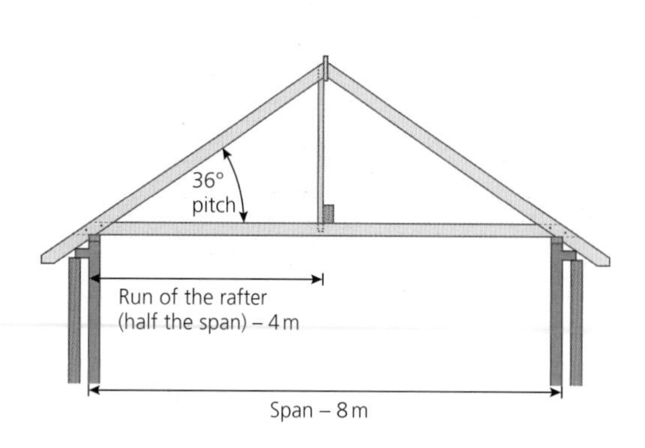

FIGURE 6.49 A roof with a 36° pitch

Referrence to roof components

Rise of common rafter

0.727 m per metre of run

Pitch 36

Run of rafter	Length of rafter	Length of hip
0.1	0.124	0.159
0.2	0.247	0.318
0.3	0.371	0.477
0.4	0.494	0.636
0.5	0.618	0.795
0.6	0.742	0.954
0.7	0.865	1.113
0.8	0.989	1.272
0.9	1.112	1.431
1.0	1.236	1.59

Bevels:		
Common rafter	Seat	36
	Ridge	54
Hip or valley	Seat	27
	Ridge	63
Jack rafter	Edge	39
Purlin	Edge	51
	Side	59.5

Jack rafter	Centres decrease
333 mm	412 mm
400 mm	494 mm
500 mm	618 mm
600 mm	742 mm
(in mm to 999 and thereafter in m)	

Example – Multiply by four to establish the lengths of the rafters and hips for a roof with a run of the rafter

FIGURE 6.50

FREQUENTLY ASKED QUESTIONS

How do you calculate the run of the rafter if it is in between the 100 mm intervals on the table?

It is quite unusual for the span of a roof to be a rounded measurement. In these cases the whole number should be multiplied by the figures in the 1 metre column and the fractions multiplied by the corresponding units.

For example, using a 36° pitch and a run of the rafter of 3.650 metres:

Run of rafter	Length of rafter	Length of hip
1.0 × 3 = 3 metres	1.236 × 3 = 3.708	1.59 × 3 = 4.77
0.6 × 1 = 600 mm	0.742 × 1 = 0.742	0.954 × 1 = 0.954
0.1 ÷ 2 = 50 mm	0.124 ÷ 2 = 0.062	0.159 ÷ 2 = 0.0795
3.650 metres	4.512 metres	5.8035 metres

Note: refer to the ready reckoner in Figure 6.50.

MARKING OUT THE WALL PLATES

The positioning of the rafters on the wall plates is determined by the weight of the roof covering, the size of the sheet materials used to cover the ceiling and the sectional size of the roof timbers. The positioning of the rafters is normally measured between the centres of each rafter; this is referred to as 'centre to centre' (c/c). Measuring between the centres of the rafters is more accurate than measuring between the rafters because of the potential discrepancy in the size of the timbers.

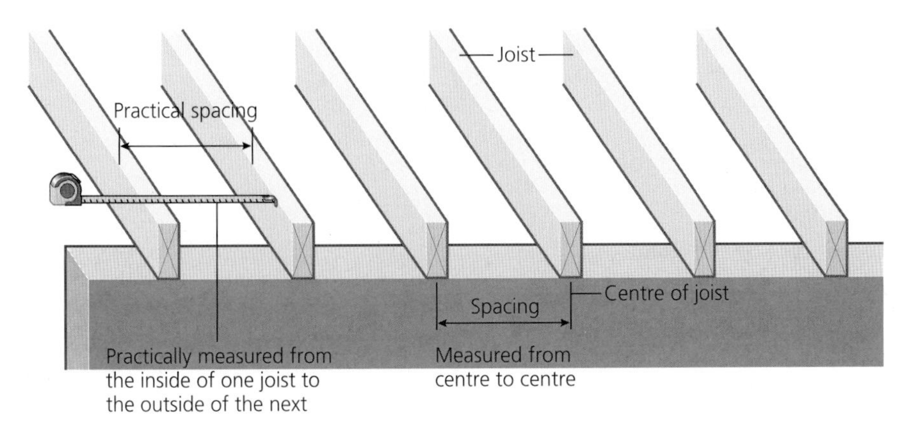

FIGURE 6.51 Potential inaccuracy

MARKING OUT HIPPED ROOFS

The positioning of the hip rafters is vitally important if the established true lengths of the timbers are going to sit in the correct position on the roof. The hips run between the corner of the wall plate and the centre of the roof at 45°. There are several methods of arranging the intersection between the hips and crown rafter; these are shown in Figure 6.52.

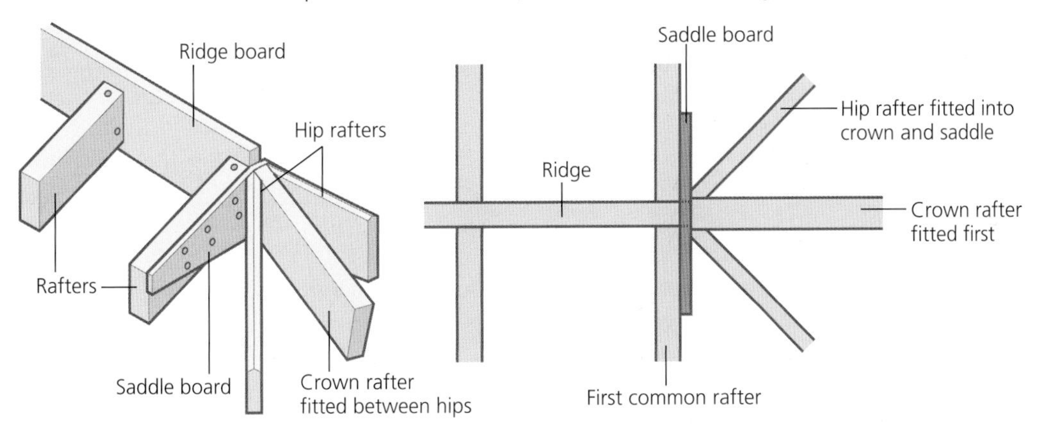

FIGURE 6.52 Joints between the crown and hip rafters

The last common rafter position is established by the setting out of the wall plate. The crown rafter should be positioned in the centre of the roof on the gable end. The distance between the outside edge of the crown rafter and the outside of the wall plate is known as the 'clear run'. The clear run is transferred along the returned edge on the wall plate; this is the position of the face of the saddle board. The thickness of the saddle board must be taken into account and marked at the clear span position; this is the starting position for the first common rafter.

FIGURE 6.53

TRUSSED RAFTER ROOFS

Trussed roofs are constructed with a series of pre-made trusses assembled in a factory and erected on site. Each truss is specifically designed and engineered using computer technology to:

- withstand the load imposed on it;
- bridge the span of the roof without sagging;
- produce accurate roof shapes.

Trussed roofs differ from cut roofs because they:

- have no ridge board;
- use 40 per cent less timber;
- are quicker to erect;
- are cheaper to purchase;
- can be constructed at ground level and craned on to the wall plates;
- are consistently accurate;
- eliminate the need for cutting complex joints.

There are many different types of roof truss, each one specifically designed to suit the roof covering, possibly accommodate living space and provide additional support for long spans.

Trusses are held together with a series of galvanised nail plates on either side of each butt joint; this intersection of the timbers within the truss is known as the 'node'.

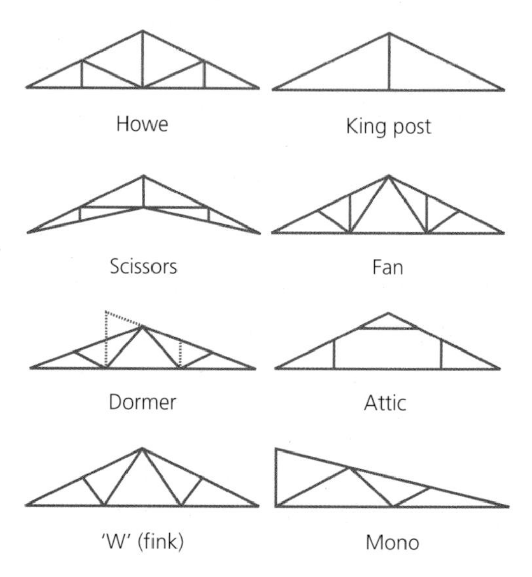

FIGURE 6.54 Roof trusses

Trusses are fairly weak until fixed into position on the wall plates and braced, so it is important that careful consideration be given to their handling and storage. Carrying large trusses horizontally will allow them to sag under their own weight, and therefore weaken the joints between the nail plates and the butt joints. Roof trusses are usually delivered to site strapped together with either plastic or metal banding; this will prevent twisting and the weakening of the trusses. Trusses should be lifted manually on to the roof in a vertical position with the banding attached. A 'spreader bar' should be used to minimise the pressure on the nodes while the lifting takes place.

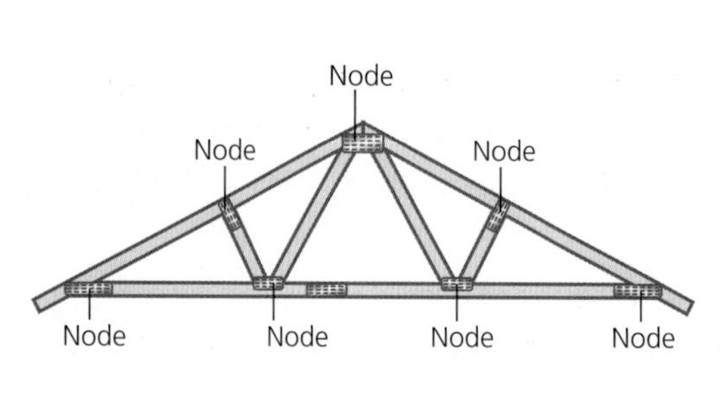

FIGURE 6.55 Nodes

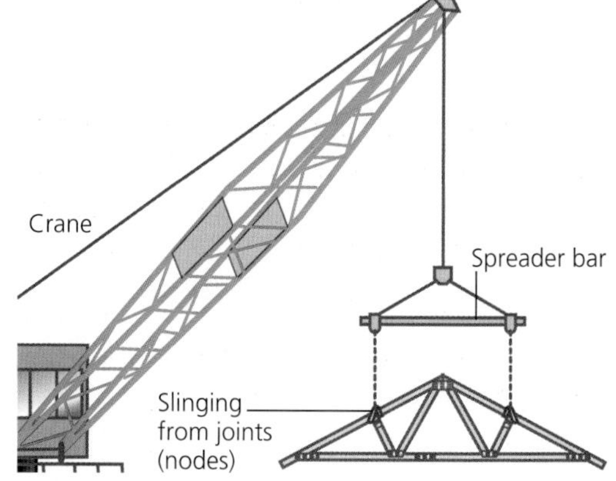

FIGURE 6.56 Spreader bar

Company method statements (a safe plan of working) should be referred to when manually handling roof trusses. In general, a site should be fully prepared for the arrival of the roof trusses.

Things to consider include:

- space to store the trusses;
- weather conditions (high winds, heavy rain, etc.);
- overhead cables;
- methods of lifting (mechanical/manual).

FIXING TRUSSED ROOFS

Traditionally, cut roof timbers are fixed into position with galvanised nails through the bird's mouth and into the wall plate. Fixing trusses using the same method may weaken the joint between the timbers and the nail plate. The preferred method of fixing roof trusses is to use galvanised truss clips; alternatively, tie-down straps can be fixed between the truss and the inner cavity wall.

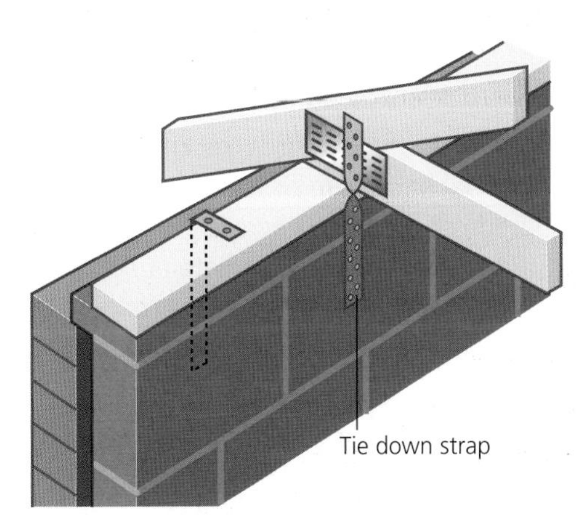

FIGURE 6.57 Galvanised tie-down strap

As the individual trusses are lifted into position on the roof they will have to be braced in position temporarily. Erect the first truss in the fifth position along the wall plate, working from the gable end, and brace off the wall plate using 22 × 97 mm battens.

FREQUENTLY ASKED QUESTIONS

How are the battens, used to brace the trusses, fixed to the trusses?

The battens should be fixed using galvanised round wire nails. All temporary bracing should be nailed with the head left slightly raised. This will allow for the easy removal of the battens at the final stages of the roof installation.

The following trusses should fill the space between the gable end and the fifth truss, with additional bracing fixed diagonally across the roof. The temporary fixing and bracing of the first five trusses provides a ridge structure to one end of the roof; this will then allow the remaining trusses to be positioned, braced and fixed.

When the final roof truss has been positioned, the remaining bracing should be fixed. Longitudinal bracing should be fixed across the trusses at every node to prevent the trusses flexing between the wall plates.

The longitudinal bracing should be kept down from the underside of the rafters a minimum of 25 mm to allow the diagonal bracing to pass through. During the construction of trussed roofs, there may be a need to lengthen the longitudinal bracing due to the restricted length of the timbers used. This can be achieved by allowing the sections of timber to run past one another over a minimum of two trusses.

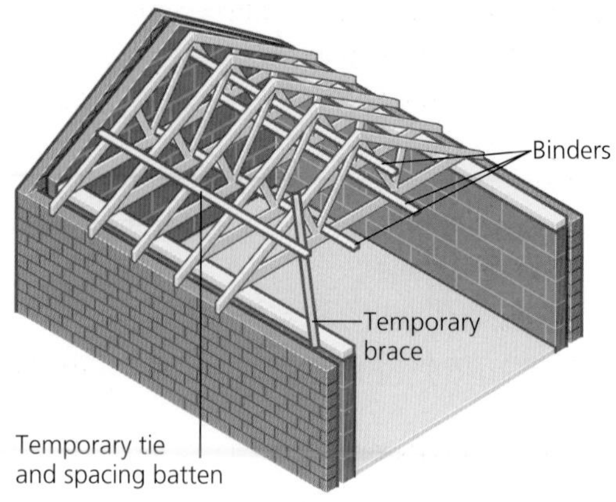

FIGURE 6.58

DIAGONAL BRACING

To prevent the trusses from 'racking', a series of diagonal braces are permanently fixed to the underside of the rafters across the whole length of the roof. Diagonal bracing should run from the ridge and over the wall plate at approximately 45°. Trussed roofs should contain a minimum of four diagonal braces running in the opposite directions on each side of the roof.

TRIMMING

Roof trusses are designed to withstand the weight of the roof covering, so it is important that they remain intact. The positioning of the roof trusses may be interrupted by a chimney; this can be overcome by installing trimmers and a section of cut rafters.

Note: Building Regulations state that all timbers should be at least 50 mm clear of the flue.

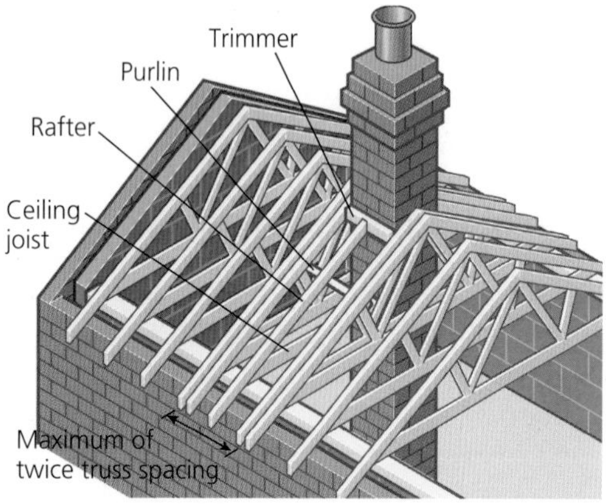

FIGURE 6.59 A chiney extending through a trussed roof

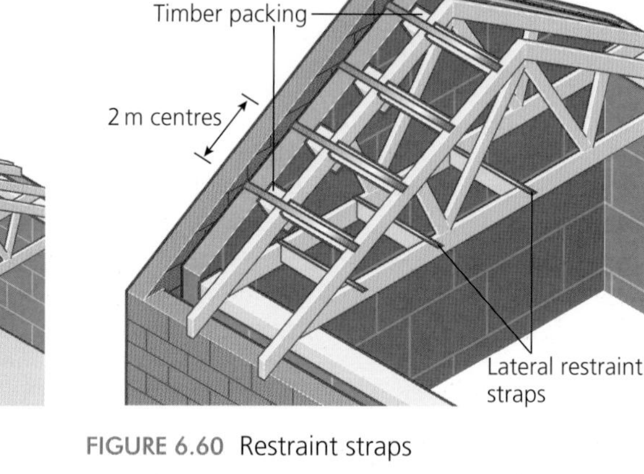

FIGURE 6.60 Restraint straps

RESTRAINT STRAPS

The brickwork that extends above wall plate height at the gable end may require additional support from the roof structure to transmit wind loads. This is normally achieved by fixing galvanised metal straps (restraint straps) to the ceiling joists and over the top of the rafters.

The restraint straps should be positioned at 2 metre centres on both the ceiling and rafters, and built into the brickwork as the build progresses. Noggings should be fixed between the trusses underneath each restraint strap; this will allow additional fixing points and strengthen the gable end.

DOS AND DON'TS

- *Always store and handle correctly* – failure to do so may twist the trusses and weaken the nodes.
- *Never cut trusses* – this will weaken the structure and may invalidate the manufacturer's warranty.
- *Always use truss clips to fix* – nailing or screwing through the nail plate may weaken the joint.

VENTILATION

Although a roof construction is designed to protect a building from the elements it must also be able to breathe.

FREQUENTLY ASKED QUESTIONS

How does a roof breathe?

Failure to ventilate a loft space will create stagnant (motionless) air, which could lead to rot in the roof timbers. To prevent this, a passage of air must be allowed to flow through the roof. This can be achieved in a number of ways, including drilling holes along the length of the soffit and covering them with soffit vents. This will prevent birds and bats from nesting in the loft area.

ROOFING UNDERFELT

After the roof structure has been formed, the roof should be covered with roofing underfelt and rows of roofing battens to which the tiles or slates will be fixed. The first row of roofing underfelt should be fixed in position along the bottom of the roof, allowing an overhang over the eaves and into the guttering. This will allow any moisture that gets trapped between the undersides of the roof tiles or slates to run directly into the guttering, and allow the remaining underfelt to overlap.

Traditional bitumen underfelt has been used on roofs for many years to help insulate the loft and prevent heat loss. The underfelt also acts as a secondary waterproof barrier to any moisture that passes through the roof tiles or slates. There are many types of underfelt available from good builders' merchants to prevent moisture passing through the roof. Bitumen underfelt is the cheapest alternative but has several disadvantages; these include those listed below.

- It is not very durable and will rot and become brittle over time, especially in areas exposed to sunlight such as the overhang into the guttering.

- Bitumen underfelt prevents water passing through and into the roof space; modern underfelts will also allow any moisture in the roof space to escape. Breathable roofing underfelts (or 'membranes') prevent the need for other forms of ventilation.
- Bitumen underfelt can tear easily when it is being nailed into position and is also very heavy compared to breathable membranes. Extreme care must be taken when fixing the felt because it can tear easily when nailed through.

ACTIVITIES

Activity 25 – Erecting truss rafter roofs

Read through the following questions and answer them as fully as you can to help you develop your underpinning knowledge of this subject area.

1. What is a 'truss'?
2. What are the advantages of using trussed roofs?
3. How can you determine the centres required for the trusses on a roof?
4. Sketch a 'node'.
5. Trusses can be cut to allow roof windows to be fitted: true or false?

FLOOR CONSTRUCTION

Floors are constructed using a number of methods and materials, depending on their location within a building. They serve a number of purposes, including:

- providing a level surface;
- prevention of moisture penetration at ground level;
- sound insulation;
- thermal insulation;
- additional space through the inclusion of upper and mezzanine floors.

Floors in general are divided into two categories:

1. ground floors (nearest to the outside floor level);
2. upper floors (any other floor above ground floor level, including the first, second, etc.).

GROUND FLOORS

SUSPENDED TIMBER HOLLOW GROUND FLOOR

These differ in the way they are constructed compared to upper floors, because they have the additional task of stopping moisture penetration from the ground rising into the building. Timber ground floors can also have intermediate support across the span, without disrupting the layout of dividing walls and room layout. Timber joists are normally supported at ground floor level by rows of low-height brick walls; these are known as 'sleeper walls'.

Sleeper walls are normally constructed with bricks or blocks, with gaps equal to half the length of a brick between them on each course (also known as 'honeycomb walls'). This will allow air to pass through the air bricks on the external walls and circulate around the timbers, reducing the risk of stale air.

FREQUENTLY ASKED QUESTIONS

What is stale air?

When air becomes stagnant or motionless around timber it will allow moisture to build up in the timber. If the moisture in the timber rises above 20 per cent, dry rot will occur.

The exact spacing of the sleeper walls will depend on the section of the timbers used for the joists, although they are normally either 400 or 600 mm centres; the smaller the section, the closer the spaces between the sleeper walls.

Ground floor joists are normally smaller in section than upper floors because they benefit from the increased support. The joists are secured in position by 'skew' nailing into the wall plates on top of the sleeper walls. The purpose of the wall plates is to spread the weight of the floor over the entire sleeper walls rather than direct loading. A damp-proof course is positioned between the sleeper walls and the wall plates to prevent moisture being drawn up into the timber.

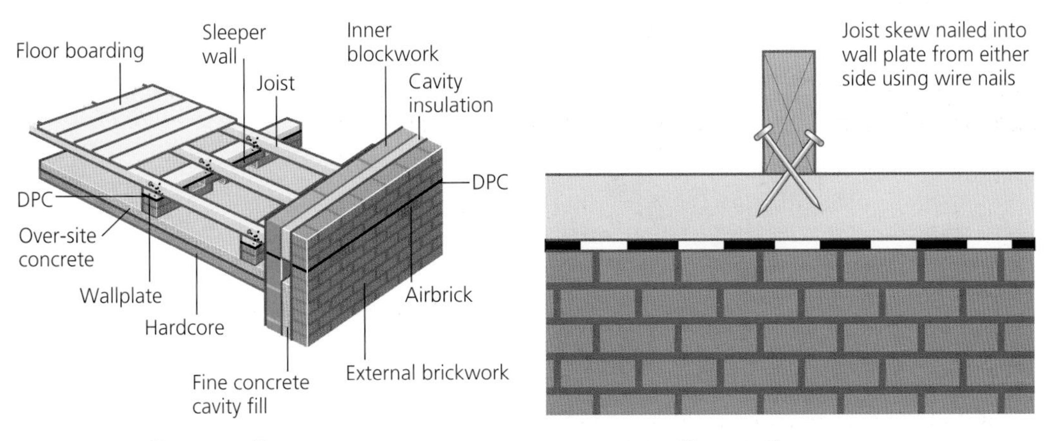

FIGURE 6.61 Sleeper walls

FIGURE 6.62 Skew nailing

FREQUENTLY ASKED QUESTIONS

What is 'skew' nailing?

Skew nailing is a method of fixing timber with the maximum amount of strength. The nails are used in pairs, and directed towards each other in a dovetail shape.

Solid floor joists – advantages:

- readily available from suppliers;
- easily cut to size;
- solid construction.

Solid floor joists – disadvantages:

- drying out will cause the timber to shrink and create gaps between solid strutting;
- joists are likely to twist or warp before fixing;
- joists will have to be drilled and notched for services (e.g. electrical cables, water pipes);
- slower to erect than manufactured joists.

WOOD-BASED PANEL-WEB 'I' JOISTS

Floor joists have progressed from being constructed of conventional solid timber, which has a number of disadvantages, to manufactured joists and beams. Wood-based panel-web 'I' joists are a surprisingly stronger and cost-effective alternative. They are manufactured under factory conditions to withstand the loads imposed upon them, and are available in the same standard sizes as solid joists.

Each wood-based panel-web 'I' joist is engineered with two parallel stress-graded timbers, usually either 72 × 47 mm or 97 × 47 mm. These are divided by V-shaped galvanised steel webs. The steel webs are fixed either side of the top and bottom timbers, with nail plates formed in each web and pressed under load into the timber.

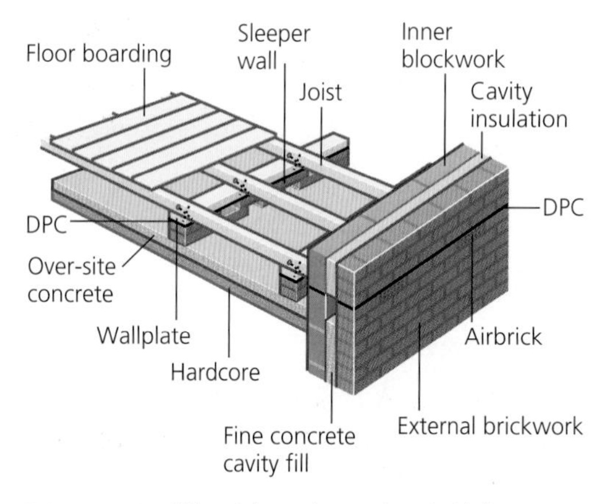

FIGURE 6.63 Wood-based panel-web I joists

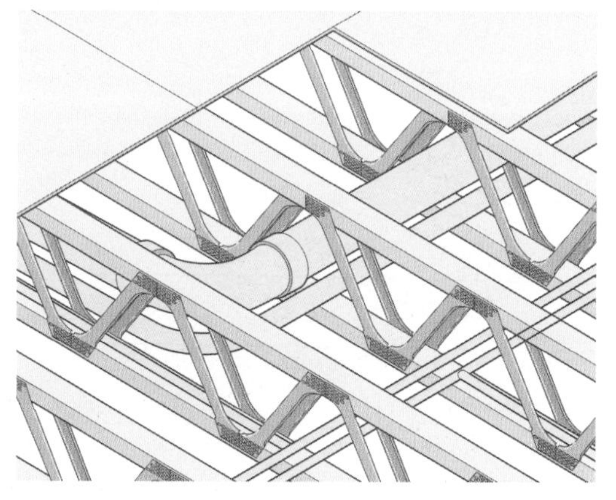

FIGURE 6.64 Services passing through joists between webs

Wood-based panel-web 'I' joists are substantially lighter than solid joists so they are easier to transport and manually handle when fixing into position. They also have the benefit of large open areas through the joists, which will allow service pipes and cables to pass through without the need for drilling and notching.

Wood-based panel-web 'I' joists are usually manufactured to order, but some suppliers will keep a limited stock of standard sizes. They are installed similarly to solid joists; they are both returned with joist hangers (although the exact shape of the hanger will differ for wood-based panel-web 'I' joists), and they both need forms of strutting. It is good practice to infill the cut ends with the same-sized section as the top and bottom timbers of the joists, to strengthen the open ends.

FREQUENTLY ASKED QUESTIONS

What is strutting used for?

All forms of manufactured and solid timber joists will either twist or flex across their length unless they are correctly supported with strutting. Strutting is the bracing fixed between floor joists to prevent any movement occurring. Although the movement of the joists is normally concealed within the floor structure, it will cause damage to the ceiling finish below.

Wood-based panel-web 'I' joists are strutted by allowing a continuous length of timber to run through them and fixing at each intersection. The continuous bridging of all the joists ensures that the load imposed on them

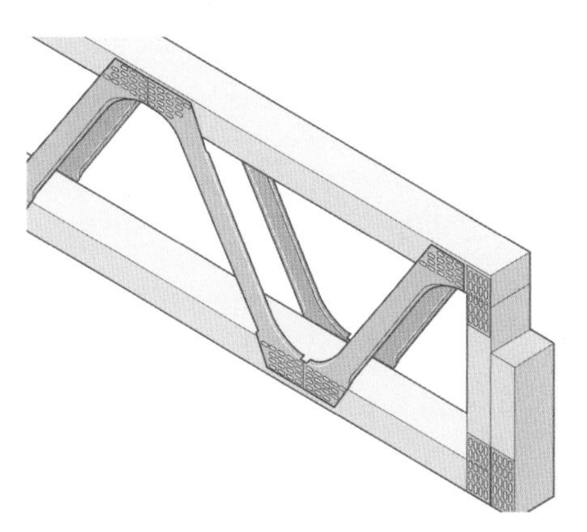

FIGURE 6.65 An infill piece between top and bottom timbers

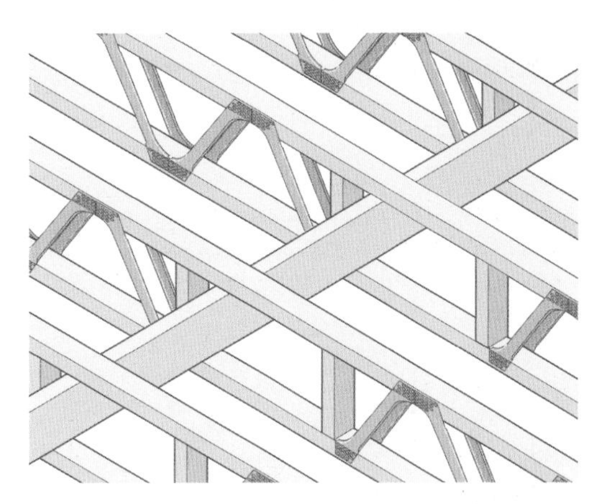

FIGURE 6.66 Strutting

Wood-based panel-web 'I' joists – advantages:

- up to 40 per cent lighter than solid joists;
- quicker to erect than solid joists;
- no notching and drilling required for services (e.g. electrical cables, water pipes);
- generally wider than solid joists, which makes fixing the floor and ceiling covering easier;
- they can span up 6.3 metres unsupported.

Wood-based panel-web 'I' joists – disadvantages:

- not readily available;
- difficult to cut to length.

'I' BEAM JOISTS

Developments in technology have enabled new adhesives, laminated timbers and man-made boards to be used to create 'I' beam joists. 'I' beam joists are factory produced, with an upper and a lower section of stress-graded timber known as 'flanges'. These are connected together with either oriented strand board (OSB) or plywood, and bonded together into grooves. 'I' beam joists are available in a range of different depths to suit varying spans and floor loads. They are also manufactured in longer lengths than solid timber joists and have the advantage of a lighter construction, making them more efficient to install. 'I' beams are a versatile building component that can also be used for walls and structural roofing members.

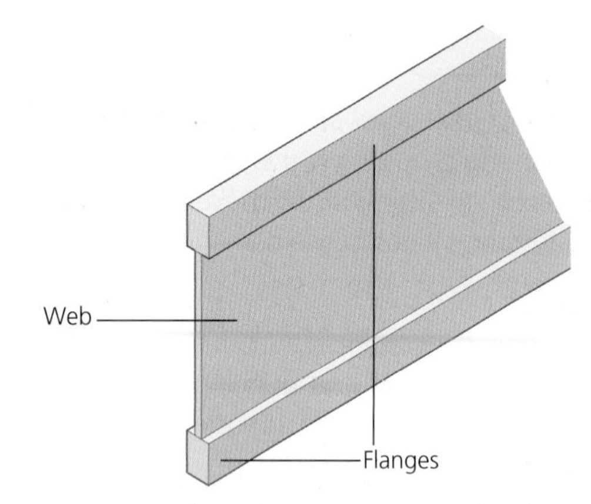

FIGURE 6.67 An 'I' beam

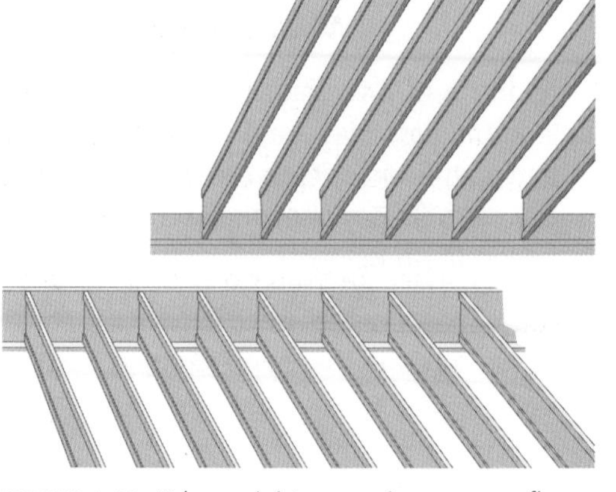

FIGURE 6.68 'I' beam joists spanning across a floor

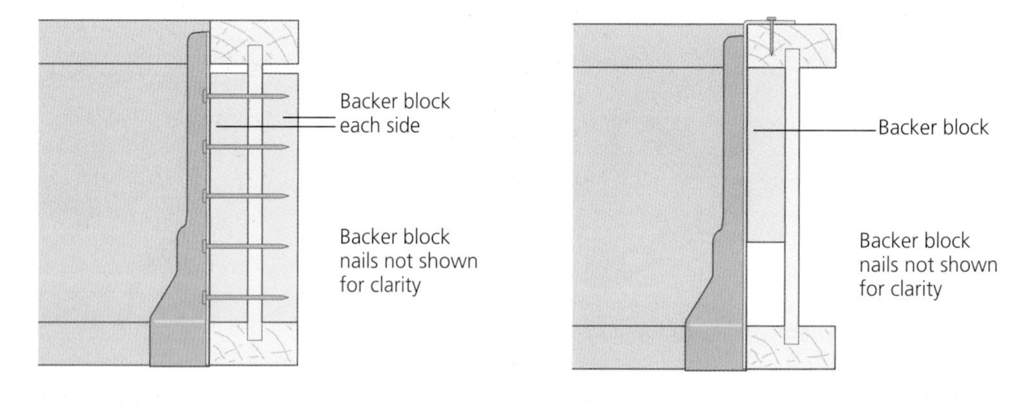

FIGURE 6.69

FIGURE 6.70

'I' beam joists – advantages:

- resist twisting, bowing and shrinking, which could lead to 'squeaky' floors;
- lightweight;
- cost effective;
- easy to install.

'I' beam joists – disadvantages:

- 'backer blocks' must be fitted behind every joist hanger to fill the void; they are produced from plywood, OSB or solid timber, and are secured in place to form a strong fixing point for the adjoining joists; face-mounted joist hangers must have backer blocks on both sides of the joist at the fixing points; top-mounted joist hangers only need to have these on the same side as the returned joists; failure to adequately support 'I' beams with backer blocks may result in the joists twisting and weakening the structure
- the solid construction of the 'I' beam means that the joists will have to be drilled for service pipes and cables.

FREQUENTLY ASKED QUESTIONS

What are joist hangers?

All the structural timbers of a building will have to be well supported to ensure a solid and safe construction. This can be achieved in a number of ways. Fixing joist hangers between connecting timbers in a floor is one method that is commonly used to support joists in a floor structure.

Joist hangers are simply galvanised sections of folded steel, each one shaped around the connecting timbers.

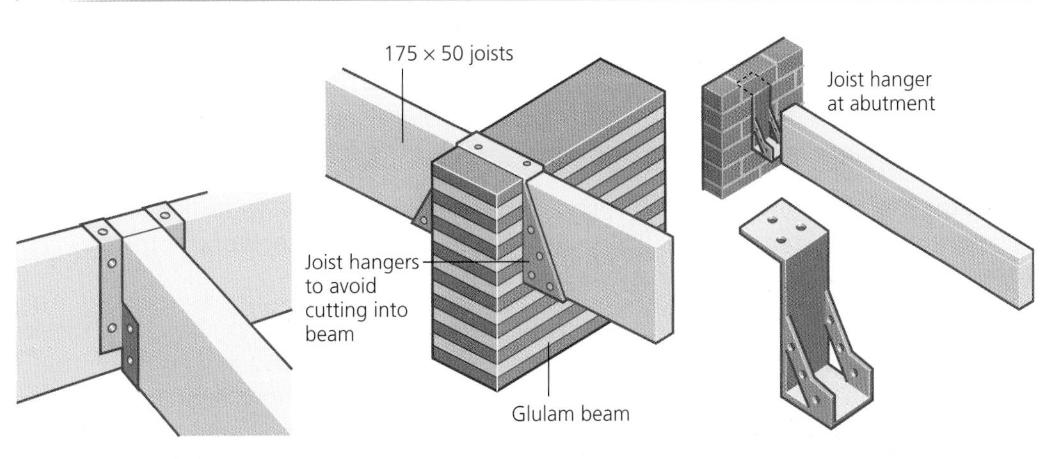

FIGURE 6.71 Joist hangers

UPPER FLOORS

The layout of the upper floor joists is similar to the ground floor layout. The main difference between the two is that upper floors normally span longer unsupported distances. Upper floor joists are usually deeper in section compared to suspended timber ground floors; this provides the additional support needed to bridge the longer unsupported spans. The span of the upper floor joists can be supported by adding load-bearing walls beneath, or with the inclusion of steel joists.

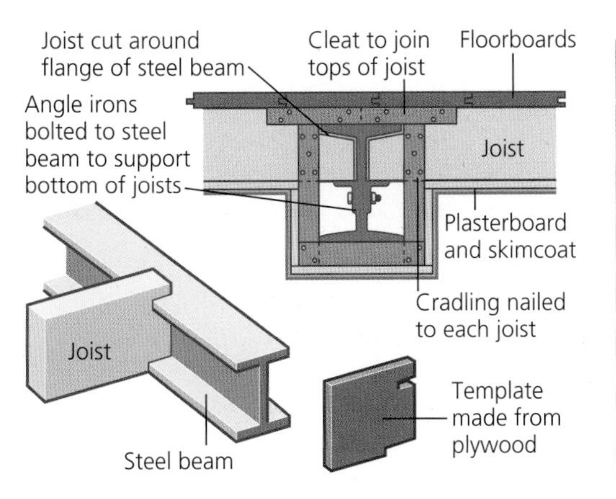

FIGURE 6.72

TRADE SECRETS

To prevent repeatedly marking each joist end to be cut around a steel joist, make a template. Templates are normally made from timber-based, man-made sheet materials (MDF, plywood, etc.). Man-made sheet materials are easy to shape with carpentry tools and do not normally misshape, shrink or expand.

STAIRWELLS

The positioning of the floor joists will vary between floors; upper floors will normally have openings for stairwells and trimming around chimney breasts/flues. The exact positioning of each stairwell should be detailed on the architect's drawings.

When positioning floor joists to form a stairwell, consideration should be given to the amount of space needed to install the staircase. If the stairwell is constructed to the exact width of the staircase it may become difficult to align, especially if the staircase is manufactured prior to fixing the joists.

FLOOR JOIST IDENTIFICATION AND LAYOUT

Floor joists should run the shortest distances between supporting walls or steel joists. This method of positioning the joists will keep the maximum depth of the joists to a minimum, and will therefore be the most cost-effective solution. All joists within the same floor should be the same depth.

SINGLE, DOUBLE AND TRIPLE FLOORS

Floor joists that span between supporting walls without an intermediate bearing (clear span) are known as single floors. 'Single' floors are the most common method of constructing upper floors in new buildings, and usually span up to 4.7 metres.

The sectional size of the floor joists can be kept to a minimum if the span of the joists is split in two with an intermediate support; this type of construction is known as a 'double' floor. The intermediate support can be built either under the floor joists or within the floor. Traditionally, large timbers would have shared the load of the floor from the underside. Although this method was very successful, it had the disadvantage of revealing the full depth of the supporting beam along the ceiling. This beam (or 'binder') is usually either covered with plasterboard and a 'skim' (thin) coat of plaster, or left exposed to create a feature.

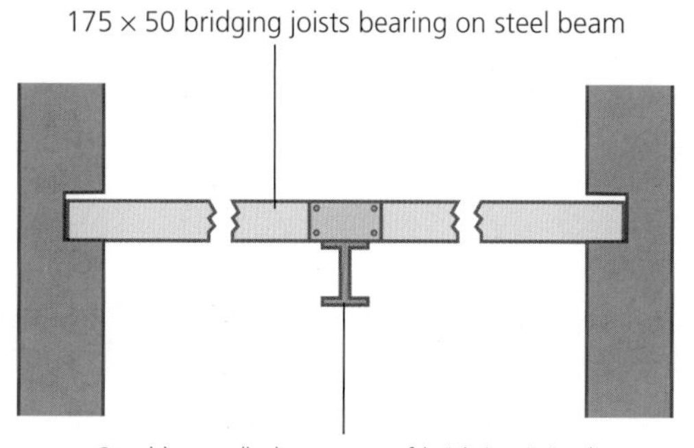

FIGURE 6.73 Floor joists running over the top of a binder

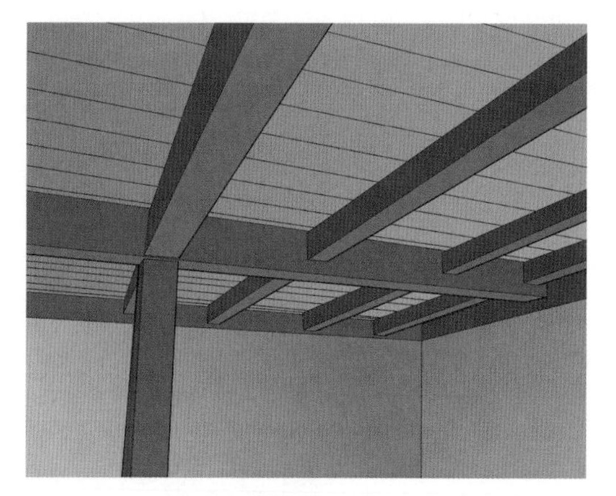

FIGURE 6.74 Exposed beams

Triple floors are rarely used in modern construction due to the cost of the heavy sections of timbers needed to span from wall to wall or beam to beam. Modern floors would usually have an intermediate support from load-bearing walls or steel joists built with the floor, or be constructed from concrete beams.

FLOOR COMPONENTS

- Bridging joists run between supporting walls, also known as common joists.
- Trimming joists are parallel to the bridging joists and line the edges of an opening in the floor.
- Trimming joists are the cut, shortened common joists with the trimmer joists fixed.
- Trimmer joists are fixed at 90° to the trimming joists and are positioned around the opening of a stairwell, chimney breast or flue.

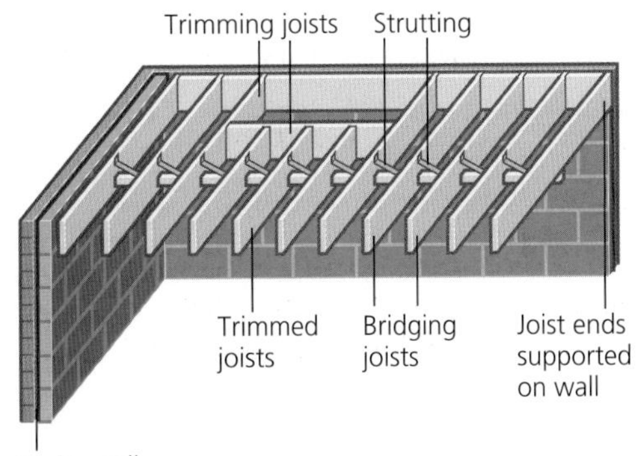

FIGURE 6.75 Floor components

Note: both the trimming joists and the trimmer joists around a stairwell will have increased loads imposed upon them due to the support they give to the staircase. These joists are usually increased in thickness by 25 mm or doubled up and bolted together.

SOLID FLOOR JOIST SPANS

Note: the following table should only be used as guidance; for exact spans and joist spacing, refer to the Building Regulations. Joist sizes may vary depending on the grade of timber used and the loads imposed upon them.

MAXIMUM CLEAR SPAN OF JOISTS			
Size of joist	Spacing of joists (mm)		
Thickness × depth (mm)	400	450	600
46 × 97	1.93	1.82	1.47
46 × 120	2.52	2.42	2.05
46 × 145	3.04	2.92	2.59
46 × 170	3.55	3.42	3.00
46 × 195	4.07	3.91	3.41
46 × 220	4.58	4.39	3.82
62 × 97	2.20	2.09	1.83
62 × 120	2.78	2.67	2.42
62 × 145	3.35	3.22	2.92
62 × 170	3.91	3.77	3.42
62 × 195	4.48	4.31	3.92
62 × 220	4.94	4.80	4.41
73 × 120	2.94	2.83	2.57
73 × 145	3.54	3.41	3.10
73 × 170	4.14	3.99	3.63
73 × 195	4.72	4.56	4.15
73 × 220	5.15	5.01	4.67

FIXING FLOOR JOISTS

Floor joists are usually fixed in position either with conventional joist hangers or by building the joists into the brickwork. Each method will provide adequate support to the floor joists, but building the joists into the brickwork is less labour intensive and a cheaper alternative. The disadvantage of building solid timber floor joists into the inner leaf of cavity walls is the potential for the ends to rot, although this risk can be dramatically reduced if the correct precautions are taken during their installation.

> **TRADE SECRETS**
>
> The cut ends of floor joists should have a liberal covering of preservative applied before building in. The end grain of timber will absorb more moisture than the faces, so it is vital that these areas are protected against decay.
> It is good practice to cut the ends of the joists to a splayed angle, making sure that their bottom edges still have a full bearing on the inner wall. Using this method allows the ends of the joists to be encased in mortar, thus increasing the protection of the ends.

Cavity walls are designed to prevent moisture passing from the outer wall to the inner wall of a building. If the gap between two walls is bridged, then moisture will travel between and into the building. Moisture that comes into contact with joists will rot the ends of the timber and potentially undermine the integrity of the floor. The following guidelines should be considered when positioning and installing built-in joists.

1. Avoid the joists overhang the cavity between the two walls. Mortar that falls down the cavity as the build progresses could build up on the joist ends and potentially bridge the gap.
2. Avoid the joists bridging the full width of the cavity. This method could draw moisture through the outer wall and promote the start of rot.
3. Floor joists should bear down on the full width of the inner wall; this is the most effective form of providing maximum support.

Bricklayers will build up the courses of bricks or blocks to the height of the underside of the floor joists. At this point carpenters will lay each of the floor joists between load-bearing walls.

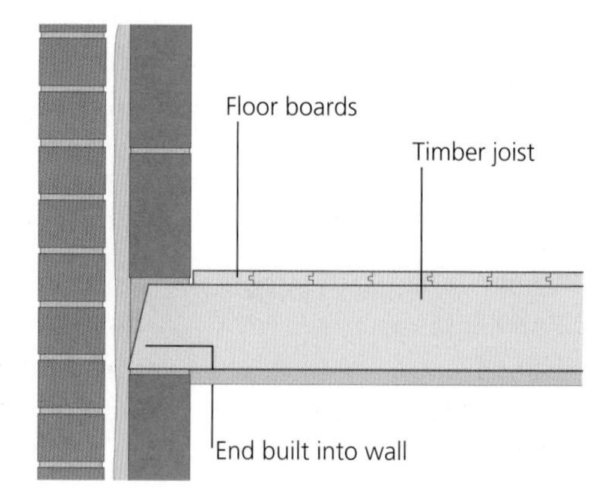

FIGURE 6.76

STRUTTING

The term 'strutting' refers to the support given between the floor joists. Strutting is sometimes confused with noggings commonly found in timber studwork, although they have similarities. Strutting is positioned between timber floor joists to prevent them from twisting, and to tie the joists together to create a solid structure. Any movement between upper floor joists may cause damage such as cracks and an uneven surface to the finished ceiling. Twisting or bowing of the joists may also distort the finished floor level over the top of the joists.

Strutting is not normally required between joists that span less than 2.5 metres, and one row will be needed for floors that span up to 4.5 metres. Floors spanning 4.5 metres or more will have two rows of strutting equally spaced along the length of the joists.

METHODS OF STRUTTING

There are several different kinds of strutting commonly available; each one prevents movement between the joists and has its own advantages and disadvantages. It should be noted that the methods of strutting described are only for solid joists; for all other types of floor joists the manufacturer's recommendations should be followed.

Solid strutting or bridging

Solid timber blocking (strutting) should be at least 38 mm thick and extend at least three-quarters the width of the joist to provide adequate support and prevent the joists from cupping. Full width strutting should be avoided as it may distort the floor above and ceiling below.

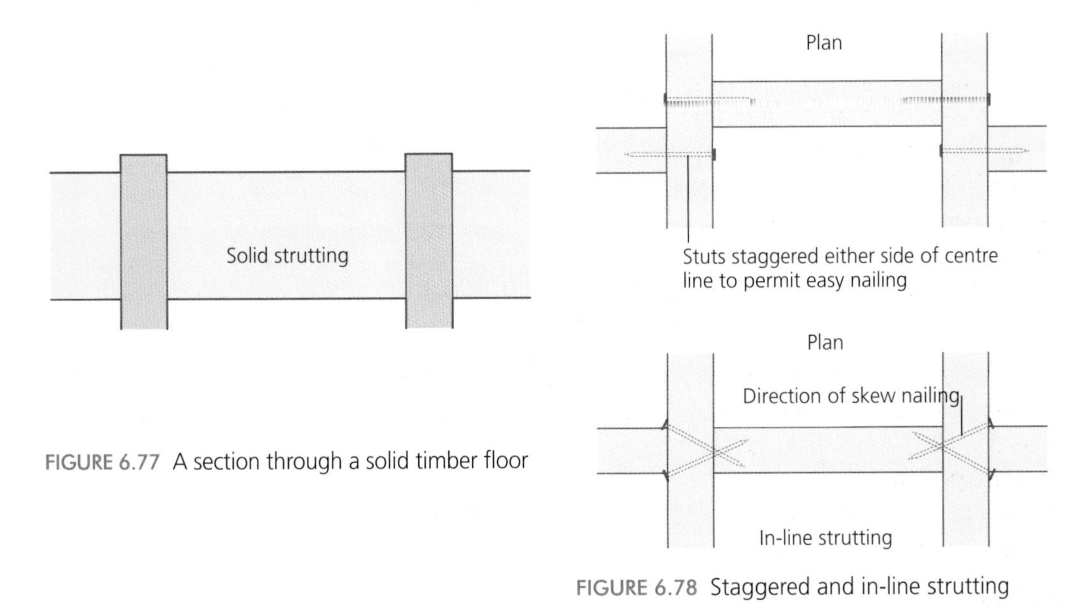

FIGURE 6.77 A section through a solid timber floor

FIGURE 6.78 Staggered and in-line strutting

FREQUENTLY ASKED QUESTIONS

What does the term 'cupping' mean?

When a tree has been converted into usable sections, the timber will go through a process of drying out known as 'seasoning'. The amount of seasoning will depend on the type of timber and the position it is going to be used. As the sections of timber are dried ready for use they may distort or misshape. In most cases the timber sections will 'cup' away from the centre of the tree (the pith).

Herringbone strutting

Although herringbone strutting is time consuming to cut and install compared with other methods, it does not have to be drilled and notched for services to pass through. The herringbone sections used should be at least 50 mm by 25 mm (although 50 mm by 32 mm is very common) and should not be in contact with each other due to noise transfer through the floor.

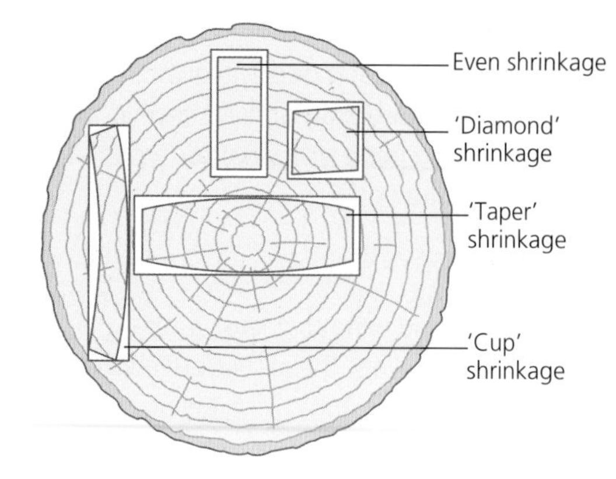

FIGURE 6.79 Cupping

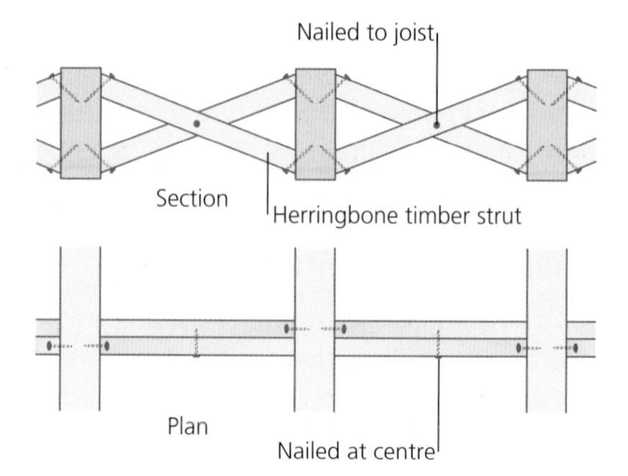

FIGURE 6.80 Herringbone strutting

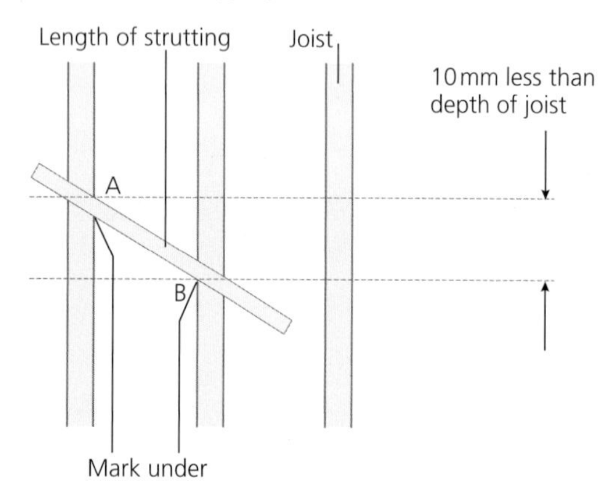

FIGURE 6.81 Marking, cutting and fixing herringbone strutting

TRADE SECRETS

Herringbone strutting is normally secured in position with clout nails, fixed directly through the top and bottom edges of each section. If saw cuts are made through the middle of both the top and bottom edges before fixing, the nails can be driven through these slots. This will prevent unnecessary splitting of the timber and a weak fixing.

Proprietary strutting

Proprietary strutting looks very similar to traditional timber herringbone strutting, the main difference being that it is made from galvanised mild steel. Proprietary strutting is approximately 1–1.5 mm in thickness and pressed into shape to provide a very light and strong restraint. There are several manufacturers of proprietary strutting, each one with a slightly different method of fixing the strutting into position. In general, the strutting can either be bent and fixed over the top and bottom edges of each joist using 30 mm clout nails, or wedged directly into the inner faces of each joist using the pointed ends of each strut to secure them in place.

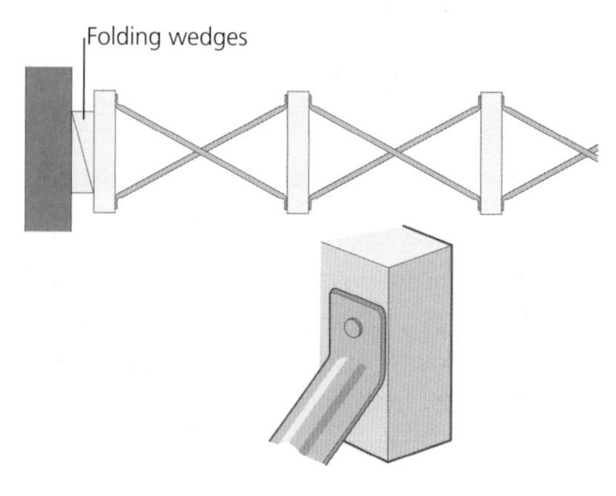

FIGURE 6.82 Proprietary strutting

Proprietary strutting is available in a variety of standard lengths to suit the depth and spacing between the joists. Herringbone strutting should always be fixed in pairs between the floor joists to form a cross, ensuring maximum strength. Care should be taken when fixing the struts to ensure that they *do not* touch, as this may lead to noise in the floor when it is under load (in use).

DRILLING AND NOTCHING JOISTS

Solid timber floor joists may have to be drilled or notched to allow services to pass through. The amount, size and positioning is strictly controlled by Building Regulations to prevent weakening of the structure. Figure 6.83 highlights the areas that are most affected.

The clear span of a joist will be under its own load as well as the weight from the floor covering and ceiling. The area through the centre of the joist is the least affected by loads; this is because the timber above the centre line is under compression, while below it is under tension. The centre line across the width of a joist is known as the 'neutral stress line'; this is the point at which all drilling should take place.

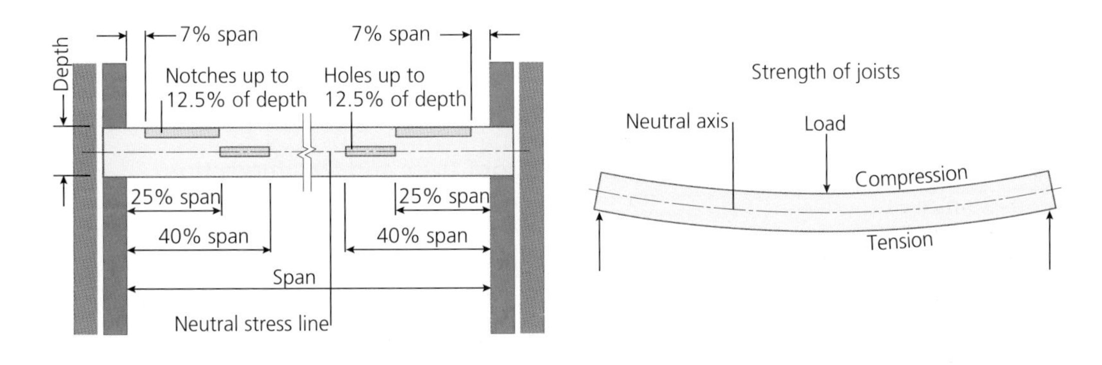

FIGURE 6.83 Positioning for drilling and notching

FIGURE 6.84 The neutral stress line

ACTIVITIES

Activity 26 – Installing floor joists

Read through the following questions and answer them as fully as you can to help you develop your underpinning knowledge of this subject area.

1. List the advantages and disadvantages of solid timber floor joists.
2. Sketch a section through a floor to demonstrate the use of 'herringbone strutting'.
3. Where should service holes be drilled in solid floor joists to prevent weakening?
4. What is the name of a joist that spans from one supporting wall to another?
5. What factors determine the depth of floor joists?

MULTIPLE-CHOICE QUESTIONS

1 The type of truss shown below is a :
 a fan
 b fink
 c king post
 d queen post

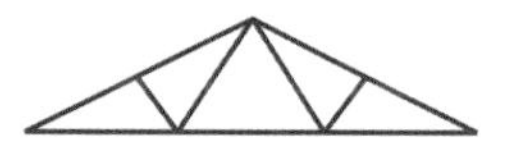

2 What is the minimum distance between a roof truss and a chimney flue?
 a 25 mm
 b 50mm
 c 75 mm
 d 100 mm

3 Which **one** of the following braces would be used to prevent trussed rafters from 'racking'?
 a Girder
 b Ledger
 c Diagonal
 d Proprietary

4 Which **one** of the following boards is the guttering fixed to?
 a Barge
 b Fascia
 c Soffit
 d Matched

5 Which **one** of the following ladders provides a fixing for the barge board?
 a Gable
 b Girder
 c Trussed
 d Extension

6 Intermediate support to joists within a hollow suspended ground floor is provided using
 a props
 b columns
 c sleeper walls
 d herringbone strutting

7 The DPC within a hollow suspended ground floor is positioned between the

a wall plate and joist
b joist and floor board
c sleeper wall and joist
d sleeper wall and wall plate

8 The illustration below shows a

a truss clip
b girder clip
c joist hanger
d joist bracket

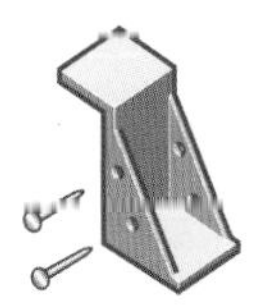

9 Upper floor joists are strengthened and prevented from twisting using:

a braces
b noggings
c restraint straps
d herringbone strutting

10 Holes bored through upper floor joists to accommodate wiring runs are best positioned along the

a top edge
b bottom edge
c neutral axis
d longitudinal axis

CHAPTER 7

MAINTENANCE

LEARNING OUTCOMES

By the end of this chapter you should have developed a knowledge and understanding of:

- repairing mouldings, windows and doors;
- replacing gutters and downpipes;
- repairing traditional sliding sash windows and replacing sash cords;
- making good plaster, paintwork and brickwork.

INTRODUCTION

The aim of this chapter is for students to be able to recognise common building products often deteriorated by their environment, and the methods used to repair and maintain them. It also identifies and explains current Building Regulations applicable to this area of study, following all the relevant health and safety law and good working practices.

REPAIRING DAMAGED, DECAYED AND INFESTED TIMBER

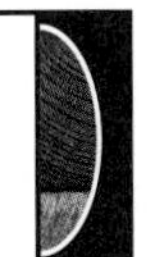

There are several reasons why timber will have to be replaced or repaired within a building; these normally include:

- damaged timber;
- wet rot;
- dry rot;
- insect infestation.

FIGURE 7.1

Note: the methods used to repair or replace damaged timber are discussed later in this chapter. Dry and wet rot can potentially be found in all areas of older buildings, causing cosmetic and structural damage amounting to thousands of pounds in repair bills. It is important to be able to recognise the different types of wood decay, treat them correctly and prevent them recurring.

WET ROT

Wet rot is a common problem found in timber with high moisture content, such as:

- the back of wooden window frames;
- along the sill or jambs on door frames;
- leaking roofs;
- along the ends of joists;
- behind toilets, sinks and baths;
- behind washing machines, dishwashers, etc.

DETECTING WET ROT

Vulnerable areas at risk of wet rot should be inspected regularly for early signs of deterioration of the timber. Damaged paintwork, discolouring and smell in the timber are all tell-tale signs and should be investigated further. This can be done simply by inserting a sharp knife into the timber; if the end of the knife disappears it is highly likely that wet rot is present.

TREATING INFECTED TIMBER

It is vital that the source of the water damage is located and made good before any repairs take place, to prevent the timber from being affected again. Voids created in the construction of buildings should be adequately ventilated to prevent stagnant air and moisture deteriorating the timber components. This is commonly discovered in voids between suspended timber ground floors in older properties. Wet rot is easily treated by cutting out the damaged areas until sound timber is discovered. Smaller areas can be spliced with new sections of timber to replace the damaged sections, and treated with preservative and several coats of wood primer, undercoat and top coat. Minor areas can be treated with an 'epoxy-based repair kit' following the manufacturer's instructions. If it is suspected that the high moisture levels may reappear, then the timber should have a wet rot treatment applied. If the problem extends to structural timbers such as the roof or floor joists, then professional advice should be sought to ensure that the maintenance does not affect the integrity of the building.

DRY ROT

Dry rot is a much more serious problem than wet rot, and is often referred to as 'building cancer' by tradespeople in the industry because of its potential to travel through walls and destroy timber structures. Unlike wet rot, dry rot is a fungus that grows in damp timber with a moisture content above 28–30 per cent, and also in dark, unventilated areas, attacking mouldings such as skirtings and flooring. Dry rot is often discovered in voids behind timber walls etc., and will often go undetected for a long period of time before the later stages of growth, when the mushroom-shaped fruit bodies start to appear.

FIGURE 7.2 Fruit bodies in dry rot

FREQUENTLY ASKED QUESTIONS

Why is 'dry rot' referred to as such if it is not dry at all?

It is referred to as 'dry rot' because the fungus is capable of transporting moisture from other sources metres away to attack dry timber.

IDENTIFYING DRY ROT

There are several stages that the rot will go through as it develops and starts to attack the timber

- 1st stage – white sheets simulating cotton wool appear on the face of the infected timber.
- 2nd stage – the sheets develop into fungal strands.
- 3rd stage – large flat mushroom-like fruiting bodies grow across the infected areas.
- 4th stage – red/brown spores are produced by the mushrooms.
- 5th stage – the infected timber will begin to develop deep cracks along and across the grain of the timber.
- 6th stage – musty damp odour starts to develop.

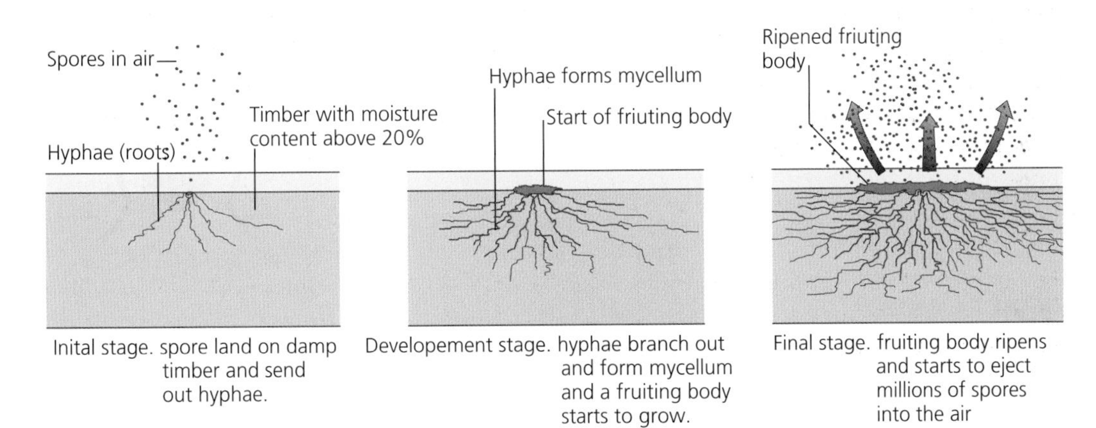

FIGURE 7.3 Stages of dry rot

TREATING INFECTED TIMBER

Generally, specialist companies are normally used to treat infected properties because of the risks associated with the materials used to eradicate the problem. They are usually better equipped and have the expertise to carry out thorough risk assessments to comply with the Control of Substances Hazardous to Health (COSHH) Regulations 1988. Wherever possible, timber lintels and infected joists etc. should be disposed of away from the site and replaced with steel to prevent new timbers being reinfected. Dry rot thrives on damp or wet timber with little or no ventilation; if these elements can be removed, then it will cease to grow, although it can remain dormant for years after. Dry rot should be cut out from the infected areas and at least 1 metre past the last signs of timber decay, and replaced with pressure-treated timber. In

areas where it is not possible to remove the damaged timber, the dry rot must be treated with a chemical or water-based fungicide. Brushing the treatment over the dry rot should be avoided, because it will only cure the surface area. Deeper penetration can be achieved by injecting the fungicides into the core of the timber. Alternatively, fungicidal paste can be used to spread over the infected areas, allowing the fungicidal oil contained in the paste to penetrate deep into the timber, provided the surface is not too wet.

INSECT INFESTATION OF TIMBER

In general, any timber discovered having dozens of tiny holes bored into its surface is usually considered to have 'woodworm'. Woodworm is not a specific species of insect, rather a generic name given to any wood-boring insect at its development stage. There are many different species of wood-boring insect; these include:

- powder post beetle;
- common furniture beetle;
- death watch beetle.

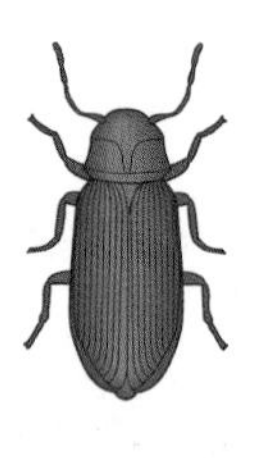

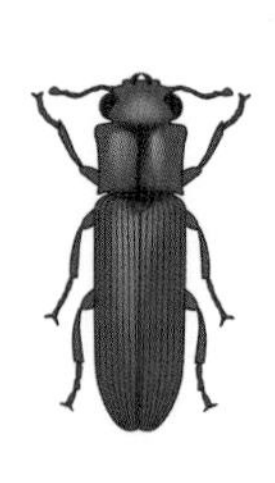

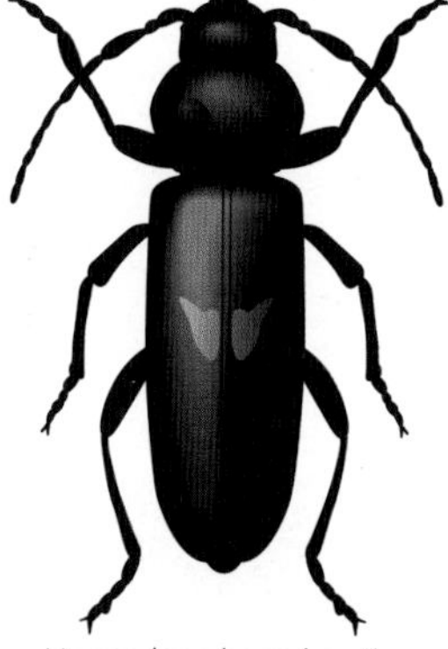

FIGURE 7.4 Beetles

Adult wood-boring beetles usually lay offspring known as 'larvae' in the surface cracks and deep grain of certain seasoned hard and softwoods. Studies have shown that timber containing 'sapwood' rather than 'heartwood' is more susceptible to widespread infestation, as sapwood is less dense and contains a ready food source. When laid, the larvae are white in colour and approximately 1 mm in length. They normally burrow down into the surface of the timber and remain there for months as they grow into fully developed adult beetles. At this stage they will resurface by boring back through the face of the timber, leaving a trail of exit holes and wood dust ('frass'). Badly

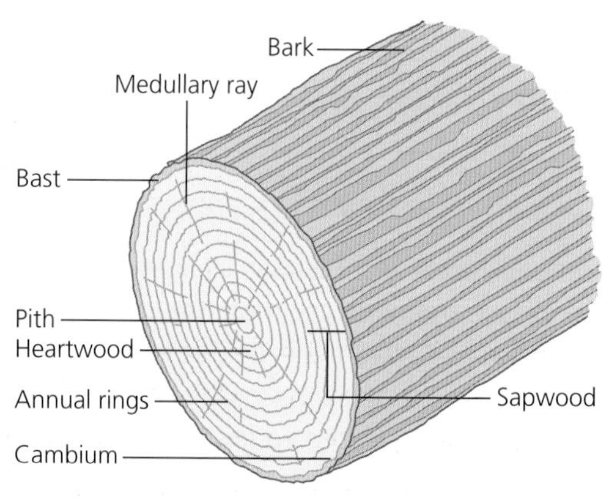

FIGURE 7.5 An insect attcking sapwood

infected areas of timber are usually destroyed by the insects, turning the sound timber into crumbling powder.

ERADICATING INFESTED TIMBERS

Wood-boring beetles are commonly discovered in furniture, flooring and the structural timbers of old buildings. Infected timbers should be replaced with treated wood, while the remaining uninfected timbers should be treated with insecticides or fumigation to control the potential for reinfection. Infected waste materials should be removed from the site immediately and correctly disposed of at a waste management depot to prevent contamination of other timbers in the building.

REPAIRING DAMAGED DOORS

Occasionally, older interior and exterior doors will need to be replaced because they have been poorly maintained or neglected, or they may require emergency repairs. (*Note*: Chapter 5, Second Fixing, should be referred to for details regarding the process of fitting new side-hung doors.) This section focuses on the effects of damaged and ill-fitting doors, and the methods used to adjust or repair them to full operational condition.

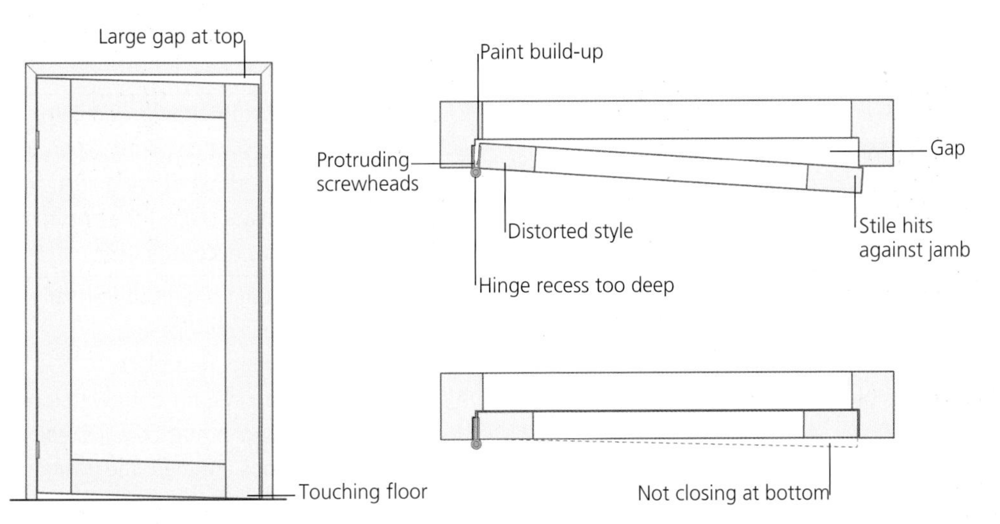

FIGURE 7.6 Potential problems with side-hung doors

ACTIVITIES

Activity 27 – Repairing mouldings and doors

Read through the following questions and answer them as fully as you can to help you develop your underpinning knowledge of this subject area.

1. Briefly explain the difference between 'dry' and 'wet' rot.
2. Explain how timber infestation could be avoided.
3. A timber door keeps springing open. Explain the possible cause and remedy.
4. The latch on a door is not engaging properly. Explain the possible cause and remedy.

Problem	Possible cause	Remedy
The door keeps swinging open	The door is 'hinge bound' or binding against the rebate on the frame	Remove the door from the frame. Recess the back edges of the hinges deeper into the door edge and frame. Re-hang the door. If the door is binding in the rebate, the hinges should be repositioned so that the face of the door sits slightly proud of the frame
The door will not close into the frame	The door may have swollen	Remove the door and ironmongery. Reduce the width of the door with either an electric plane or jack plane. Make sure the leading edge of the door is removed before re-hanging
The door is difficult to close	The bottom or top of the door is rubbing against the floor/frame	Remove the door from the frame. Look for signs of rubbing on the top and bottom edges. Reduce the door along these edges and re-hang. Check that a suitable margin has been achieved along the top and bottom of the door
The door is dragging across the floor surface	The door has 'dropped' on the lock side	Remove the door. Dismantle the components of the door and re-glue together. Re-hang the door in the frame
The door swings into the frame correctly, but the latch will not hold the door closed	The striking plate is incorrectly aligned	Close the door against the frame, and mark either side of the latch with a pencil. Open the door and check the position of the marks against the striking plate. Adjust the position of the striking plate as necessary
The door will not close flush in the frame; one corner of the door sticks out at the top	The door is twisted	Once a door is twisted it is difficult to rectify the problem and usually results in a replacement being hung in its place. Alternatively, if it is an internal door with loose door stops in the lining, it may be adjusted. Remove the door stops. Re-mark the position of the top hinge so that it sits deeper in the lining. Recess the hinge, re-hang the door and replace the loose stops
The door handles fail to spring back to their original positions	The spring mechanism in the handles or latch is either damaged/worn or rubbing against the door	Remove the ironmongery; check the operation of the components and apply some light oil. Check for signs of the ironmongery rubbing against the door, and chisel the necessary areas. Replace damaged parts and re-hang the door

REPAIRING AND REPLACING ROOF GUTTERING AND DOWNPIPES

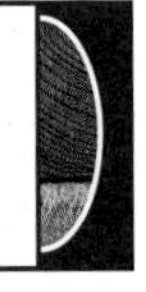

Roof guttering is a system of drainage channels used at the eaves level of a roof surface to collect and distribute rainwater to the drains. It is normally secured with fixing brackets along the edge of the fascia boards or rafter ends (the eaves) and drained through a number of vertical downpipes. Guttering and downpipe products are manufactured in a number of different materials, including:

- cast iron;
- copper;
- aluminium;
- uPVC (plastic)

CAST IRON

Rainwater goods have been used on roofs since the early nineteenth century to replace lead, as a cheaper alternative. Cast iron guttering is extremely strong and durable, with a long life expectancy of around 100 years. It is rarely used these days because it is expensive, extremely heavy and needs to be painted and regularly maintained. The insides of the guttering should be primed with zinc chromate paint, before a finishing coat of bituminous paint. It is especially important to ensure that the brackets and fixings are checked regularly for rust or weakness.

Guttering and downpipes are usually cut to length on site, with an electric disc cutter or manually with a tungsten-tipped hacksaw. The joints between the components were originally connected together with 'plumber's putty' and gutter bolts. The putty between the joints would often dry out over the years and break down before the guttering, allowing water to leak. Replacement gutters are now sealed with an improved 'jointing sealing tape' or 'compound' to provide a watertight joint that is cleaner to install and can be painted over.

Victorian cast iron guttering and downpipes were usually enhanced with decorative mouldings and detailed design features. Some of these profiles can still be seen today on some older, listed or historic buildings. The traditional architectural details found on Victorian rainwater goods have been replicated on other materials (e.g. uPVC) to add character to modern buildings at a fraction of the cost.

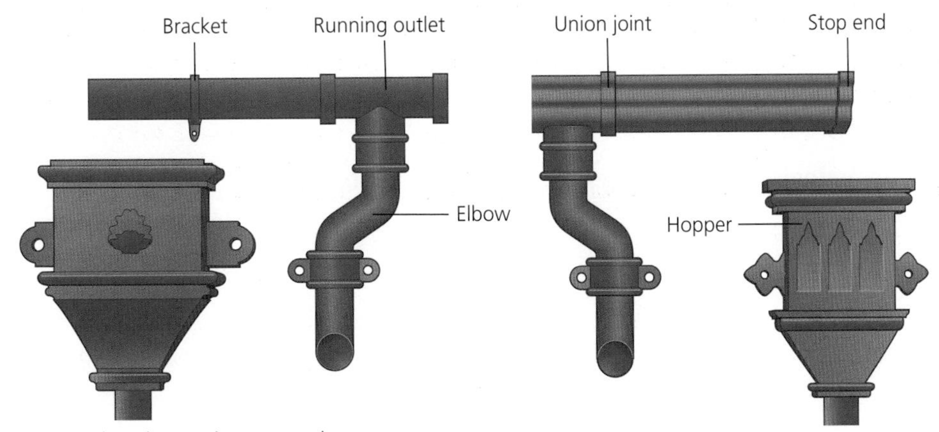

FIGURE 7.7 Victoria cast iron guttering

COPPER

Copper is a natural, non-toxic material that is environmentally friendly because 100 per cent of the product can be recycled. It is also substantially lighter than cast iron and easier to install. The copper can simply be cut to length with a pair of 'sheet metal snips', and jointed together by applying a bead of silicone between overlapping sections of guttering by approximately 50 mm. The joints are secured together with pop rivets either side of the gutter; a further bead of silicone is applied along the inside of the joints to complete a watertight seal. Copper guttering is relatively maintenance free, durable and resistant to corrosion. New copper components are generally supplied highly polished; their exposure to moisture and the elements when installed causes them to oxidise and, in time, turn green; this is known as 'verdigris'. Care should be taken to avoid leaning ladders etc. against thin-gauge copper guttering while accessing the roof, as this will cause damage.

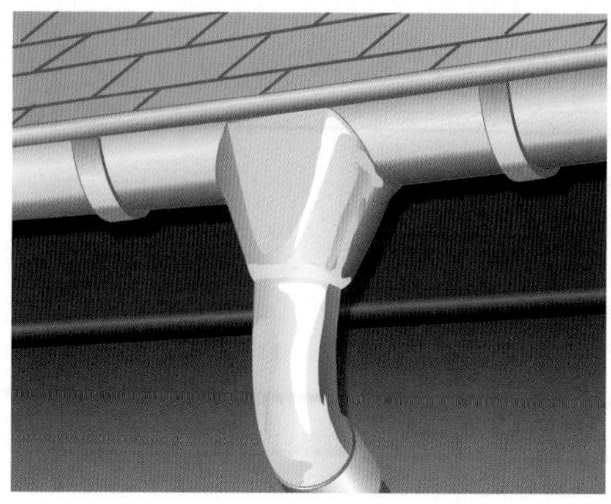

FIGURE 7.8 Joint between sections of copper guttering

FIGURE 7.9 New copper guttering versus discoloured sections

ALUMINIUM

Aluminium guttering and downpipes can be constructed using two different methods:

1. sectional gutters;
2. seamless gutters.

'Sectional' gutters are constructed in a similar way to other types of guttering, with relatively short component parts jointed together to build the required lengths. They are commonly available in a range of over 1900 standard colours.

As the name suggests, 'seamless' gutters are manufactured in lengths of up to 30 metres, making the need for joints in the gutter virtually non-existent. Aluminium guttering is approximately 40 per cent more expensive than uPVC (plastic), but is expected to last more than 20 years longer. Similar to other modern types of guttering, aluminium is low maintenance, resistant to atmospheric corrosion and, as a recyclable material, it is environmentally friendly. Larger roof surfaces will require deeper and wider sections of guttering to move the flow of water away from the eaves and prevent it overflowing. Standard sections of guttering are normally available in 100 mm widths; aluminium profiles are produced in sections up to 150 mm.

FREQUENTLY ASKED QUESTIONS

How are the long lengths of aluminium guttering transported to site?

Aluminium is delivered to site in long rolls of sheet metal, in the back of relatively small, customised vans. The flat sheet metal is 'pressed' into shape on the vans as it is passed through a moulding machine. As the guttering exits the machine, it is cut to the exact dimensions required, which results in no waste material. The machine can easily be adjusted on site to produce different profiles, including 'torus' and 'ogee'.

UPVC

uPVC (unplasticised polyvinyl chloride) guttering was developed in the 1950s as an alternative to cast iron. It is now commonly used in modern house construction because it is very light, inexpensive and quick to install. The components are available in a limited range of colours; these include:

- black;
- white;
- brown/caramel;
- grey.

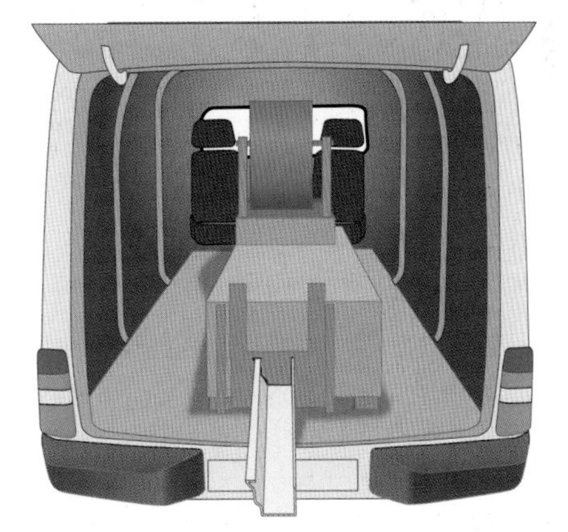

FIGURE 7.10

Low-maintenance uPVC guttering does not require painting to protect it from the elements, although the ultraviolet (UV) rays in natural sunlight will eventually weaken the plastic after approximately 10–12 years. After this period the plastic may begin to discolour and the joints creep apart, causing leaking. (*Note*: some cheaper products may begin to fade earlier, although some companies will guarantee against this happening.)

Products that have been independently tested and meet the required standards will be entitled to carry the BSI kitemark

Plastic guttering is available in maximum lengths of 4 metres, with downpipes up to 5.5 metres. The guttering is connected to other sections by simply clipping the system together 'dry'. Rubber gutter seals already fitted to the joints allow them to become watertight without the need for silicone.

FIGURE 7.11 uPVC guttering

FIGURE 7.12 British Standards kitemark

TRADE SECRETS

Plastic guttering is capable of connecting to other products and metal systems with the use of extensive ranges of adaptors. This may be especially useful when replacement gutters are required to connect to existing ones – for example, on a semi-detached or terraced property

PROFILES (UPVC)

Plastic guttering is commonly available in the following standard profiles:

- half round;
- square;
- ogee;
- large capacity half round.

Some drainage systems may have a combination of two profiles – for example, a half round gutter with square downpipes. When two different profiles meet, special adaptors are required for the transition between component parts.

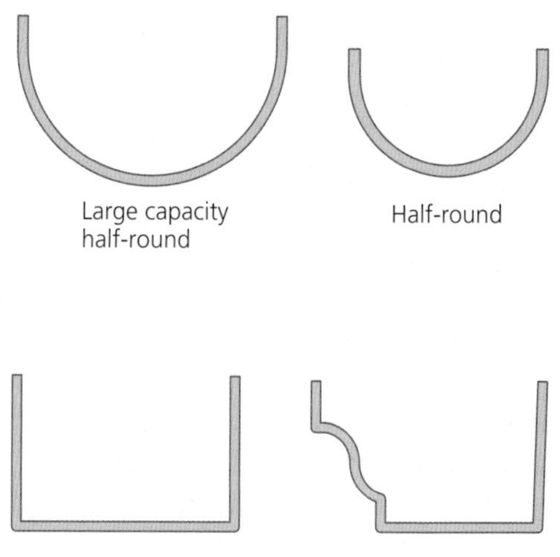

FIGURE 7.13 Guttering profiles

GUTTERING AND DOWNPIPE COMPONENTS

- **90° and 135° gutter angles** – connected to straight lengths of guttering, allowing it to return around square and angled corners at roof level.
- **92.5° and 112.5° offsets** – these joints allow the downpipes to be jointed and redirected over the shape of the building to the drain holes.

- **Branch** – these are used to connect two lengths of downpipe together, allowing the rainwater to be directed to one drain.
- **Downpipes** – are jointed into the underside of guttering with an 'outlet' to direct the flow of water down the face of the building into the drains. It is important that downpipes are well supported up against the building with pipes clips to prevent any movement and weakening of the joints.
- **Fascia brackets** – are used to support and secure the guttering across the fascia and eaves. They are usually spaced no more than 1 metre apart, and within 150 mm of any fittings or joints. If the fascia brackets are spaced wider apart than 1 metre, the guttering may 'sag' and hold the water in hollows along the channel, which could promote weed growth. The spacing of the brackets may be reduced if heavy snow fall is experienced, and may also vary between manufacturers; the installation instructions should be referred to for the exact positioning. The brackets should be positioned with a slight fall (6:1000 or 'six in one thousand') along the length of the guttering to allow water to flow to the outlets.
- **Gutter** – is fixed around buildings to drain the flow of rainwater away from the surfaces of the roof to the drains at ground level. This will protect the faces of the building from being saturated due to excessive amounts of water travelling down the outer walls, which could eventually lead to damp.
- **Half round to square adaptor** – as mentioned previously, there are numerous adaptors that allow uPVC guttering to be connected to various other types of material. This particular adaptor will connect half round and square uPVC guttering together.
- **Hopper** – these components are used to allow several downpipes to join one downpipe, midway down a wall, before continuing to the drain.
- **Stop ends** – can be found at the highest points on lengths of guttering. The components can be jointed either with rubber seals on the outside of the guttering (externally) or internally with silicone sealant. External stop ends are the preferred method of finishing the guttering because they provide a stronger watertight joint, although internal stop ends provide a seamless finish.
- **Pipe clips** – are wrapped around the downpipes and screwed back to the face of a building to secure them in position. They should be fixed no more than 1.8 metres apart, unless the manufacturer's installation instructions state otherwise.

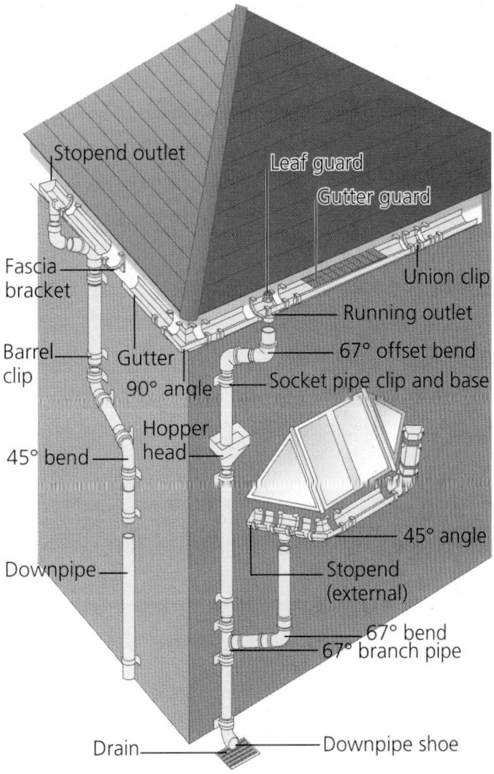

FIGURE 7.14 uPVC guttering components

FIGURE 7.15 Fitting fascia brackets

- **Pipe connectors** – are used to increase the length of downpipes.
- **Running outlets** – are normally used midway along the length of gutters to connect downpipes to the guttering channel.
- **Shoes** – are fitted to the feet of downpipes, approximately 40 mm from ground level, to direct the waste water into the drain. If 'shoes' are not used, the water will splash on the edge of the drain, causing a damp patch on the outer wall.
- **Square to round adaptors** – are usually connected to the underside of square outlets to allow a transition between two different profiles.
- **Stop end outlets** – are a combination of two components: stop ends and running outlets. These are particularly useful when the drain position is directly below the end of a run of guttering.
- **Unions** – are used to extend the length of guttering by simply clipping the sections together.

FREQUENTLY ASKED QUESTIONS

I sometimes have difficulty fitting the parts together. How can I make this easier?

Expansion and contraction in the plastic sometimes cause difficulty in fitting the joints together. Silicone spray lubricant should be used between the joints to allow the components to slide together easier. The guttering is best fitted into the fascia brackets by tilting it slightly to locate it against the back of the bracket, then straightening it to the front clip and 'snapping' it in place.

IDENTIFYING AND REPAIRING DAMAGED PARTS

Poorly maintained or blocked guttering could lead to water penetrating through 'porous' bricks and pointing at roof level; it could also cause water damage to timber fascias and soffits. Moisture damage and damp can be costly to repair and will devalue a property unless repaired correctly by professionals. The following points should be considered when inspecting existing guttering.

- Gutters free of leaves, weed growth and moss. (*Note*: 'leaf guards' can be inserted into new or existing guttering to prevent build-up).
- Make sure the joints between the guttering, downpipes, etc. are watertight and firmly together.
- Inspect the fascia brackets and pipe clips for signs of cracking, splits or corrosion on metal components.
- The paint on some gutters and downpipes will crack, peel and flake if not regularly sanded down and repainted. Check around the joints between components to look for early signs.

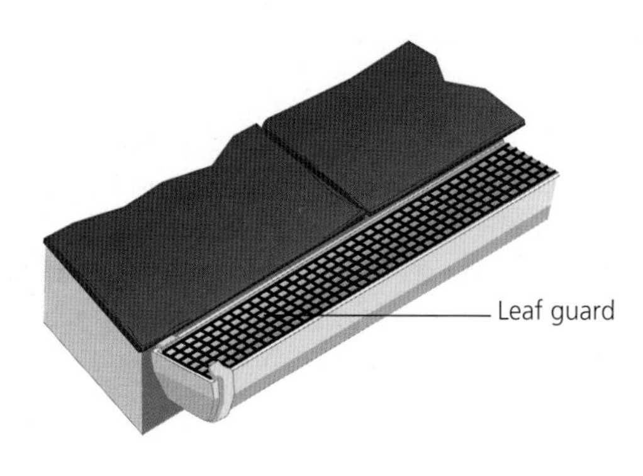

FIGURE 7.16 A leaf guard

- uPVC guttering should be regularly wiped clean with a soft, damp cloth to maintain its appearance and prevent early deterioration.

REMOVING EXISTING GUTTERING

In most cases, removing or installing guttering will require working at height with some form of access equipment. Falls from height are the major cause of accidents in the construction industry, so it is vital that consideration is given to the type of equipment used, the length of time it will be needed and how many people will be using it at any one time. (Further information on 'Basic working platforms' is available in Chapter 1, Safe Working Practices.)

Removing uPVC guttering is fairly straightforward and it can normally be replaced by one person. Heavier systems, such as cast iron, will need at least two people to take the weight of the larger sections as they are being lowered. Care should be taken to ensure that no damage occurs to the face of the building, including protruding window sills and doors. Old cast iron guttering usually has a large build-up of paint over the screws securing the fascia brackets and the bolts in the guttering, etc. The build-up of paint, together with corrosion around the fixings, can make the task of removing the guttering extremely difficult. These fixings are normally removed by cutting the heads off with either a hacksaw or a small electric disc cutter (also known as an 'angle grinder'). It may be necessary to cut the fixings on the gutter from the underside so it is important that the correct personal protective and access equipment is used. This will include safety goggles, together with builders' gloves used to protect your hands from the sharp edges found on cast iron, especially after cutting.

ASBESTOS GUTTERING

'White asbestos' is sometimes encountered when replacing old guttering and downpipes. It was commonly used in domestic house building up to the late 1960s to strengthen the cement-based products, which included roof guttering. There are several types of asbestos:

- white – used in domestic appliances and building;
- brown – used in thermal insulation;
- blue – used for insulation lagging.

The fibres in the asbestos can enter the body if they become airborne and are inhaled, which can cause damage to the delicate cells in the lungs. Brown and blue asbestos are considered to be more dangerous because they have finer fibres and will therefore travel deeper into the lungs. There can be a delay of 15–60 years between exposure to the fibres and the onset of asbestos-related diseases . The effects of damage caused to the lungs as a result of exposure to asbestos are permanent and irreversible. (*Remember* – asbestos can be found in a variety of other building materials besides guttering; these include insulation and fire-resistant products.)

TRADE SECRETS

When replacing existing guttering, it is worthwhile investigating and assessing the condition of the fascia boards and soffits. These items can easily be renewed or repaired at this stage, thus avoiding the need to remove newly installed work to access the areas at a later date.

All types of asbestos are a major health hazard if cut, drilled, sanded or broken.

If you are unsure whether or not a product contains asbestos, you should always treat it as if it does. Asbestos cannot be identified visually because it is normally mixed with other materials; the only way it can be confirmed is by laboratory analysis under a microscope. It is recommended that guttering containing asbestos should be removed by licensed asbestos removal contractors. It is now illegal to use products containing asbestos in new building or refurbishment of existing properties.

Employers have a legal duty to manage and control the potential risks caused to employees and the general public by hazardous materials, in line with the Control of Substances Hazardous to Health (COSHH) Regulations.

Further information regarding managing the risks of asbestos can be sourced from the Health and Safety Executive (HSE) (see http://www.hse.gov.uk).

INSTALLING PLASTIC GUTTERING

It is important to ascertain the positions of the drains before producing a sketch detailing the layout of the building, with the overall dimensions of the roof perimeter. Refer to the job specification before choosing an appropriate range of guttering, to check the variety of component parts manufactured and available. Some ranges may be limited to standard components, potentially making it difficult to joint to existing rainwater systems on adjoining properties.

Use the sketched plan of the guttering layout to produce a list detailing the lengths of guttering, downpipes and components needed to complete the project, before ordering the materials. The following sequence details a method of installing standard uPVC guttering, although the basic principles are the same for other materials.

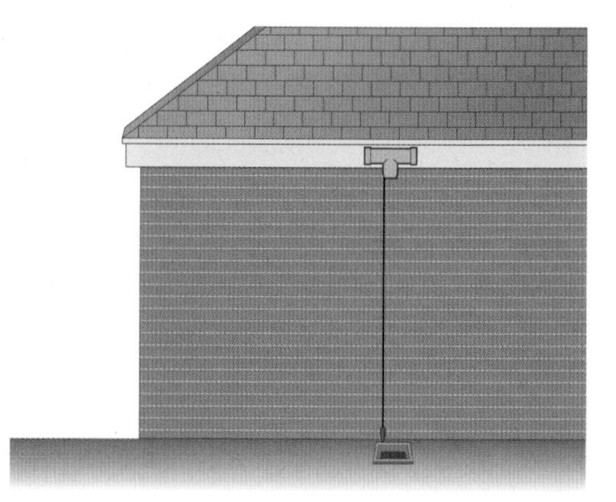

FIGURE 7.17 Use a plumb bob to align the running outlet above the drain

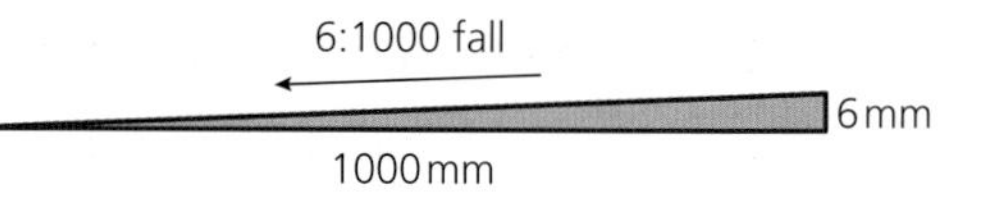

FIGURE 7.18 A 6:1000 fall

- Stage 1 – Use a 'plumb bob' between the fascia and ground level to centralise the running outlet above the drain.
- Stage 2 – Measure and fix one fascia bracket 100 mm away from the edge of the running outlet. The height of the bracket will be determined by the depth of the fascia and the total length of the guttering. In general, the guttering should be central to the depth of the fascia board, taking into account the angle or 'fall' of 6:1000.
- Stage 3 – Measure and fix another fascia bracket 100 mm in from the opposite end of the fascia, having calculated the fall; use two 25 mm screws to secure it in position. Ensure that the screws are not over-tightened, as this will cause the plastic to split.

- Stage 4 – Tie a 'string line' between the two fascia brackets, and equally space and secure further brackets between the two ends. It may be necessary to support the string line in the middle with a small screw in the fascia to prevent sagging over longer distances. 'Sighting' along the length of the string line is a simple and effective method of checking that a string line is straight. Ensure that the brackets are no more than 1 metre apart and any joints between unions etc. are supported with further brackets no more than 150 mm away from either side of the components.
- Stage 5 – Measure and cut to length the guttering with a hacksaw. Lubricate the seals around the fittings with either silicone spray or washing-up liquid. These lubricants will make it easier to slot the components together, and allow the plastic to contract and expand freely between the joints without damaging the seals.
- Stage 6 – Starting at the outlet, tilt the guttering into the back of the fascia brackets, making sure that it fits tight against the 'lugs'. Bring the front edge of the guttering down until it is level and snaps into place behind the 'lugs' on the front of the brackets. Any further lengths of guttering should be fitted into the unions and fascia brackets at this stage.
- Stage 7 – Slide the outlet on to the end of the guttering over the drain and install the remaining smaller section of guttering on the opposite side.
- Stage 8 – Fit the stop ends to each end of the channel.
- Stage 9 – As it usually requires two lengths of downpipe to reach the gulley, fit the upper first then measure for the lower section. If the eaves project, two offset bends are required to allow the downpipe to fit back to the wall. Hold the shoe in position, approximately 40 mm above the ground, and measure the length of the downpipe.
- Stage 10 – Connect the downpipe into the running outlet, making sure that the socket or square to round adaptor is fitted to the top. Fix one pipe clip around the top of the downpipe to support it against the wall while it is levelled, then fit the rest of the clips to the wall using pipe connectors between any joints.
- Stage 11 – Finally, fit the shoe to the bottom of the downpipe.

ACTIVITIES

Activity 28 – Replacing gutters and downpipes

Read through the following questions and answer them as fully as you can to help you develop your underpinning knowledge of this subject area.

1. List four materials used to manufacture gutters and downpipes.
2. Sketch three different profiles commonly found on guttering.
3. What is the purpose of a running outlet?
4. 'Shoes' are normally connected to what part of a rainwater drainage system?
5. List the potential hazards associated with removing old guttering.

VERTICAL SLIDING SASH WINDOWS

There are two methods of constructing sliding sash windows:

1. traditional box frames with counterbalancing weights;
2. modern, solid frames with spiral spring balances.

BOXED FRAMES

The boxed frame method of construction is commonly used to replicate and replace existing windows in historical/listed buildings and conservation areas. The sliding sashes are suspended by sash cords, running over a pulley and connected to counterbalancing weights that run in the box of the jamb and through the top of the jamb through a pulley. The cord then runs down the other side of the pulley stile and is tied to cylindrical weights; the weight of each completed sash should be equal to one pair of weights. This will allow the sash to slide freely up and down the frame with ease.

> The weight of each pair of cylindrical weights = the weight of each sash + the glass + the glazing bead or putty

(*Note*: the weights for the top sash should be 5 per cent heavier and 5 per cent lighter for the bottom sash. This ensures the sashes close tightly to the frame naturally.)

Timber boxed sash windows will need to be serviced regularly to maintain their performance; part of the maintenance may include the replacement of damaged or broken sash cords. The only access to the weights while the windows are in situ is through the pockets cut into the pulley stiles.

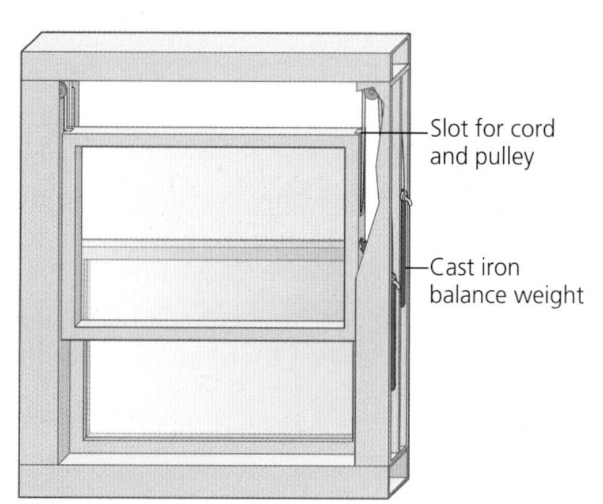

FIGURE 7.19 Sash weight

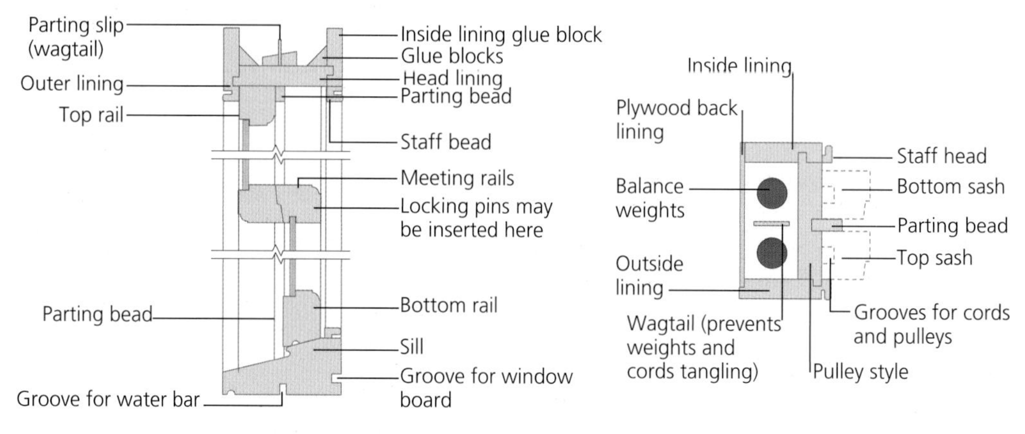

FIGURE 7.20 A boxed sash

DEFINITIONS

- The *wagtail* is a strip of timber loosely held in position at the head lining and falls between the weights. Its function is to prevent the sash weights and cords tangling up.
- The *parting bead* is the moulded section of timber dry fixed into a groove in the pulley stile. It is dry fixed to allow easy removal when the sashes need to be removed from the window frame.
- *Staff beads* are dry fixed into position using galvanised nails on the inside of the boxed frame; they are used to hold the inner sliding sash in position.
- *Glue blocks* are fixed to the head of the boxed window frame to provide additional support.
- *Meeting rails* provide a section of rail for the glazing and window lock to attach. They differ from the other rails in the sash because they have a splayed rebate. When the sashes are fixed in position the splayed rebates bridge the gap between the two sliding sashes.
- The *Sill* is the large section of moulded timber position at the bottom of the window frame. It has weathering on the outer edge and a rebated step to prevent water passing through the window frame.
- *Inside and outside lining* provides sections of the boxing that make the main frame to construct the window.

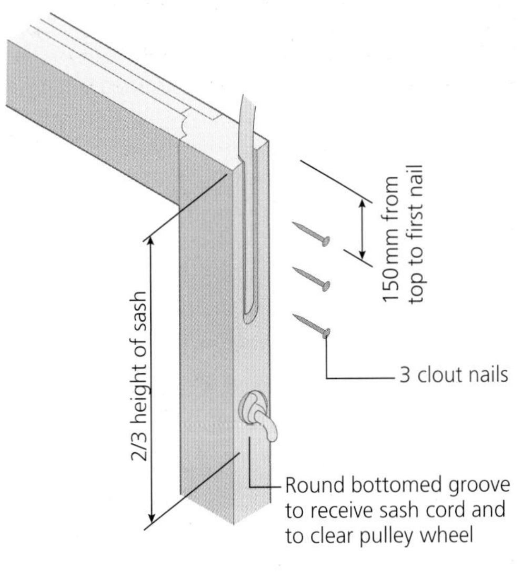

FIGURE 7.21 A pocket in the side of a pulley stile

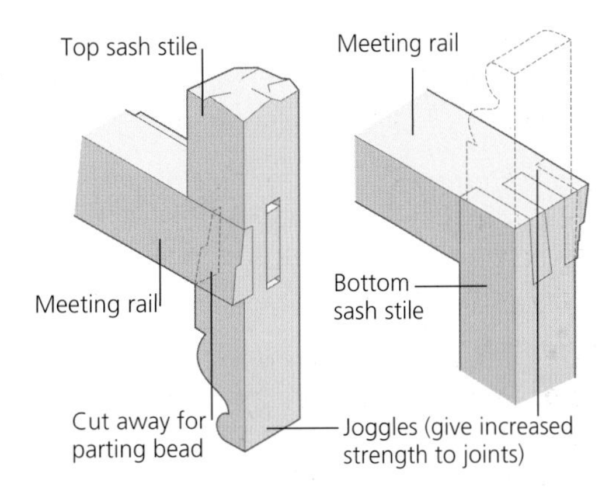

FIGURE 7.22 A box frame

SPRING-BALANCED FRAMES

This type of frame assembly uses a less complex method of construction. The outer frame is made of solid components rather than a boxed section, and uses spring balances to counterbalance the weight of the sliding sash.

Spring balances are available in a variety of lengths to suit different-sized windows and can be finely adjusted by twisting the coiled spring to increase or decrease the tension. If the springs are too tight, the sash will be difficult

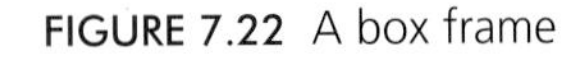

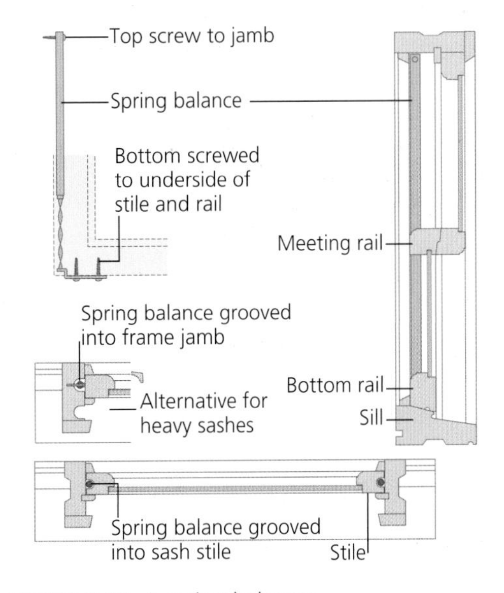

FIGURE 7.23 A spring balancer

to pull closed and will fly open. If the springs are too loose, the sash will close under its own weight and will require increased effort to open. Correctly adjusted balances should allow the sash to slide up and down the window frame with ease.

The advantage of spring-balanced windows is the simple construction method, making this the preferred method and the most cost-effective when mass produced. There are two ways to accommodate the spring balances into the window frame:

1. spring balance grooved into the frame (this is a stronger option but has the disadvantage of seeing the grooves in the jambs);
2. spring balance grooved into the sash (this method removes a lot of timber, weakening the stiles of the sash, but the grooves are not seen)

FREQUENTLY ASKED QUESTIONS

How are the spring balances fixed to the window?

The spring balances are screwed into the frame at the top of the jamb. The bottom of the spring balance has a flange that is recessed and screwed to the stile and bottom rail of the sash.

SLIDING SASH IRONMONGERY

FIGURE 7.24 Thumb latch

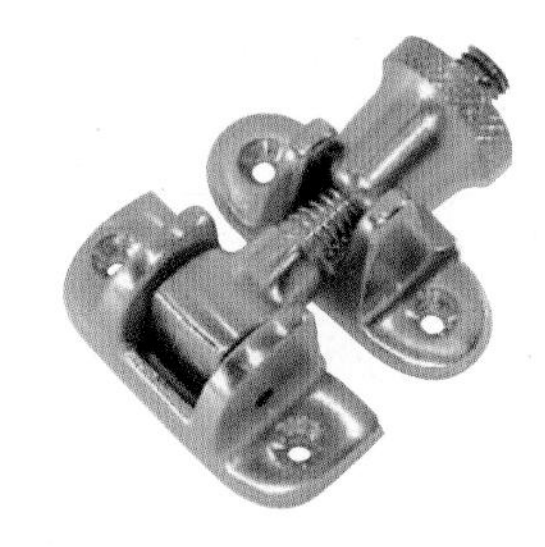

FIGURE 7.25 Spring screw fastener

BOXED FRAME ASSEMBLY

It is important that any framework assembly is carried out on a clean, dry and level surface. You cannot always assume that the bench you are assembling the joinery frame on is level; even the slightest bit of twist in the bench will cause difficulties later on when installing. The best way to check your bench is not in twist is to use bearers sighted through and packed up accordingly to overcome any twist.

The bearers are used in pairs and normally nailed in position temporarily. It may be necessary to use additional bearers to prevent sagging between bearers; this may depend upon the length and thickness of the frame.

FIGURE 7.26 Bearers

- Step 1 – Before assembling any frame it is important to 'clean up' all the parts that are difficult to 'clean up' when the frame is assembled. Glue and screw the head lining to the pulley stiles across the bench bearers, position the pulley stiles in the housings in the sill, and glue, wedge and screw into position.
- Step 2 – The inner lining should be glued and nailed on to the soffit and pulley lining. Turn the frame over and repeat the process using the outer lining. Punch all the nails below the surface of the frame.
- Step 3 – Fix the glue blocks to the head of the boxed frame. Check the frame is square, using a squaring rod.
- Step 4 – Insert the wagtails and fix the back lining, leaving the fixings proud (this will allow for easy removal when installing the sash weights).
- Step 5 – Insert the outer sash and check that it moves freely up and down the frame, with a 3 mm gap on either side.
- Step 6 – Dry fix the parting bead into position (this will allow for easy removal when removing the sash).
- Step 7 – Insert the inner sash and check that it moves freely up and down the frame with a 3 mm gap on either side.
- Step 8 – Dry fix the staff beads into position with oval nails. Leave the heads slightly proud, so that they can easily be withdrawn to remove the sashes. Fit the window ironmongery.
- Step 9 – 'Clean up' both sides of the assembled frame and remove all the arrises (an 'arris' is a sharp edge or corner).

(*Note*: the pocket will be cut but not knocked out until after assembly.)

FREQUENTLY ASKED QUESTIONS

What does the term 'clean up' mean?

The term 'clean up' means to remove all the pencil and machining marks, etc. from the timber to achieve a smooth surface ready to receive a paint finish. This can be achieved by using a smoothing plane, belt sander and finishing sanders. Large quantities of timber could be 'cleaned up' with a machine known as a 'drum sander'.

SASH ASSEMBLY

1. The inner edges of the frame must be 'cleaned up' prior to gluing together. This will prevent difficulty at a later stage when trying to sand the internal corners after the frame has been assembled.

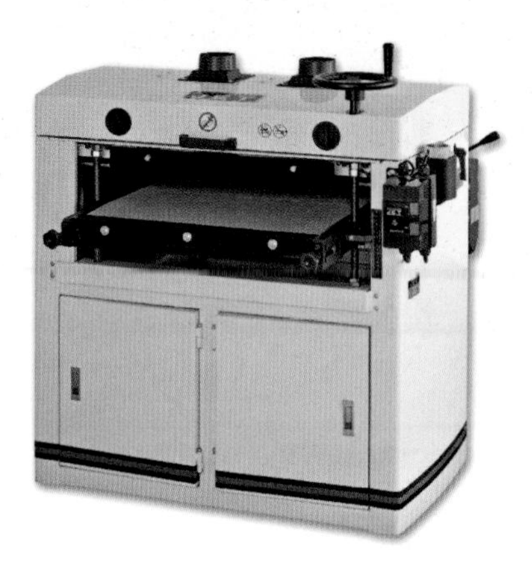

FIGURE 7.27 Drum sander

TRADE SECRETS

Do not plane or sand across any joints when preparing a disassembled frame for assembly. This will alter the accuracy of the joint by rounding over and reducing the material thickness. This mistake may result in poorly fitting joints and a reduction in the overall size when the frames are assembled.

2. Apply waterproof adhesive to the joints, including the shoulders, and assemble the frame.
3. Lay the sash flat on bearers across a joiner's bench.
4. Use a pair of sash clamps to squeeze each joint up tight.

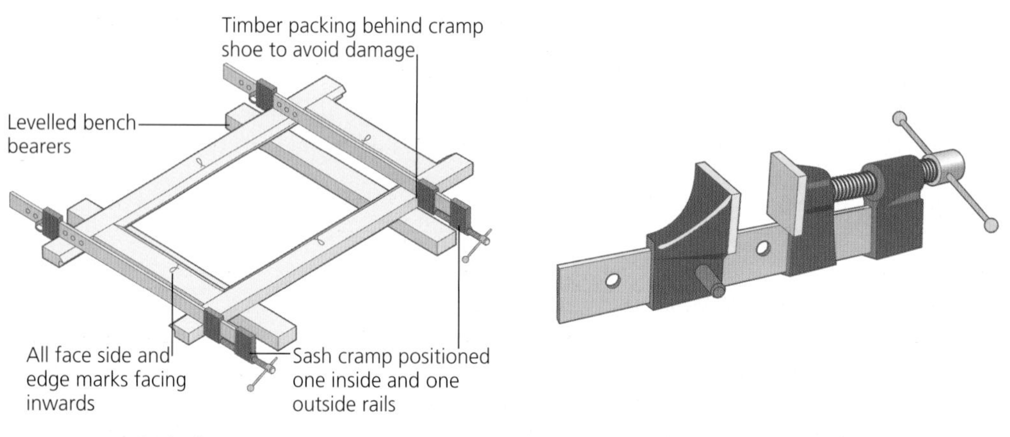

FIGURE 7.28 Sash clamp

FIGURE 7.29 T-bar clamp

5. The joints should be wedged and for additional strength a star dowel is driven through the face of the sash.
6. The sash clamps should now be removed. Using a squaring rod, measure from corner to corner to check that the frame is square. If the sash is not square, remove it from the bench and drop one corner of the sash onto the floor and recheck. (*Note*: the horns on the corners of the sash will protect it from damage during the squaring process.)
7. Check the frame is not 'in twist' by sighting through. If the sash is twisted, gentle pressure must be applied in the opposite direction to realign the frame.

8. Fix the sash to the joiner's bench by nailing through the end gain of the horns. Leave the top of the round head nails raised for easy removal. Securing the frame using this method allows the whole of the sash surface to be cleaned up without any obstructions.
9. Finally, remove any steps in the joints with a sharp smoothing plane and prepare the surface with an orbital sander ready for finishing. Remove the horns with a tenon saw and prepare to hang the sash.

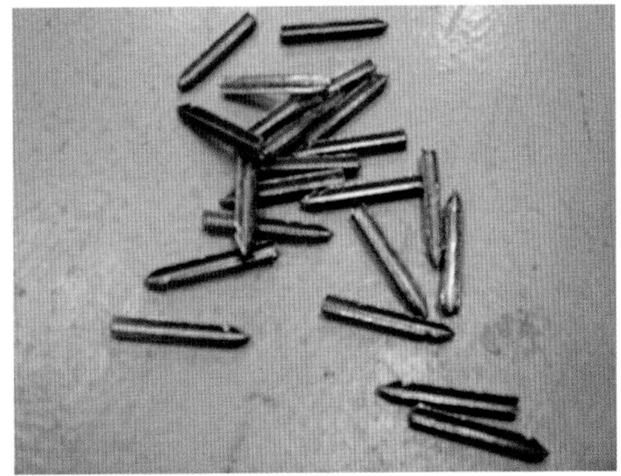

FIGURE 7.30 Alunium star dowels

MAINTAINING VERTICAL SLIDING SASH WINDOWS

COMMON DEFECTS FOUND IN TIMBER WINDOWS

If the windows are not maintained regularly the sashes may begin to stick and jam in the frames, and this will prevent them sliding freely. This is normally the result of the sashes not being suitably protected with paint, and swelling against the channels in the frame. Over many years, a build-up of paint between the frame and sashes normally seals them firmly in the closed position. Timber sliding sash windows can be expensive to replace and install, so in many cases the frame has to be serviced and maintained to its original condition. There are several ways that the sashes may be 'freed up' and made operational again; these include:

- using a putty knife to cut along the joint between the sash and the frame; avoid using a sharp knife as this may cause damage to the timber;
- carefully strike the meeting rail with a hammer and block of timber to prevent damage, in the opening direction;
- remove the layers of paint on the faces of the sashes with a heat gun before sanding down; if the width of the sashes needs to be reduced, use either a jack plane or an electric planer to remove large quantities of waste; reapply several coats of good-quality primer, undercoat and top coat before refitting;
- it is advisable to remove and replace both the parting and staff beads because they also usually have a build-up of paint, causing the sashes to stick; this also avoids having to strip the old paint and sand relatively small inexpensive mouldings;
- apply candle wax or talcum powder along the outer edges of the sliding sashes to encourage the smooth operation of the sashes.

Older, traditionally made vertical sash windows usually rattle in the wind and cause draughts, heat loss and increased noise levels travelling through the windows. This problem is normally overcome by installing draught excluders around the joints between the frame and sashes.

REPAIRING ROTTEN AND DAMAGED WINDOW COMPONENTS

Rot is a common problem with old, poorly maintained timber windows. Simply filling the rotten area and covering with numerous coats of paint will only mask the problem for a very short period of time. To remedy the problem, the rotten area must be completely removed until sound timber is discovered and a new section must then be 'spliced' in its place. This method of repair is known as 'scarfing'. Smaller sections of replacement timber are usually moulded on site. Larger pieces are normally replicated in the joinery workshop with the use of samples supplied by the site carpenter. It is sometimes more productive to send large quantities of damaged sashes to a joiner's workshop to be repaired, or in some cases replaced.

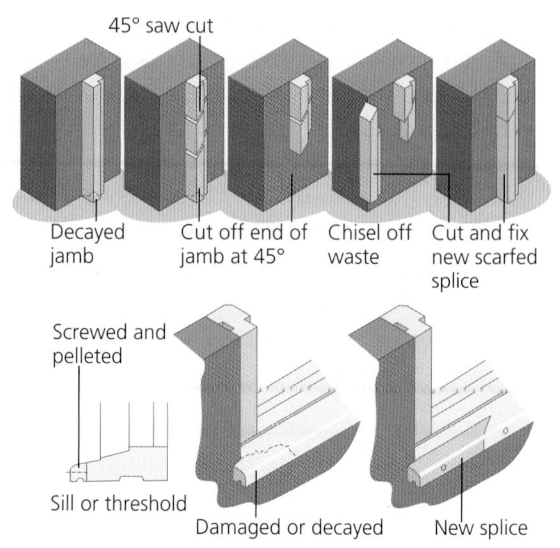

FIGURE 7.31 Scarfing

REPLACING BROKEN SASH CORDS

Over a period of years the movement of the sash cords over the pulleys and the weight of the sashes wear and cause them to break. Even if only one of the cords breaks, it is likely that the other cords will be in a similar condition, so it makes sense to replace them all at the same time. This can be achieved by the following steps.

- Stage 1 – Remove the staff beads from the window frame.
- Stage 2 – Insert wedges into the gap between the pulleys and the cord to prevent the weights dropping in the boxed frame.
- Stage 3 – Lift the lower sash out from the frame and cut the remaining intact sash cords.
- Stage 4 – Unscrew and remove the pockets from the pulley stile before removing the pair of weights belonging to the lower sash from the frame.
- Stage 5 – Run a knife around the edge of the parting beads to break the paint holding them in position. Carefully prise the parting beads out of the grooves in the pulley stiles with a chisel.
- Stage 6 – Lift the top sash out from the window frame and cut the remaining intact sash cords.
- Stage 7 – Feed one length of sash cord through the pulleys and pockets using the method illustrated in Figure 7.32; this will ensure that there is no wasted cord after cutting to length.

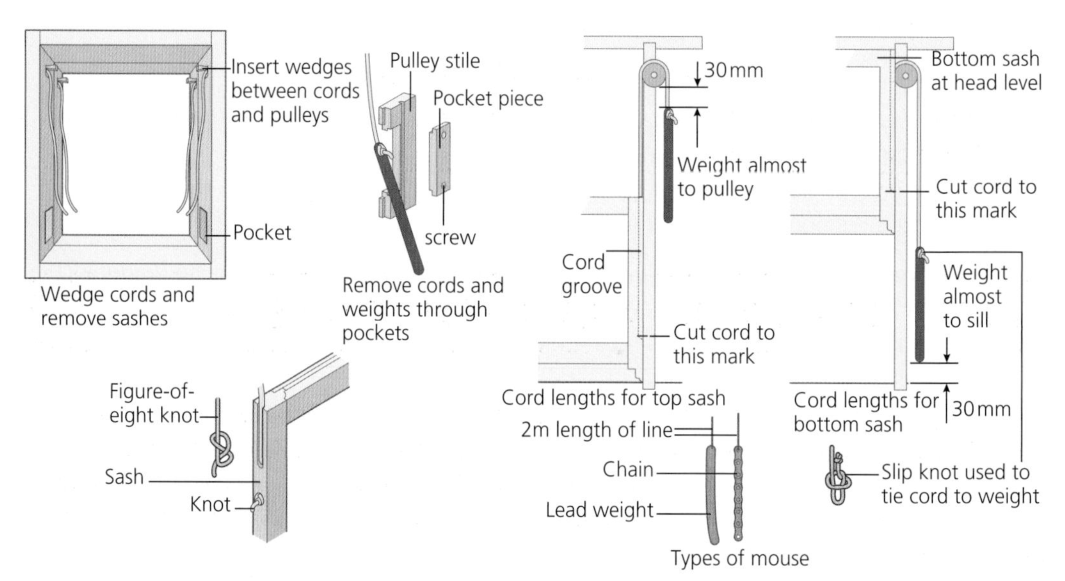

FIGURE 7.32 How to thread a sash cord through a window frame

- Stage 8 – Tie the end of the sash cord projecting from the pocket to the outer sash weights. Lift the weight to the top of the pulley stile and hold it in position temporarily by nailing through the cord into the pulley stile. Measure the length of cord required down from the top of the pulley and cut to size. Repeat this process on the remaining three sash cords, (*Note*: the outside sash is always hung first, followed by the inside sash.)
- Stage 9 – Feed the sash cords through the sashes and tie a firm knot to the ends. Insert several clout nails through the knots in the cords and into the side of the sashes to prevent the knots from coming undone.
- Stage 10 – Reinsert the parting beads back into the grooves in the pulley stiles and reposition the top sash. Insert the staff beads into the frame and secure them in position using oval head nails, leaving them slightly raised. Slide the sashes up and down the channels several times to check they move freely before finally knocking the heads of the nails below the surface of the staff beads.

TRADE SECRETS

Threading the sash cords through the window frame can be made easier with the use of a 'mouse'. A mouse is simply a small weight, such as a piece of lead, attached to a length of string and the sash cord. The weight of the mouse helps to feed the sash cord through the boxed window frame.

ACTIVITIES

Activity 29 – Repairing traditional sliding sash windows

Read through the following questions and answer them as fully as you can to help you develop your underpinning knowledge of this subject area.

1. The weights in vertical sliding sash windows should be equal to what?
2. How can access be gained to the weights in a vertical sliding sash window?
3. What is the purpose of the staff bead?
4. Explain the process of repairing a sash that has been painted closed.
5. Weights are traditionally used to balance the sashes in a vertical sliding sash window. Name one other method.

MAKING GOOD PLASTERED WALLS

Site carpenters will be expected to make good small areas of damaged walls and reveals, caused while replacing linings, frames and mouldings. This is more common on smaller construction sites where plasterers may not be available. 'Patching' small areas of damaged walls is fairly straightforward, provided that the surfaces are well prepared and the correct materials and tools are used.

PLASTERING TOOLS

HAWKS

Hawks are used by plasterers to hold small quantities of 'mix', before spreading over the wall.

FLOATS

Floats are used to spread plaster and render on walls from the hawk. They are available in a range of different lengths from 330 mm up to 508 mm for larger areas.

FIGURE 7.33 Plaster's hawk

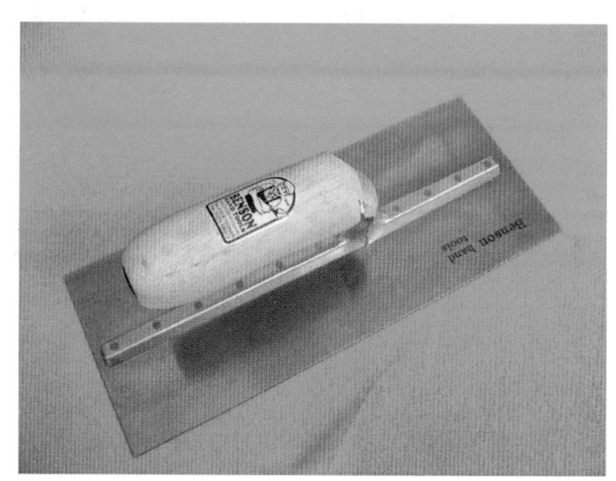

FIGURE 7.34 Plaster's float

BUCKET TROWELS

Bucket trowels are used to mix small quantities of plaster and dry wall adhesive in a bucket. The square noses on the trowels allow them to get into the corners of the bucket, ensuring that all the parts of the mixture blend together evenly.

FIGURE 7.35 Bucket trowel

TRADE SECRETS

Experienced plasterers will use trowels that they have owned for many years because they are 'worn in'. This simply means that the sharp corners and edges have been removed or shaped through continued use. Over a period of time the plaster trowels will become shaped to suit the individual tradesperson. Good-quality trowels can be purchased already 'worn in'.

FINISHING TROWELS

Finishing trowels are usually slightly 'dish' shaped across the width of their bases to prevent them 'digging in' as they pass over the plastered wall.

SCARIFIERS

Scarifiers are comb-shaped tools, used to scratch the first coat of render applied to brick, block or stone walls. This provides a rough surface, or 'key', that allows the following layer of plaster to bond.

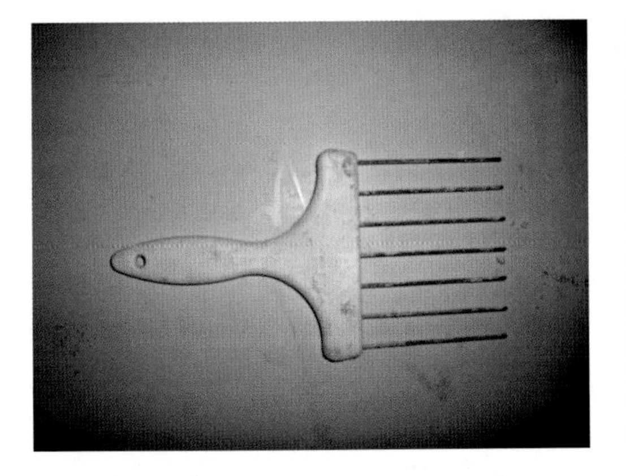

FIGURE 7.36 Scarifier

FIGURE 7.37 Spot board

SPOT BOARDS

Spot boards are usually constructed from a piece of moisture-resistant manufactured board, approximately 1 metre square. They are normally used by a labourer to load mixed plaster or render on to; the plasterer will then gather a manageable portion of the plaster to use from the hawk. Spot boards should be elevated off the ground to prevent dirt getting into the plaster.

DERBY

A derby is normally manufactured with a bevelled edge on one side. This is used to check that the surface of the wall is flat, and can also be used to scrape the semi-dry render from the wall to even out any high points.

MIXING WHEELS

Mixing wheels are used to manually mix small quantities of plaster; larger amounts of plaster should be mixed with a 'mixing paddle' attachment that can be fitted to an electric drill.

TYPES OF PLASTER

LIME PLASTER

Lime plaster can be found on the internal, non-load-bearing walls of older buildings, where it was used in conjunction with timber 'laths' to cover the surface of hollow walls. Laths are strips of 25 × 6 mm timber nailed horizontally across timber stud walls, with a gap of approximately 6 mm left in between. Plasterboard was not developed until the middle of the twentieth century so this method of 'lath and plaster' was the norm instead. The lime plaster was usually mixed with animal hair to improve its strength. Lime plaster is softer than modern gypsum products, but can be 'worked' for longer periods of time before setting. The disadvantage of lime plaster is the extremely long periods of time needed between coats to allow it to dry fully; in some cases this could be up to three to four weeks. Lime plaster is sometimes used to construct new environmentally friendly buildings because it is a natural mineral mined from the ground.

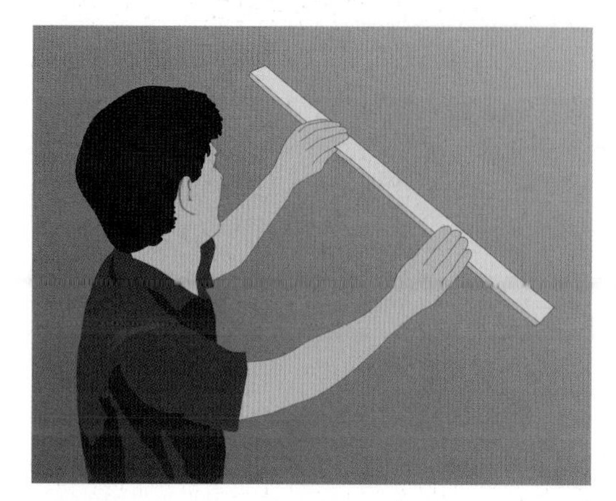

FIGURE 7.38 Steel straight edge

FIGURE 7.39 Lath and plaster wall

GYPSUM

Gypsum is a natural product mined underground. It has been used in the construction industry since the nineteenth century, when it was added to lime-based plaster to reduce the setting times between coats. Gypsum has good hardening qualities that make it naturally resistant to fire and heat. It is now commonly used in plasterboard and as a replacement to lime traditionally used in plaster because of its quick drying times.

BROWNING AND BONDING

Modern plaster is normally applied to walls and ceilings in two stages; these are:

1. render layer;
2. finishing layer, or 'skim'.

Browning and bonding are normally grey or pink types of plaster, suitable for the first render layer; these are known as the 'backing' plasters. Bonding coats of plaster are required over absorbent surfaces such as brick or block walls to prepare them to take the thin layer of finishing plaster (skim). Applying the render between 9 and 12 mm thick to large areas requires a lot of practice and skill to achieve a flat surface.

FREQUENTLY ASKED QUESTIONS

What is the purpose of 'polishing' the surface of the plaster?

As the plaster starts to dry, any imperfections in it will be revealed. At this point it is still possible to smooth them out with a finishing float before polishing the area to produce a flawless surface. If the plaster dries too quickly it may have to be dampened with a wet brush, while following behind with the float.

FINISHING PLASTER

Finishing plaster can be applied over browning or bonding, or directly on to plasterboard. It is usually spread approximately 2 mm thick over the surface area; any thicker than this may result in cracking as it dries. When the plaster starts to dry it should be 'polished' with a float to complete a shiny surface ready to decorate.

FIGURE 7.40 Polishing a plastered wall

MULTI-FINISH PLASTER

This is a finish plaster that can be applied to a range of backgrounds, including stone.

ONE-COAT PLASTER

One-coat plaster is an ideal product for 'patching' or repairing damaged walls. This type of plaster prevents the need to build up the thickness of the wall with several coats, thus speeding up completion times.

METHODS OF FINISHING AN INTERNAL WALL SURFACE

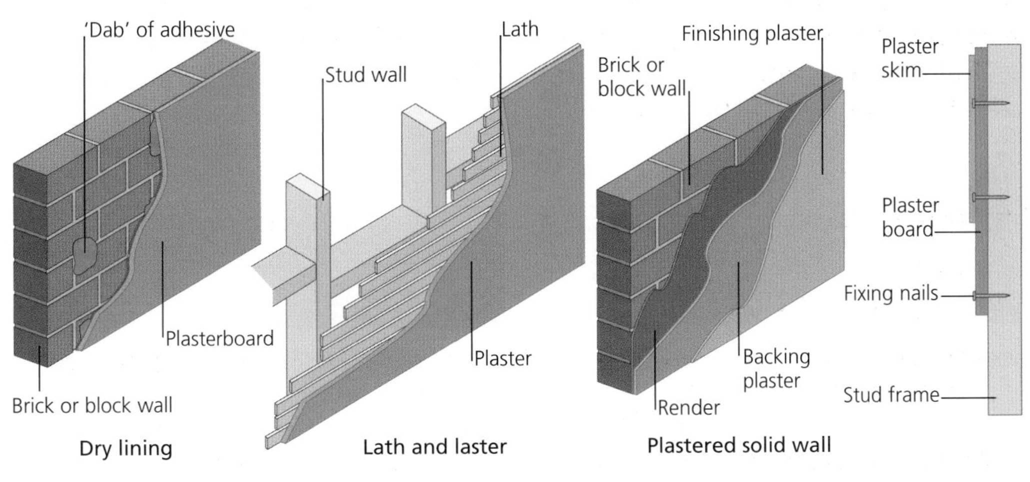

FIGURE 7.41 Finishing an internal wall

PREPARING OLD SURFACES FOR REMEDIAL WORK

Damage caused to plastered surfaces should be repaired using the same methods as originally used; this will prevent cracking appearing around the edges of the repaired area. It is important to make sure that the surface is free of loose plaster and dust, as this will prevent the new plaster from sticking and will probably result in it dropping off on to the floor. Wetting the area and sealing porous walls with polyvinyl acetate (PVA) adhesive will prevent the moisture being sucked out of the plaster before it has bonded correctly.

1. Use a club hammer and bolster to remove the loose or 'blown' plaster from the damaged area (the term 'blown' means the plaster is no longer bonding or attached to the wall behind).

2. Dust the area with a stiff brush, before damping down with clean cold water.
3. Dilute the PVA sealing agent following the manufacturer's instructions and brush on to the whole of the damaged area.
4. Add a small quantity of 'backing plaster' to half a bucket of clean cold water and mix with a 'paddle'. Keep adding the plaster until it is completely smooth and creamy.

FIGURE 7.42 Removal of loose plaster

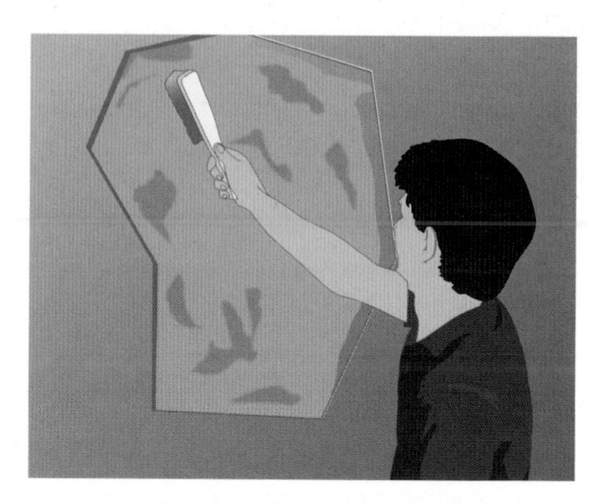

FIGURE 7.43 Dusting down with a stiff brush

FIGURE 7.44 Mixing and applying backing plaster

FIGURE 7.45 Scooping plaster from the hawk onto a float

5. Empty the plaster on to the spot board and scoop up a manageable quantity on to the hawk.
6. Tilt the hawk towards the damaged area and use the float to scoop a small amount of plaster on to its edge.
7. Starting at the bottom of the wall, spread the plaster up over the damaged area. Angle the float slightly to allow the plaster to slide down to its bottom and on to the wall.
8. Keep applying the plaster to the damaged area before levelling it out so that it is 2–3 mm under the finished wall surface. Take care not to run the float continually over the area as this will cause the water in the plaster to rise to the surface of the mix and could result in the plaster falling off the wall.
9. As the backing plaster begins to dry, use the 'scarifier' to scratch the surface of the plaster to provide a 'key' for the finishing coat.
10. The backing plaster should be left for a few hours until it begins to go hard. Brush down the damaged area to ensure there is no loose plaster. Mix up a further batch of finishing plaster in a clean bucket, following the manufacturer's ratio instructions. The top coat of plaster should be slightly thinner than the backing plaster.

FIGURE 7.46 Using a scarifier

FIGURE 7.47 Levelling plaster

11. Spread the finishing plaster over the surface of the damaged area. Allow the plaster to dry for approximately 35–45 minutes before using a steel straight edge to remove any raised excess plaster from the area. The straight edge should be used on its edge, moving it in a backwards and forwards motion across the plastered area, until it is completely flat.
12. Apply further amounts of finishing plaster to any hollowed areas. Leave the plaster to dry for approximately half a hour before using the float to smooth out the plastered area, making sure that the float is held at a slight angle.
13. Leave the plaster to dry for a further 10–15 minutes before using a clean wet float to smooth over the surface of the wall. Keep the surface damp with a wet painter's brush as the process is repeated until any imperfections are removed and it becomes polished. Make sure that the float remains clean and wet at all times while polishing the plaster, to prevent it being pulled off the wall.

FREQUENTLY ASKED QUESTIONS

Why should the float be held at an angle?

Using a float at a slight angle moves surplus plaster over the wall, filling any low spots. If a float is moved flat over the newly plastered area it could result in the plaster sticking to the float and dropping off the wall.

REPAIRING PLASTERED CORNERS

All external corners on plastered walls are prone to potential damage caused by general bumps and knocks. 'Galvanised angle beads' should be cut to length and bedded into plaster on the external corners to provide protection and a clean, sharp edge to plaster up to; uPVC angle beads should be used in areas of high moisture or damp, to prevent deterioration of the metal.

FIGURE 7.48

FIGURE 7.49

The adjoining internal surfaces of a wall should be plastered in two stages, making sure that one surface is completed before the other. Plastering the returning surfaces of an internal wall at the same time will cause difficulties when trying to achieve a straight corner.

LAYING AND REPAIRING BLOCK AND BRICKWORK WALLS

Tradespeople who can offer their employers other skills as well as their own are highly valued and sought after. 'Making good' damaged areas of block or brickwork while undertaking maintenance on a property could potentially avoid the expense of having to send a bricklayer to a site for a short period of time. Before repairing or replacing bricks or blocks it is important to source an identical match. Matching new bricks with the existing ones can sometimes be difficult, especially if they are old and 'weathered'; alternatively, reclaimed bricks from a reclamation yard may be a better option. (The term 'weathered' means the material has been shaped by its environment, e.g. old bricks may discolour and become slightly less uniform on their surfaces compared with new.)

TYPES OF COMMONLY USED BRICKS

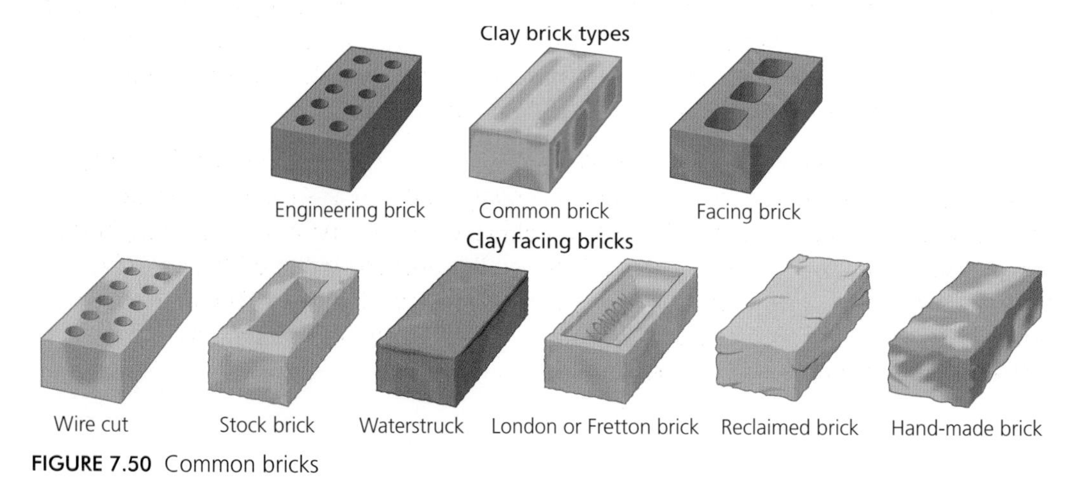

FIGURE 7.50 Common bricks

TYPES OF BOND

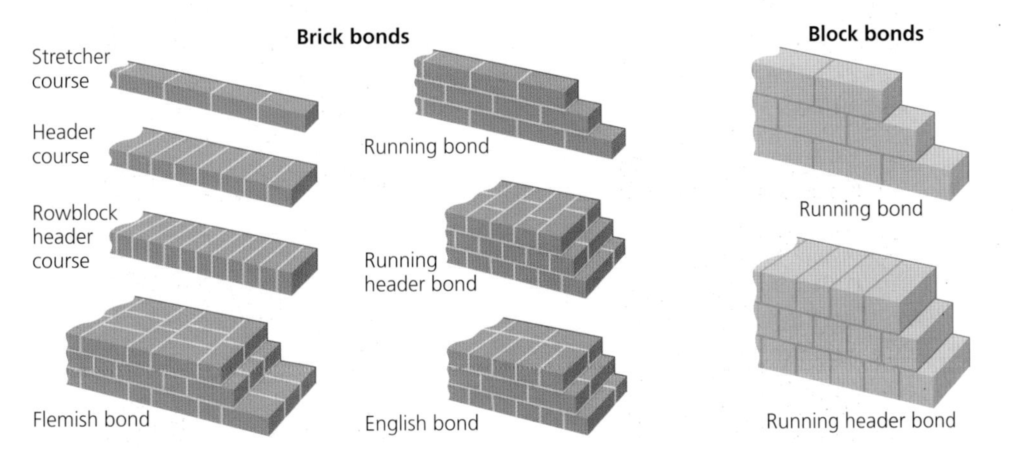

FIGURE 7.51

BOND ARRANGEMENTS

There are many ways that bricks can be arranged to provide strength and a uniform appearance to a wall. In general, the joints between bricks should be staggered to avoid weak points in the wall; this method of overlapping the bricks is referred to as the 'bond'. There are two ways of creating a bond; these are:

1. quarter bond;
2. half bond.

LAYING BRICKS IN HOT AND COLD WEATHER CONDITIONS

Brick and blockwork should not be laid in conditions below 4°C, as this will cause the water in the mortar to freeze before it sets. As the mortar dries and the conditions become warmer, the previously frozen water will cause cracks between the bonds and potentially weaken the whole structure. Any brickwork that has been newly laid and become exposed to freezing conditions as it dries should be taken down and rebuilt when the temperature reaches at least a few degrees above freezing. It is important to check that any bricks or blocks stored in freezing conditions have been thoroughly inspected to ensure that they are not frozen, especially if they come from the centre of a pile. If there is a risk of temperatures falling to a critical level after the brickwork has been laid, then it should be protected with hessian and clean plastic sheeting. There should ideally be a gap left between the protective sheeting and the brickwork to maintain air flow and prevent staining.

Hot weather conditions may also potentially cause problems while bricklaying. If porous bricks or blocks are exposed to extremely hot conditions, they may suck the moisture out of the mortar as they are being laid. This may result in a poor bond between the bricks, blocks and courses, and weaken the structure. To prevent this happening, the bricks should be dampened with a hose immediately before laying. (*Note*: avoid soaking.) This method should not be used for engineering bricks because they may slide across the mortar joints and higher courses may become unstable.

HEALTH AND SAFETY

The Construction (Design and Management) Regulations 2007 cover safe working temperatures in the workplace. Extreme heat and cold can cause short- and long-term ill effects; these include:

- heat/direct sunlight – stress, dehydration, sunburn, blistering, skin cancer;
- cold – stress, discomfort.

The Health and Safety Executive reports that: 'Skin cancer is the one of the most common forms of cancer in the United Kingdom, with over 50,000 new cases every year.' The risk to health and safety through exposure to extreme weather conditions can be reduced by following the employer's health and safety policy. Remember that employers have a legal duty to provide safe working conditions for all employees while at their place of work. Personal risks can be reduced further by:

- wearing high-factor sun cream;
- covering up bare skin;
- wearing a hard hat with a peak;
- taking regular breaks to warm up, or cool down in the shade;
- avoiding working in the midday sun.

Remember – a tan is not healthy sign; it is a sign of damaged skin.

REPLACING DAMAGED BRICKS

The difficulties encountered while replacing old bricks include matching the existing size, texture and colour; even the thickness and colour of the mortar joint will have to be replicated.

(*Note*: cutting and breaking bricks while repairing damaged walls will result in particles flying into the air, so it is vital that the correct personal protective equipment (PPE) is worn. This includes goggles and leather gloves, etc.)

FIGURE 7.52 Coloured mortar

FREQUENTLY ASKED QUESTIONS

How can you match coloured mortar?

Coloured mortar is achieved by using coloured building sand or by adding 'mortar pigment' to the mix. If coloured mortar is required, it is important to experiment with the pigment beforehand and allow it to dry fully before a true comparison can be made between the new and the existing.

- Stage 1 – Any damaged bricks should be removed from the wall by carefully breaking the mortar joint with a 'cold chisel' and club hammer. Drilling a series of holes along the joints (stitch drilling) will loosen the mortar and allow the bricks to be removed more easily. Make sure that other courses or individual bricks are not disturbed, and the old and loose mortar is completely cleaned off.

(*Note*: try to avoid using half bricks to repair damaged walls as this may result in the vertical joints, known as the 'perps' (short for perpendicular), aligning. This will cause a weak point in the brickwork and make it obvious where the repairs have taken place.)

FIGURE 7.53 Cutting bricks with a bolster and disc cutter

FIGURE 7.54 Removal of damaged bricks with a cold chisel

FIGURE 7.55 A poorly built wall

- Stage 2 – Remove the waste material, and dust the area with a stiff hand brush.
- Stage 3 – Brush water on to the area where the new bricks are going to be laid.
- Stage 4 – Prepare a mix of one part cement to six parts builder's soft (fine particle) sand (this is referred to as a 1:6 mix), thoroughly mixed dry before adding water. Keep adding

small quantities of water to the mix until it starts to stick together; try to avoid adding too much water as this will cause difficulties when laying the bricks. Experienced bricklayers have different preferences for the exact consistency of their mixes, and for this reason they often prefer to work with labourers that have mixed for them previously.

- Stage 5 – An additive known as 'plasticiser' should be added to the mix, at this stage to improve the fluidity of the mortar. The manufacturer's instructions should be referred to for the exact quantities of plasticiser to be added to the mix.
- Stage 6 – Use a 'pointing trowel' to apply a bed of mortar approximately 1 metre along the middle of the wall, about 25 mm thick. Run the point of the trowel along the centre of the bed of mortar to even it out along the width of the wall.

FIGURE 7.56 Applying mortar using a pointing trowel

FIGURE 7.57 Positioning the brick and clearing off mortar

- Stage 7 – Apply mortar to the end of the first brick; this process is known as 'buttering up'. Place the brick in position with the 'frog' face up, and squeeze the mortar until a joint of 10 mm is achieved (or more, or less, to match the existing joint). Check the brick is level and correctly aligned with the face of the wall with a spirit level; slight adjustments can be made by tapping the brick with the 'heel' of the trowel. Never lay the spirit level on the brickwork and strike it to straighten or level it as this will reduce the accuracy of the tool.

FREQUENTLY ASKED QUESTIONS

What part of a brick is the 'frog'?

The 'frog' is the shaped hollow section found on some faced bricks. It is used to improve the strength of the joints between the courses of bricks.

The bricks must be laid with the frogs uppermost so that they can be completely filled with mortar as the build continues. If the frogs were laid face down, they would create voids in the brickwork that would weaken the structure and increase sound transmission through the wall.

FIGURE 7.58 A Face brick

FIGURE 7.59

- Stage 8 – Remove the excess mortar by using the edge of the trowel, held at a slight angle, and cutting the excess mortar away, leaving the facework free of mortar smudging. Repeat the processes of buttering up and laying the bricks until the damaged area is completed.
- Stage 9 – Once the mortar has slightly hardened, the face mortar can be either 'weather struck pointed' or 'jointed' (to match the existing finish). This compresses the mortar and makes it resist moisture penetration more effectively later. The perp joints would be done first, followed by the bed joints. The joints between the courses should be completed by running a 'brick jointer' or piece of bent copper tube along before the mortar dries. Alternatively, the pointing trowel can be used to bevel the joints downwards to allow water to run off; repeat this method on the short vertical joints to complete the wall.

ACTIVITIES

Activity 30 – Making good plaster, paintwork and brickwork

Read through the following questions and answer them as fully as you can to help you develop your underpinning knowledge of this subject area.

1. What is the purpose of a 'hawk'?
2. Sketch a 'scarifier' and explain its purpose.
3. Name two different types of plaster.
4. What component should be used to seal a damaged wall prior to plastering?
5. Why should you not lay bricks in freezing conditions?

MULTIPLE-CHOICE QUESTIONS

1 Which **one** of the following is a wood-boring beetle?
- a Rhino
- b Colorado
- c Stag
- d Death-watch

2 Dry rot will attack timber when the moisture content is above:
- a 5 per cent
- b 10 per cent
- c 15 per cent
- d 20 per cent

3 Which **one** of the following materials is not used in the fabrication of guttering?
- a uPVC
- b Pewter
- c Copper
- d Aluminium

4 Plastic guttering is joined in its length using a
- a union
- b collar
- c spigot
- d compound

5 In a sliding sash window the counterbalancing weight is accessed through the
- a hatch
- b mouth
- c pocket
- d window

6 Plastic guttering is fixed to a slight
- a fall
- b rise
- c fascia
- d indent

7 The component that separates the sashes within a pulley stile is called the
- a wagtail
- b divider
- c staff bead
- d parting bead

8 To assist in replacing a sash cord, which **one** of the following aids would be used?

a Jig
b Rat
c Mouse
d Spanner

9 Mortar is made from a mix of

a putty and paint
b sand and cement
c glue and sawdust
d unibond and plaster

10 Plaster is applied to a wall using a

a float
b derby
c trowel
d scarifier

CIRCULAR SAWS

CHAPTER 8

LEARNING OUTCOMES

By the end of this chapter you should have developed a knowledge and understanding of:

- setting up fixed and transportable circular saws;
- changing saw blades;
- cutting timber and sheet materials.

INTRODUCTION

The aim of this chapter is for students to be able to recognise the different types of circular saw and their component parts. This chapter also provides practical guidance on the safe methods used to set up, use and service/maintain them. Throughout this chapter references are made and explained regarding current machinery regulations applicable to this area of study, following all the relevant health and safety law and good working practices.

HEALTH AND SAFETY

Employers have a duty under the Health and Safety at Work Act 1974 (HASAWA) to provide employees with equipment and machinery that is safe to use. They must also ensure that employees are suitably trained and competent in the use of the machinery; this should be done by:

- providing information;
- training;
- instruction;
- supervision.

FIGURE 8.1 Supervision

New employees, even if suitably qualified, may need to undergo a period of refresher training, instruction and familiarisation. You cannot expect them to be competent in the use of all makes and models of woodworking machinery, so every employee must be assessed to establish their training needs.

FREQUENTLY ASKED QUESTIONS

If somebody is trained, doesn't that mean that they are competent?

No, not necessarily. A person that has received training has to then prove their competence through testing. For example, a person may have been shown how to set up and use a circular saw, but that does not mean that they can set up and use the saw.

This system of training and testing will be repeated throughout Level 2 Site Carpentry.

PROVISION AND USE OF WORK EQUIPMENT REGULATIONS 1998 (PUWER)

PUWER gives practical guidance for the safe use of *all* manual and powered work equipment; these items may range from a microwave oven to a forklift truck. The Provision and Use of Work Equipment Regulations cover all sectors of work, not just construction. The aim of the Regulations is to ensure that *employers* provide equipment to their employees that:

- is regularly serviced;
- is maintained;
- is safe to use for its intended purpose;
- will not put people at risk.

RISK ASSESSMENTS

Employers have a statutory duty to carry out a thorough 'risk assessment' of all equipment in the workplace before use. All risk assessments should be carried out by a competent person(s) appointed by the employer; alternatively, an outside agency may be employed (risk assessments are covered in more detail in Chapter 1, Safe Working Practices). A risk assessment will highlight the risks or potential risks associated with a particular task, the people it will affect and measures needed to remove or control the risk. Once the control measures have been implemented they should be reviewed periodically, or if significant changes have occurred in the workplace. If a company employs five or more people, the risk assessment should be recorded in writing.

The Provision and Use of Work Equipment Regulations state that the following areas of a circular saw could potentially pose significant risks to operatives' health.

Noise

Noise is the level of sound, measured in decibels (dB), transmitted by the saw. Measures are required to be in place to control an operative's exposure where noise levels exceed 80 dB (e.g. ear protection, signage).

Wiring

It is essential that the saw has been correctly earthed by a qualified electrician and an isolation switch mounted independently from the machine. The mains power to the circular saw will run through the isolation switch; disengaging this switch allows the saw to be maintained, serviced or adjusted without risk of electrocution (any inspections carried out should be recorded in the machine's maintenance log).

FIGURE 8.2 Operating a circular saw

FIGURE 8.3 Good practice

Dust

Most woodworking machines will create a certain amount of dust when they are in use. Exposure to dust for long periods of time may cause skin disorders, asthma, nasal or lung cancer unless the correct personal protective equipment (PPE) is worn (dust mask, eye protection, gloves). In some situations fine concentrations of airborne wood dust can potentially explode if ignited (termed 'instantaneous combustion'), so it is important that the working area is kept clear at all times, especially around electrical equipment. Wood dust should, where ever practical, be collected with an extractor; this will prevent static dust becoming airborne again when disturbed.

FIGURE 8.4 Bad practice

The Control of Substances Hazardous to Health Regulations (COSHH) recognise the hazard posed by wood dust and suggest extraction should be used as an effective tool to remove the waste from the area, therefore minimising the danger. (*Note*: the maximum exposure limit to wood dust is 5 mg/m³.) A dust extraction system will maintain its effectiveness only if it is regularly maintained and serviced. If the 'ducting' from an extractor becomes blocked due to it being either full or because there are foreign objects trapped within the system, it will lose suction. Any blockage within a dust extraction system is a potential fire risk and will have to be removed immediately.

There is a wide range of different kinds of dust extractor available for both static and transportable machines. Each system is specifically designed and tested by specialists to ensure it performs at optimum efficiency when one or more machines are in use at any one time. The working environment can be improved further with the use of an 'air filtration system'. Air filtration units are usually mounted at ceiling height in workshops and factories to extract airborne dust from the areas. The size, quantity and position of the units is determined and calculated by the volume of space requiring cleaning.

FREQUENTLY ASKED QUESTIONS

What is a dust extraction 'system'?

Dust extraction can be as simple as a single-bag collection unit through to an extensive system of 'ducting', serving a dozen machines or more. The term 'ducting' refers to the lengths of pipes between the main extractor unit and the individual machines. The size of the ducting may be reduced between the main pipework and smaller machines to prevent a loss of suction and effectively remove the dust from the working area. Ducting is usually manufactured from galvanised metal and aluminium fittings; it can also range in size between 80 and 355 mm.

Dust extraction systems will normally have a number of 'blast gates' distributed between the end of the ducts and the machines that they are serving. Some dust extraction systems will lose their suction if all the blast gates are left open; generally only the machines in use should need extraction.

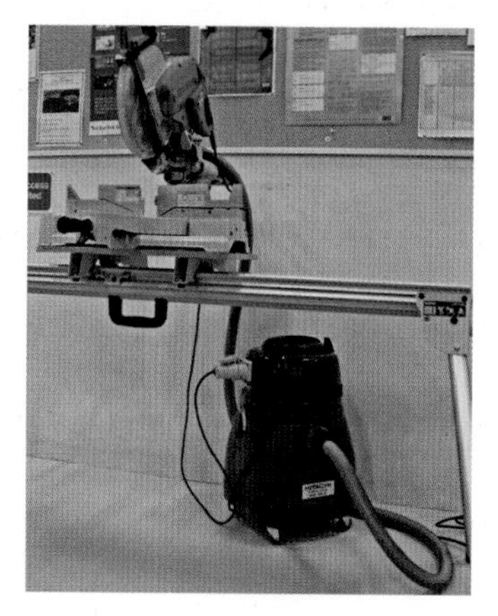

FIGURE 8.5 A mobile dust extraction

FIGURE 8.6 A Dust extractor

FIGURE 8.7 Blast gates

TRADE SECRETS

When you are choosing an extraction system, consider the environment and how you will dispose of your waste. There are some waste extraction systems that compress the dust and wood chippings into small square briquettes. The briquettes are then used as fuel for heating systems for the factory or workshop. Waste material can also be used for:

- animal bedding (no treated timber, some hardwoods or manufactured board material);
- composting;
- industrial spillage absorbent;
- man-made timber-based sheet materials.

STOPPING TIMES

Once a woodworking machine has been turned off, the moving parts generally continue rotating for a short period of time. The Supply of Machinery (Safety) Regulations 1992 state that the length of time a machine should take to stop should be no more than 10 seconds for new machines. New woodworking machines normally have an electronic braking device built into them; older machinery may require 'retrofitting' to bring the machine to a safe condition within the same period of time. 'Retrofitting' simply means adapting the machine to conform to current regulations; in this case the braking device is normally mounted on the outer body of the machine.

Some older machines may have a manual braking system; this is also an acceptable form of stopping the moving parts, provided it does so within the governed time.

WORKSHOP HAZARDS

It is essential that the machine shop/area:

- is free from trip and slip hazards, e.g. offcuts, sawdust, grease and dirt;
- has flat, level and non-slip floors directly around the machine;
- is planned so that machines have adequate space around them, ensuring the operator will not be knocked or distracted;
- has good natural or artificial lighting (poor artificial light may cast shadows on the machine and cause glare, which could be a potential hazard); at some high speeds saw blades can appear to become motionless – this is known as the 'stroboscopic' effect; twin tube light fittings are recommended to prevent this happening;
- has adequate working space and provision to store materials in use;
- is temperature controlled; the temperature in a machine shop must not fall below 16°C, and 10°C in a saw mill;
- is supplied with the *correct* fire extinguishers (e.g. carbon dioxide for electrical fires).

FREQUENTLY ASKED QUESTIONS

What is the difference between a 'machine shop' and a 'saw mill'?

After a tree has been 'felled' (cut down) it usually has to be sawn into smaller usable sections, a process known as 'timber conversion'. Timber conversion is carried out in a 'saw mill'. A saw mill is usually a lot larger than a machine shop simply because of the lengths of the materials being cut and the volume.

A 'machine shop' is normally paired with a joinery workshop. Its purpose is to prepare the sections supplied by the saw mill into planed and shaped sections. A machine shop usually contains some or all of the following:

- mortiser;
- tenoner;
- four-sided planer or through moulder (commonly a router table);
- surface planer and thicknesser;
- spindle moulder
- computer numerical control (CNC) machinery;
- circular saw;
- bandsaw;
- radial arm cross-cut saw;
- drum sander.

GUARDING

Poorly maintained guards or guarding that does not conform to PUWER pose a potential risk to the user. The HSE has the power to prosecute employers that provide machinery for their employees (including guarding) that is considered dangerous. The correct positioning of all the circular saw guards is stated in the HSE's Information Sheets and also in the HSE Safe Use of Woodworking Machines 'Code of Practice'. The main guards on a circular saw are depicted in Figure 8.8:

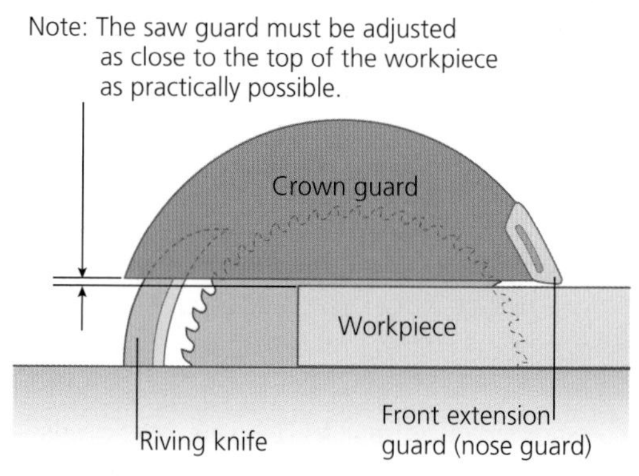

FIGURE 8.8 Crown guard

Both the *crown and nose guard* should be adjusted to a maximum of 10 mm above the top of the timber being fed; this will prevent the user's fingers passing between the timbers and guard.

The *riving knife* maintains the separation of the cut and prevents the timber from closing in on the back of the saw, causing it to bind, and possibly throw the timber back at the operator. It also acts as a protection device to prevent the person 'pulling off' the back of the saw and coming into contact with the back of the saw blade. The circular saw should also be totally enclosed below the table to prevent exposure of the saw blade.

BRITISH STANDARD EMERGENCY STOP BUTTONS

In the event of a dangerous situation occurring while a machine is in use, an emergency stop button must be available to isolate the machine from the power source. Careful consideration must be given to the position of the emergency stop button on the machine. It should be within easy reach of the operator, but not where it could accidentally isolate the machine from the power. *Emergency stop buttons are designed for light use and should therefore only be used as last resort*; this will prevent unnecessary wear and ensure their safety and reliability. Emergency stop buttons should comply with British Standards; this means that they will be easily recognisable, with a red mushroom head on a yellow background. BS emergency stop buttons are operated by depressing the mushroom-style button to isolate the machine; the button is then reset by twisting the spring-loaded button until it rises. Emergency stop buttons are available with a locking mechanism that prevents the machine from being started without a key; this prevents unauthorised use.

FIGURE 8.9 British Standard stop button

FIGURE 8.10 Stop button with a security key

SAFETY MARKING LABELS

Employers must ensure that every woodworking machine displays the necessary markings to ensure employees' health and safety; this will include:

- the safe working speed of the machine (displayed on the machine if possible);
- the safe working speed of the saw blade (displayed on the body of each saw blade);
- the minimum diameter blades allowed to be used *must* be displayed on circular saws; this will ensure that the speed of the blade (the 'peripheral' speed) does not exceed safety limits.

Note: the manufacturer's information and service schedule should be referred to if there is any doubt due to unclear markings.

CODES OF PRACTICE

Codes of practice are approved documents available from the Health and Safety Executive. There are many different codes of practice available for different regulations. Each one will give practical advice on how to comply with the Regulations – for example, 'Safe Use of Woodworking Machinery – Approved Code of Practice and Guidance'.

The codes should not be confused with the law – they only given advice on ways to comply with the law; alternative safe methods can be used provided that they meet with PUWER. Further details are available on 'Information Sheets' for specific machinery from the Health and Safety Executive.

RECAP

HEALTH AND SAFETY AT WORK ACT (HASAWA)

'Secure the health, safety and welfare of all persons at work'

PROVISION AND USE OF WORK EQUIPMENT REGULATIONS (PUWER)

'An employer that provides work equipment must ensure that it is safe to use'

APPROVED CODES OF PRACTICE (ACOP)

'Gives practice advice on how to comply with the law'

FIGURE 8.11 HASAWA flowchart

CIRCULAR SAW DEFINITION

The term 'circular saw' refers to a round metal disc with shaped teeth around its perimeter. A circular saw can be mounted and powered in either a portable power tool or a static machine. The size, shape and amount of teeth will be determined by the machine or power tool it is used in, its intended use and the material being cut.

FIGURE 8.12 Portable power tools with circular saw blades

Static circular saws are usually driven by larger motors than portable powered tools; therefore they are capable of cutting larger sections of solid timber and sheet materials with ease. Static circular saws are usually used only in saw mills and machine shops for production work; their weight and overall size make them impractical to transport to site. There are several different types of static circular saw, each one designed to either cut along the grain (ripping) or across the grain (cross-cutting). Other machines are specifically designed to cut man-made boards or timber-based sheet materials. These machines may have a smaller secondary circular saw blade in alignment with the main blade; this is known as a 'scoring'. A scoring saw can usually be raised and lowered so that it just protrudes above the surface of the saw table (also known as the bed), so that it cuts the surface of veneered, laminated or melamine sheet materials. Scoring saws are normally only used on panel, dimension and wall saws, to prevent the expensive finished surfaces of the sheet materials from breaking out as the circular saw blade passes through the underside of the board.

FIGURE 8.13 Panel saw

FIGURE 8.14 Damage to the underside of the board

TYPES OF CIRCULAR SAW

Table or fixed bed rip saw

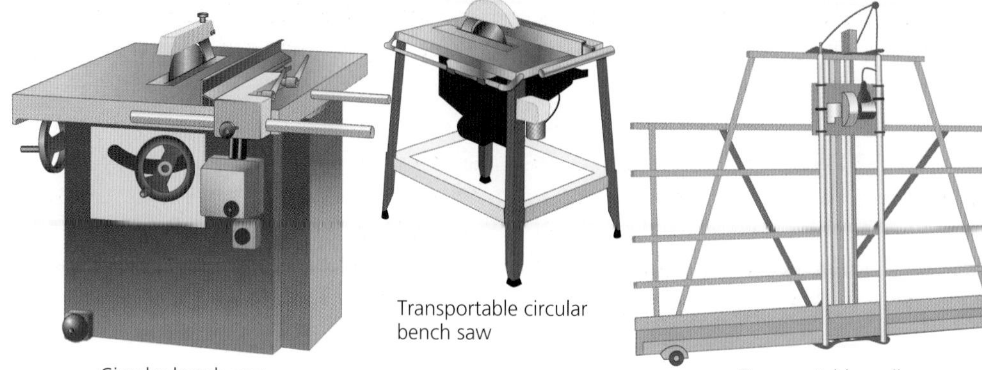

FIGURE 8.15

This type of saw, the main focus of this chapter, can be used for a variety of tasks, including:

- ripping;
- cutting to length (with a sliding fence attachment);
- angled cutting with the use of jigs and tilting the blade;
- cutting wedges and firrings (used to create the fall on flat roofs).

FIGURE 8.16

Circular saw blades can range in size from 100 mm for a scoring blade, up to 2.97 m (9 feet) for a rip saw used to convert trees in a saw mill. Generally, the most common sizes used in the construction industry range from 350 to 600 mm.

Panel/dimension saw

These types of machine are designed to cut timber-based sheet materials to size. They each have a large lightweight 'bed' attached to the left-hand side of the circular saw; this enables the weight of the material to be balanced across the machine table. The bed is able to slide along the length of the machine, thus allowing the sheet material to be trimmed and cut to width. Panel/dimension saws are available in four standard sizes: 1600 mm, 2200 mm, 3200 mm and 3800 mm. This relates to the length the bed can travel along the side of the main saw table. They also have the facility of an additional scoring saw blade and the ability to tilt to 45°. In general, this is a very versatile machine that is normally used in the production of joinery and shopfitting items.

Wall saw

A wall saw is capable of all the aforementioned operations but has the advantage of taking up less floor space; this is particularly useful for smaller workshops. A wall saw is normally only used to cut timber-based sheet materials. It functions by simply positioning the panels on a wall rack, before adjusting the circular saw to the desired position. The saw is then pulled over the face of the sheet material to cut it to the required dimension.

Cross-cut saw

As the name suggests, this saw is used to cut across the grain of timber to the required length. Some machines are capable of cutting timber up to 125 mm in thickness and 700 mm in width. Most machines have the ability to raise and lower the height of the saw blade over the timber; this action allows the saw to be used for cutting housing joints. It is normally used in conjunction with a fence with adjustable length stops.

FREQUENTLY ASKED QUESTIONS

What is a 'fence and stop'?

A *fence* is a fixed guide used to support an item of timber. Fences provide a solid platform to rest the timber against while it is being machined or cut. They are commonly used on all items of woodworking machinery, including rip and cross-cut saws. It is common practice to attach a tape measure along the length of the fence to use as a guide; this prevents repetitive measuring.

A *stop* is simply an adjustable block that can be moved along the length of a fence. An 'offcut' of timber, securely fixed in position with a 'G' clamp, can be used to create a stop. Alternatively, manufactured metal stops can be purchased; these are usually more accurate because they interlock with the fence and can easily be flipped over when they are not in use.

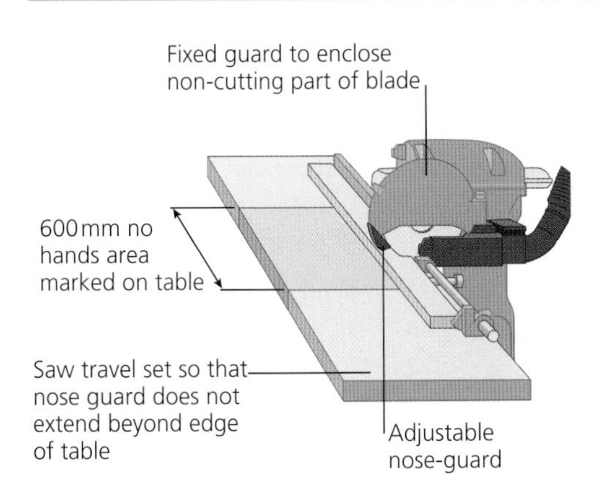

FIGURE 8.17 Cross-cut saw

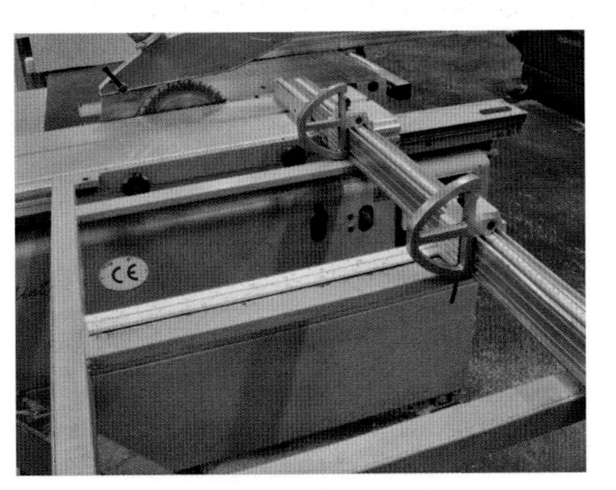

FIGURE 8.18 Stops used on a panel saw

Radial arm saw

This type of saw is similar to a cross-cut saw. Although they are both used to cut across the grain of timber, a radial arm saw is more versatile. Radial arm saws are normally used to cut smaller sections than cross-cut saws and therefore tend to be slightly smaller and lighter in weight. In simple terms, it is best described as a portable circular saw mounted on a long 'arm' that overhangs the base of the machine that it is mounted on. Timber is then fed between the radial arm saw and the base on the machine to the required position, and then the saw is drawn across the timber by the operator to cut the timber to size. The saw is controlled with the use of a handle attached to the motor and saw assembly. Some radial arm saws have the start and stop switches mounted on the handle of the saw for easy use. They may also have a spring coil attached to the cast iron arm; this allows the saw to return automatically to its starting position.

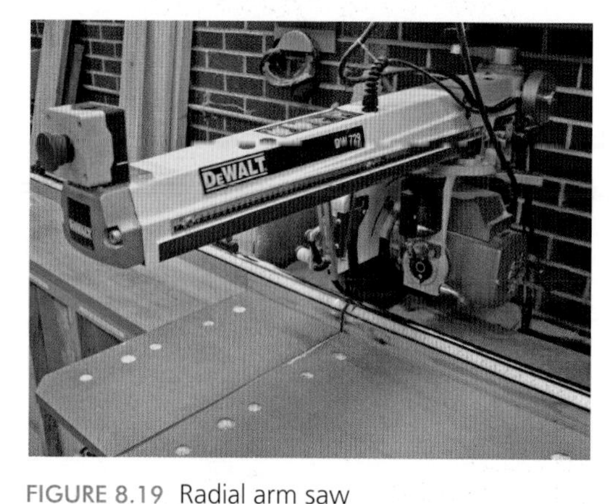

FIGURE 8.19 Radial arm saw

FIGURE 8.20 Spring coil

Radial arm saw features

As mentioned previously, this saw has many features besides straightforward cross-cutting. Radial arm saws are capable of cutting up to 610 mm in width, 110 mm in depth and 920 mm in length when repositioned to rip timber to width. When the saw has been adjusted to cut across the grain at an angle, the capacity to cut wide boards is reduced. The saw can be easily and quickly adjusted with the use of fixed levers and stops to make the cuts shown in Figure 8.21. (*Note*: The saw blade will have to be changed to suit the operation undertaken.)

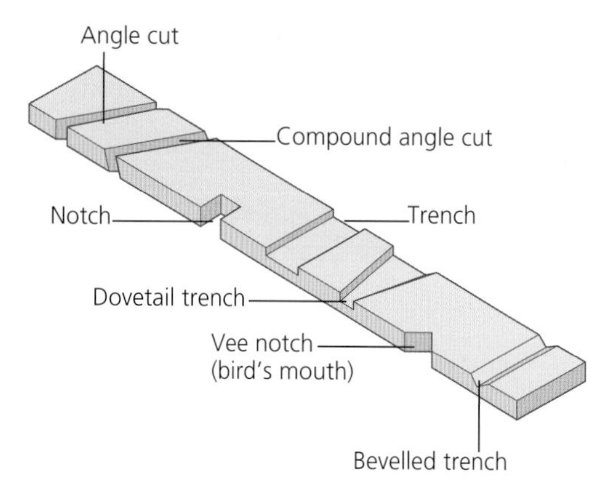

FIGURE 8.21

TRADE SECRETS

A study by the HSE of 1000 accidents occurring on woodworking machines showed that 35 per cent happened on circular saws. The majority of these accidents could have been avoided if the guards provided were used and positioned correctly. The majority of these accidents resulted in the loss of fingers as the timber was being 'hand fed' into the machines (Figure 8.23); these types of machine are considered to be 'high risk'. The risk of accidents occurring on woodworking machines can be greatly reduced if 'powered feed rollers' are used. Once they are set up, they eliminate the need for the machinist to position their fingers anywhere near the cutting edges. Feed rollers can also improve the quality of the machined timber by running the material through the machine at a consistent speed.

Mitre cutting

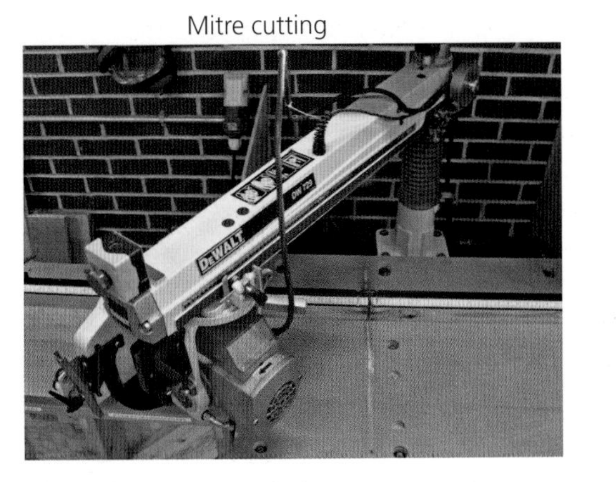

Square cutting

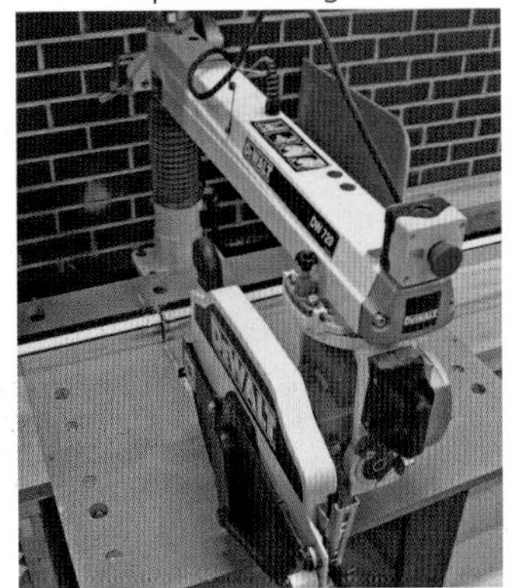

FIGURE 8.22

FIGURE 8.23 Hand-fed timber

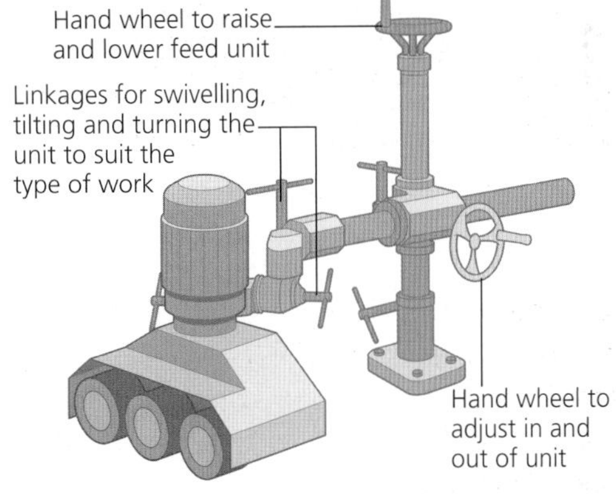

FIGURE 8.24 Powered feed rollers

Circular bench saw – component identification

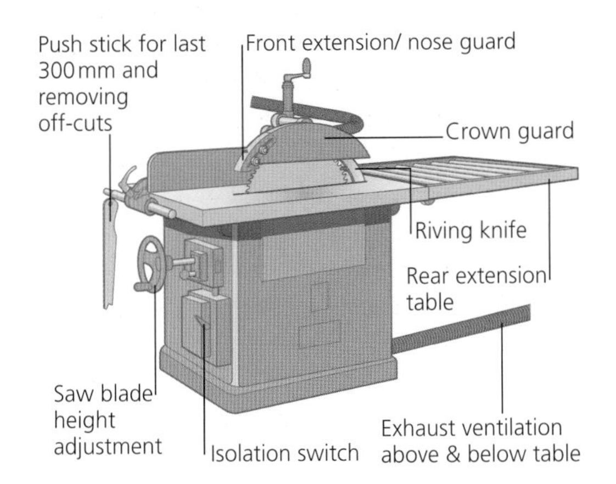

FIGURE 8.25 Circular saw and outfeed table

Crown guard

A crown guard is positioned over the circular saw to minimise the amount of the blade visible; it also reduces the risk to the operator. Crown guards on modern machines are designed with an extraction outlet built in; this is a very effective point to remove the sawdust. The crown guards on transportable circular saws are normally attached to the riving knife, and adjusted by raising and lowering the height of the saw blade.

Note: 'under no circumstances should the safety guards provided on a machine be removed or altered' (PUWER 1998).

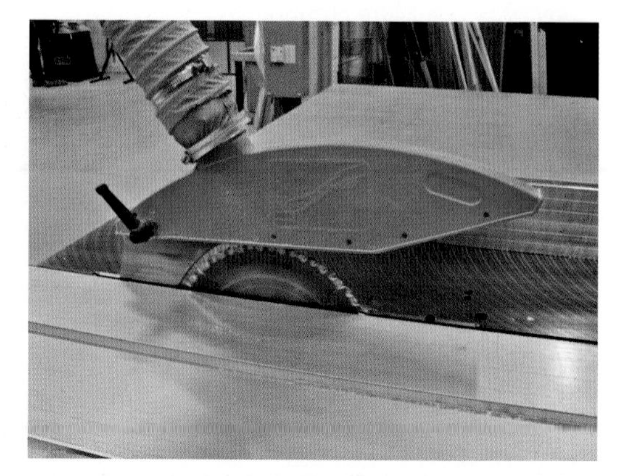

FIGURE 8.26 Crown guard

Extraction outlets

Extraction outlets are positioned at strategic points on a machine to remove the maximum amount of waste possible. It is vital that these points are checked regularly and kept clear to maintain an effective extraction system.

FIGURE 8.27 Extraction outlet

Fine adjustment screw

This allows accurate alignment of the fence position.

Finger plate

The finger plate is loosely fitted within the bed of the machine along one side of the saw blade. The finger holes in the finger plate allow it to be removed without the need for any service tools. When it is removed, it allows easy access into the body of the machine to adjust, maintain and service the blade and riving knife.

FIGURE 8.28 Removal of the finger plate

Mouth piece

The mouth piece is positioned directly in front of the saw blade, at table-top level. Its purpose is to narrow the gap between the circular saw blade and the metal table, and prevent damage to both parts. As circular saws are started, they begin to build up their speed until they reach the maximum revolutions per minute (RPM). During this start-up period, large-diameter saw blades are likely to wobble slightly before reaching their maximum speed and perfect balance.

Nose/front extension guard

The nose guard is mounted over the front edge of the crown guard with an adjustment wing nut. It is used in addition to the main protection devices as a lightweight fine adjustment to minimise the gap between the underside of the crown guard and the top of the timber being cut.

Outfeed table

This is used to provide additional support to the timber or sheet material as it exits the back of the circular saw after cutting. Outfeed tables should extend at least 1200 mm in length beyond the centre of the circular saw blade. Any person employed to 'take off' (remove) material at the back of the saw should remain 1200 mm away from the blade and avoid reaching over the table. The outfeed table also provides rear support for the timber being cut.

Packing pieces

Packing pieces are very similar to the 'mouth piece' in the front of the saw bed, only they are positioned along both sides of the blade. Over a period of time the packing pieces will become worn and will have to be replaced. Excessive wear will cause the gap between the side of the saw and the packing pieces to grow; this may lead to smaller 'offcuts' slipping between the gap during use. A build-up of material slipping through the table bed at this point may cause the dust extraction outlet to clog up; larger offcuts may act as a wedge against the blade, forcing it to jam to a halt.

FIGURE 8.29 Mouth piece

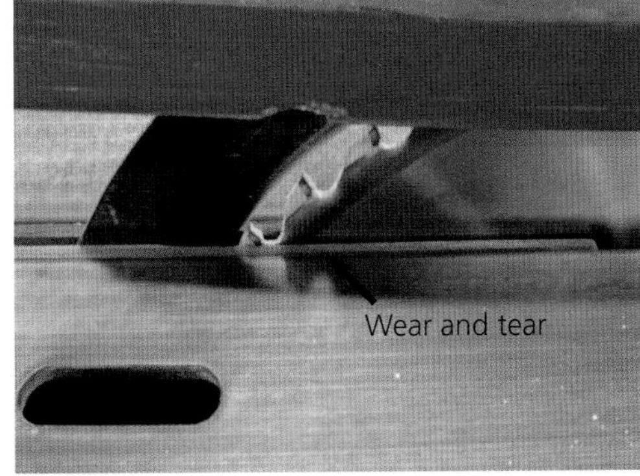

FIGURE 8.30 Wear and tear on packing pieces

Push sticks

Push sticks are used to apply pressure and firmly hold the material in position as it is being pushed through the circular saw. They prevent the need for the machine operator to put their fingers too close to the saw blade while it is rotating. Push sticks should be stored on the machine on which they are going to be used so that they are within easy reach of the machine operator. They should be used to feed timber through a circular saw that is shorter than 300 mm, or for the last 300 mm of longer pieces. They should also be used to remove the cut timber between the saw blade and the rip fence, unless the timber is 150 mm in width or bigger. Push sticks should be a minimum length of 300 mm plus the handle; this usually amounts to 450 mm over all. (*Note*: the machinist's hands should not be in line with the circular saw blade as material is being fed through the saw. This reduces the risk of the operator's hands slipping towards the teeth on the blade.)

Rip fence

The rip fence is used to guide and control the width of the timber being cut. The correct positioning of the rip fence is essential to prevent the timber jamming and resulting in a poor finish to the sawn edge as it is fed in to the saw blade.

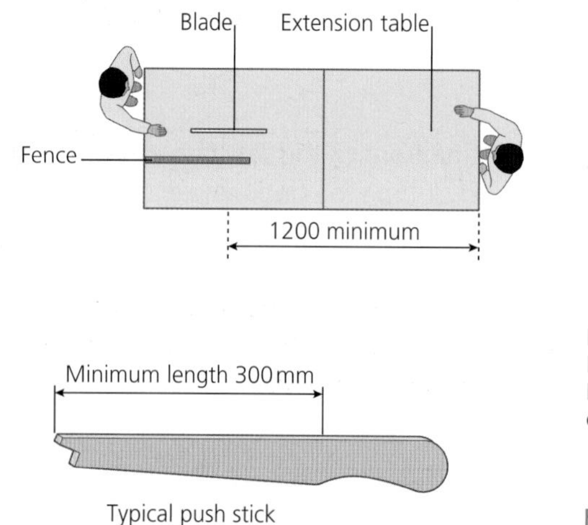

FIGURE 8.31 Push sticks

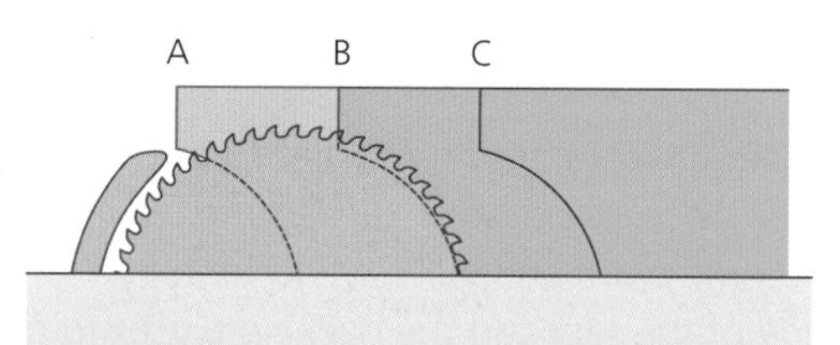

FIGURE 8.32 Rip fence

Riving knife

The riving knife should be made of ridged steel construction and accurately aligned directly behind the circular saw blade. The purpose of the riving knife is to prevent timber 'binded on the saw blade as it is being cut; this is a result of the width of the saw cut (the 'kerf') being reduced as fresh

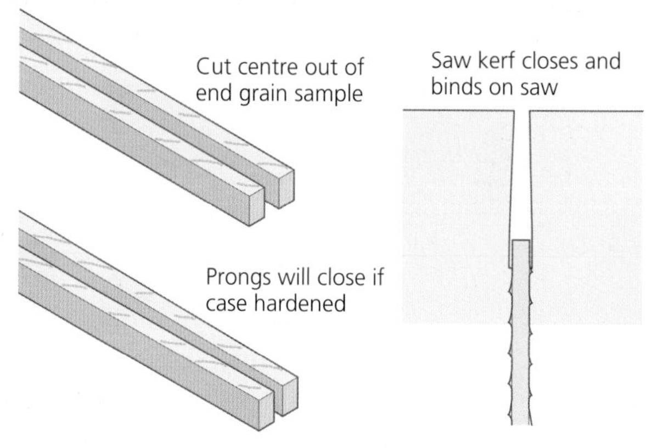

FIGURE 8.33 The kerf closing behind the riving knife

timber exposed by the saw cut moves. A riving knife should be 10 per cent thicker than the width of the kerf, and have a shaped/rounded front edge to guide it through the saw cut and prevent it from 'binding'. The riving knife must be adjusted to suit the diameter of the circular saw blade being used. The Approved Code of Practice states that the maximum distance between the riving knife and the saw blade should not exceed 8 mm at the table-top level. It also states that, with blades greater than 600 mm in diameter, the riving knife should be at least 225 mm above the table surface. For blades less than 60mm diameter the top of the riving knife should not be more than 25mm below the top of the saw.

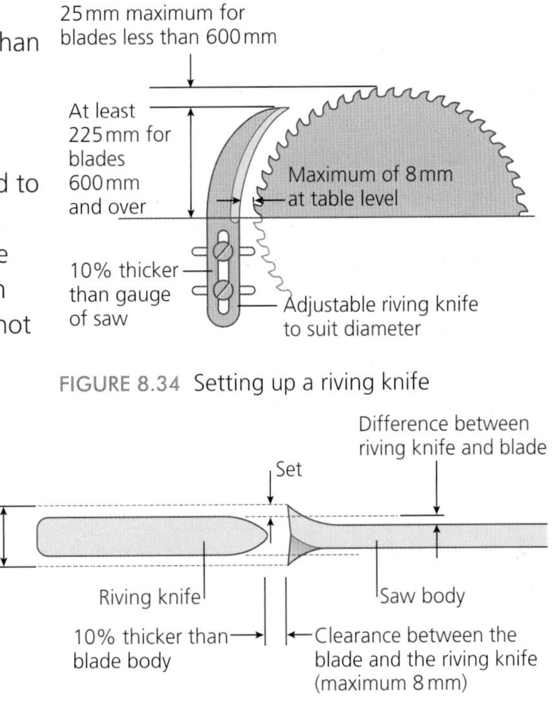

FIGURE 8.34 Setting up a riving knife

FIGURE 8.35

FREQUENTLY ASKED QUESTIONS

What is the 'kerf'?

The 'kerf' is a term given to the width of a saw cut and should not be confused as being the thickness of the saw blade. The teeth on a circular saw blade will protrude on each side, giving a clearance for the 'plate' and preventing jamming.

YOUNG PEOPLE AT WORK

All types of industrial machinery are considered to have an element of risk to the user. The level of risk depends on the following factors:

- young person (under the age of 18);
- supervision;
- instruction;
- experience;
- training;
- competence;
- familiarity with the machinery;
- authorisation.

Young people are permitted to use low- and high-risk machinery provided that they do so as part of their training, and thereafter under supervision. There is a greater risk of injury to younger users, because they may not be at a suitable level of maturity and may lack experience; they may also not be as physically able as a mature adult. (*Note*: maturity must be demonstrated and not just achieved in age.) Employers have a duty under the Health and Safety at Work Act (HASAWA) to ensure that young people are not employed to carry out tasks that are beyond their physical and psychological capabilities.

INSTRUCTION AND TRAINING

The Provision and Use of Work Equipment Regulations (PUWER) require employers to provide all the necessary instructions, information and training for the safe use of machinery in the workplace. Where appropriate, written information should be available for employees regarding all matters of health and safety; this may be in the form of 'method statements'.

Examples of training records

✓: Sheet 1 – List of authorised machine operators

The authorised trainer of ________________ is ________________

(the company) (name of trainer)

Date ________________

I certify that:

(a) I have carried out training, as indicated on the machines listed.

(b) I am satisfied that the people named below have demonstrated competence in the operation of the machines listed and have met all the training objectives for those machines, including:

(i) correct selection of machine for type of work to be done;

(ii) purpose and adjustment of guards and safeguards;

(iii) correct selection and use of safety devices – push-sticks, push spike, jigs and work-holders;

(iv) practical understanding and application of legal requirements;

(v) safe working practices to include feeding, setting, cleaning and taking off.

Operator's name	Circular rip saw	Cross-cut saw	Dimension saw	Surface planing machine	Thickness planing machine moulder	Single-ended tenoner	Spindle moulder	High-speed router	Four-sided planer/	Narrow band saw	Band re-saw	etc
J Brown	✓	✓	✓	✓	✓	✓						
D Smith	✓	✓	✓	✓	✓					✓		
C White	✓	✓										

FIGURE 8.36 Training records

Everyone involved in the machining process must be adequately trained and informed in all matters of health and safety; these people include:

- the machine operator;
- the person 'taking off' material from the back of the machine;
- the person who services or maintains the machinery.

Employers that provide woodworking machinery for their employees to use must ensure that a suitable process for implementing and recording training is in place. The 'Safe Use of Woodworking Machinery – Approved Code of Practice and Guidance' gives examples of ways to record the level of training given and the range of machinery.

It is essential that people employed to use machinery at work have completed a period of training and demonstrated competence to satisfy their employers. A person that has demonstrated competence will be able to:

- Select the safest machine and tooling for the task;
- Demonstrate safe methods of working;
- Correctly adjust guarding to minimise the risk;
- Demonstrate strong knowledge and understanding of their legal responsibilities.

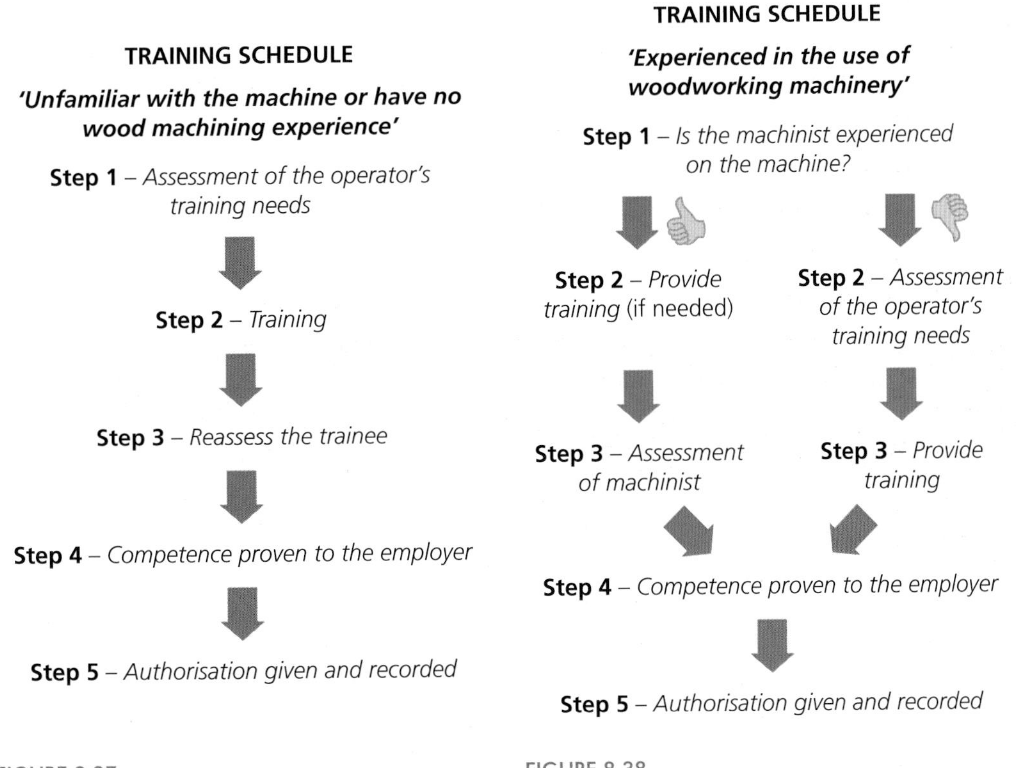

FIGURE 8.37

FIGURE 8.38

MACHINE MAINTENANCE AND SERVICING

It is essential that machinery is regularly maintained and serviced to ensure an efficient state and prevent unnecessary wear. If a machine breaks down, it could be costly in terms of damage to the machine and unused labour, due to repairs and lost production time. Poorly maintained machines may deteriorate to an unsafe condition and put the users and people in the immediate area at risk; this is against the law and could lead to the HSE prosecuting the employer.

The exact amount of servicing and maintenance necessary will vary between machines and the amount of use. Woodworking machinery should be maintained in accordance with the manufacturer's maintenance instructions and schedule; these documents will detail the type of maintenance work required and the frequency. Service schedules should be kept up to date and any maintenance recorded in a log. Both items should be kept close to the machine or maintenance area.

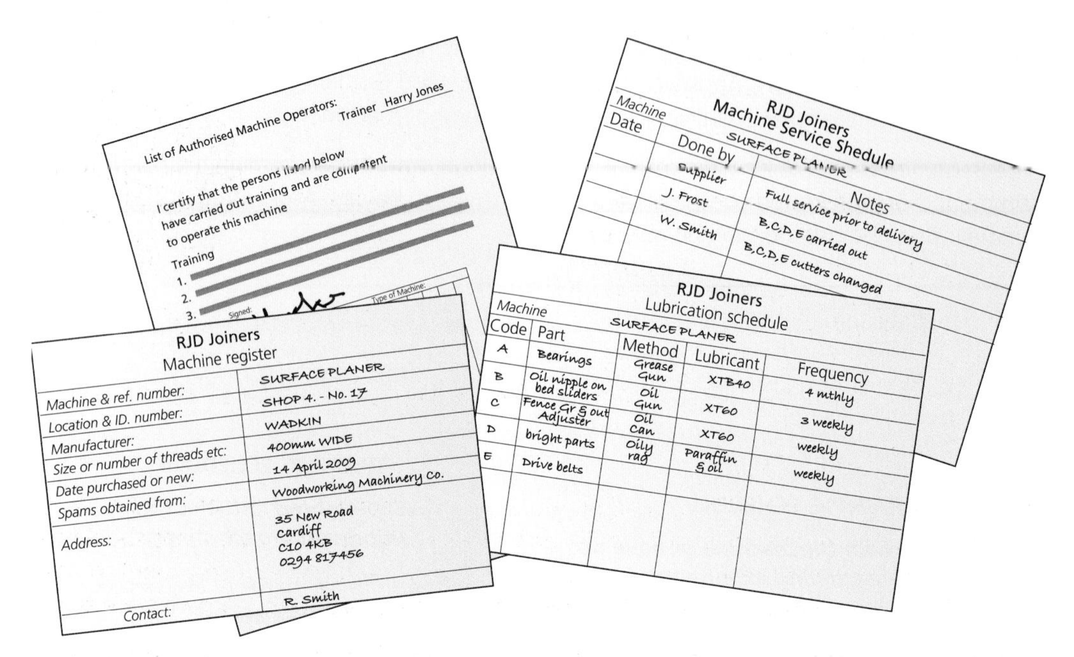

List of Authorised Machine Operators:

Trainer Harry Jones

I certify that the persons listed below have carried out training and are competent to operate this machine

Training

1.

2.

3.

RJD Joiners
Machine Service Shedule

Machine	SURFACE PLANER	
Date	Done by	Notes
	Supplier	Full service prior to delivery
	J. Frost	B,C,D,E carried out
	W. Smith	B,C,D,E cutters changed

RJD Joiners
Machine register

Machine & ref. number:	SURFACE PLANER
Location & ID. number:	SHOP 4. - No. 17
Manufacturer:	WADKIN
Size or number of threads etc:	400mm WIDE
Date purchased or new:	14 April 2009
Spams obtained from:	Woodworking Machinery Co.
Address:	35 New Road Cardiff C10 4KB 0294 817456
Contact:	R. Smith

RJD Joiners
Lubrication schedule

Machine: SURFACE PLANER

Code	Part	Method	Lubricant	Frequency
A	Bearings	Grease Gun	XTB40	4 mthly
B	Oil nipple on bed sliders	Oil Gun	XT60	3 weekly
C	Fence Gr & out Adjuster	Oil Can	XT60	weekly
D	bright parts	Oily rag	Paraffin & oil	weekly
E	Drive belts			weekly

FIGURE 8.39 Machine service schedule

FREQUENTLY ASKED QUESTIONS

Why do you have to record all maintenance and servicing carried out?

You are *not* required by law to keep a maintenance log, but advised to do so if one is available. A maintenance log will provide details to other users of the machine of the type of work carried out and future actions required.

MAINTENANCE SCHEDULES

'Planned'

The manufacturer's handbook will provide details of the machine's maintenance schedule, together with the period of time between services and routine inspections. A planned service would include:

- greasing and oiling of moving parts;
- checking, adjusting and replacing worn or damaged parts;

FIGURE 8.40 Servicing a circular saw

- checking guards and safety devices are working correctly over their full range;
- general cleaning.

'Condition-based'

Worn or damaged parts should be noted and ordered during the planned maintenance; these parts are then replaced at a more convenient time (e.g. changing saw blades).

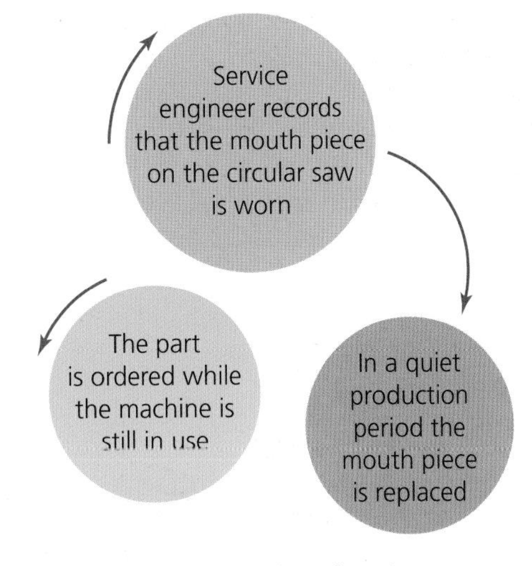

FIGURE 8.41 'Condition-based' maintenance

'Emergency'

Regular servicing and maintenance will reduce the likelihood of an emergency breakdown occurring. Unplanned stoppages are costly to employers through lost labour time, emergency engineers' expenses and missed deadlines or delivery dates. Poorly serviced machines will function less accurately and reduce productivity; they could also put the user at risk.

ROUTINE INSPECTIONS

Routine visual checks should be carried out on all woodworking machines before use to ensure that they are in good working order and a safe condition. The inspections should be carried out by a competent and authorised person, and preferably the person about to use the machine. During this time particular attention should be paid to the following areas:

- extraction ducting – should be free from splits, dust build-up and blockages, and kinks;
- dust extraction bins or bags – below maximum capacity and capable of collecting the waste from the next job;
- floor area around the machine – should be free from obstructions and dust, and should be non-slip;
- start, stop and emergency stop buttons – securely fixed to the machine, operate correctly and undamaged;
- guards, push sticks and safety devices – undamaged, fully functional and safe to use;
- rip fence – moves freely along its full range;
- table surface – clean and clear of waste material and offcuts;
- saw blade faults – as illustrated below.

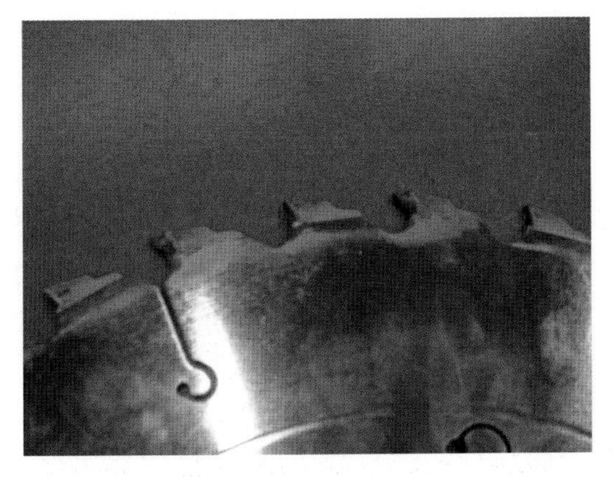

FIGURE 8.42 Saw blade with damaged or missing teeth

FIGURE 8.43 Build-up of resin

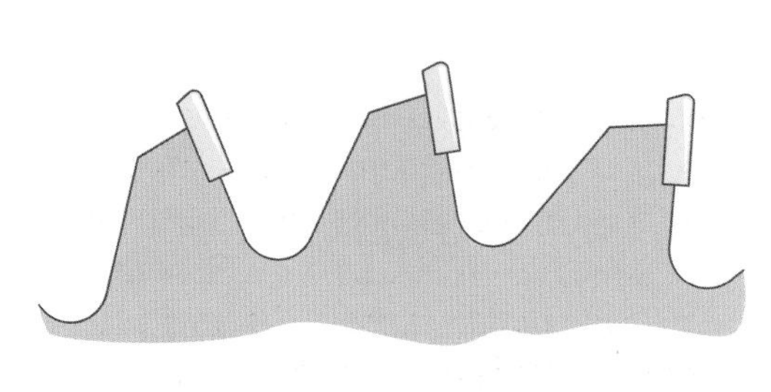

FIGURE 8.44 Blunt or dull cutting edge

FIGURE 8.45 Black eyes

TRADE SECRETS

Machine and hand tools with blunt cutting edges are referred to in the trade as 'losing their edge'. It is not always easy to identify tooling that has 'lost its edge', especially if you are inexperienced or a trainee.

If increased effort is needed while pushing material through a saw, this is normally a result of the circular saw blade being blunt; other signs are:

- *increased breakout on the underside of the material being cut;*
- *during the ripping operation, the material drifts away from the rip fence;*
- *the saw blade becomes unbalanced and wobbles due to overheating;*
- *poor-quality saw cut;*
- *burning or scorching of the material during cutting.*

ACTIVITIES

Activity 31 – Setting up fixed and transportable circular saws

Read through the following questions and answer them as fully as you can to help you develop your underpinning knowledge of this subject area.

1. What do the initials PUWER stand for?
2. What is an 'approved code of practice'?
3. Name one effective method of removing dust from a working area.
4. All moving parts on a circular saw are required by the law to stop within how many seconds?
5. What should the minimum temperature be in a machining area?

The exact process of changing a circular saw blade will differ between manufactured machines, although the principles are very similar. Circular saw blades are manufactured with either one central hole that locates over the drive shaft or a secondary hole just off the centre. The secondary hole is positioned over the drive pin to secure the blade in position and prevent the saw from slipping while it is in motion. When a saw blade is changed and relocated over the drive shaft, it is important to pull the saw blade back against the drive pin to ensure it is correctly positioned while tightening the nut.

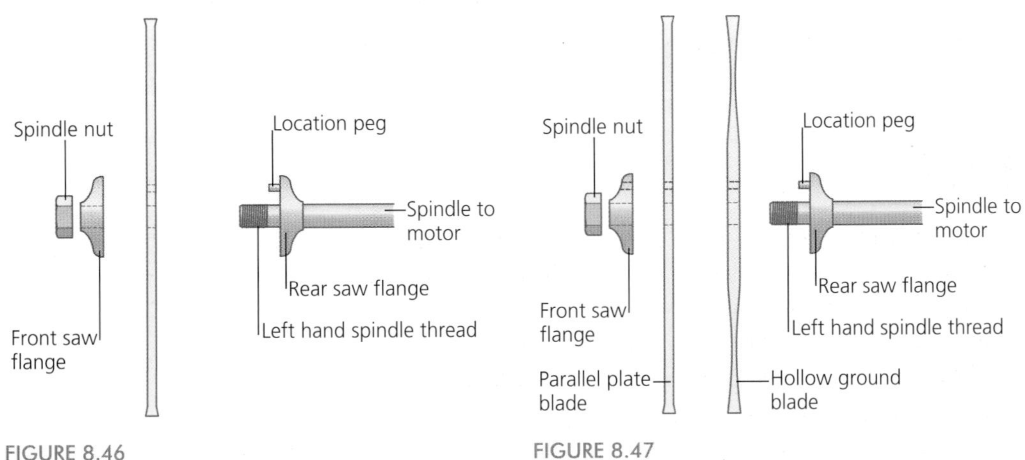

FIGURE 8.46

FIGURE 8.47

The location holes through the circular saw blades will vary in diameter between machines; the size of the hole is referred to as the 'bore' size. It is important that the location hole suits the spindle or drive shaft perfectly. If the bore size is too small it will not fit over the spindle; if it is too big, the blade will not rotate centrally while it is in motion, possibly resulting in a poor finish and increased stress on the machine parts. Bore sizes will be stated in the machine manufacturer's handbook and on the 'plate' of the circular saw blade.

FREQUENTLY ASKED QUESTIONS

▶ Older machines seem to have an unusual 'bore size'; can you still get blades to fit these machines?

It is quite common for some older or imported machines to have imperial-sized parts (e.g. ¼ inch, ½ inch). The most common metric bore sizes are 16 mm, 20 mm and 30 mm. Saw blades with metric holes can be adapted to suit imperial machines; this is normally achieved by inserting metal 'bushes'.

SERVICE TOOLS

All woodworking machinery is usually purchased and supplied with a selection of basic service tools. Each tool is designed specifically to fit the component parts of each machine. Using other tooling may result in poor fitting and could potentially damage the parts; this would also be a potential hazard for the service engineer.

CIRCULAR SAW BLADE SELECTION

Circular saw blades are specifically designed and shaped to cut through timber and timber-based sheet materials, both effectively and efficiently. It is important to consider the following factors before selecting the most suitable circular saw blade:

- type of material being cut (e.g. hardwood, softwood, veneered sheet material);
- operation (cross-cutting, ripping, multifunction);
- scoring (panel, dimension and wall saws only);
- finish required (general purpose, high-quality/extra-thin kerf).

Step 1

Step 2

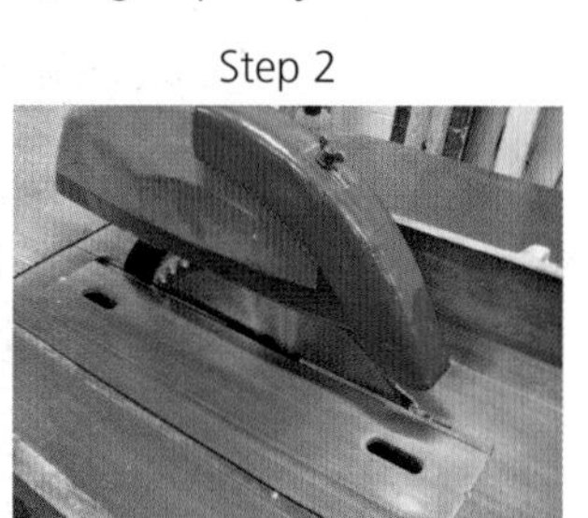

Step 3

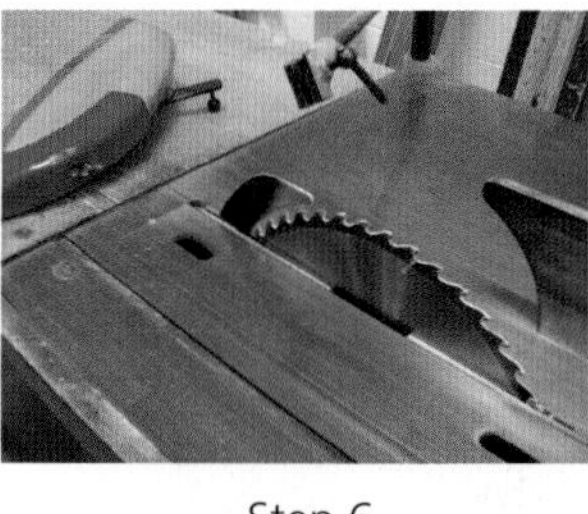

Step 4

Step 5

Step 6

FIGURE 8.48 Changing a circular saw blade

IDENTIFICATION OF PARTS

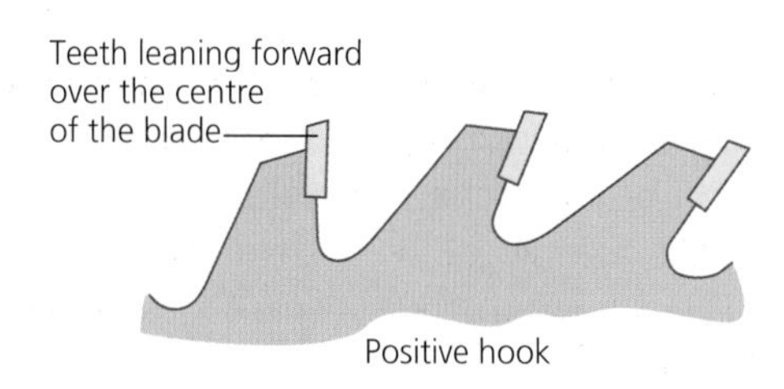

FIGURE 8.49 Circular saw blade with positive rake

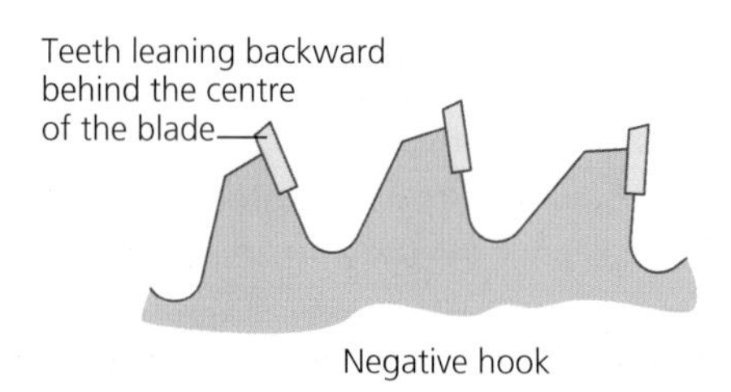

FIGURE 8.50 Circular saw blade with negative rake

TYPES OF SAW BLADE

Generally, circular saw blades have been manufactured with a parallel plate with teeth shaped and bent on either side of the body; this clearance is known as the 'set'. The set on the saw gives clearance to the body of the blade as it cuts through the material; this prevents the saw from binding and overheating as it cuts. The amount of set on a blade will vary depending on the operation and materials being cut. If the set is too much, the saw blade will produce a poor finish, require increased effort to feed the material through and produce a wider 'kerf'. If the set is too little, it will result in the material jamming and being forced back towards the operator; this is known as 'kick back'.

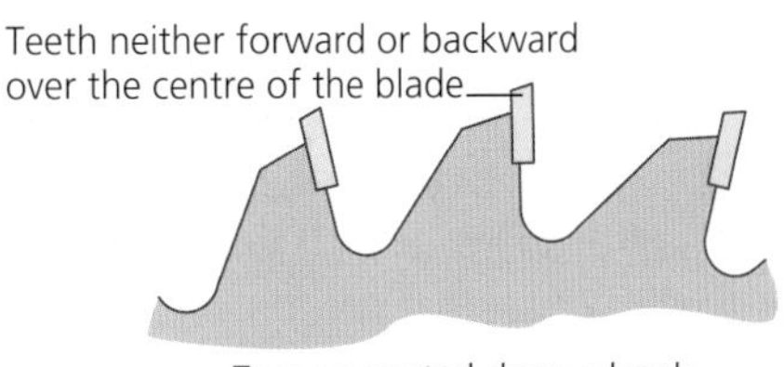

FIGURE 8.51 Circular saw blade with zero degree rake

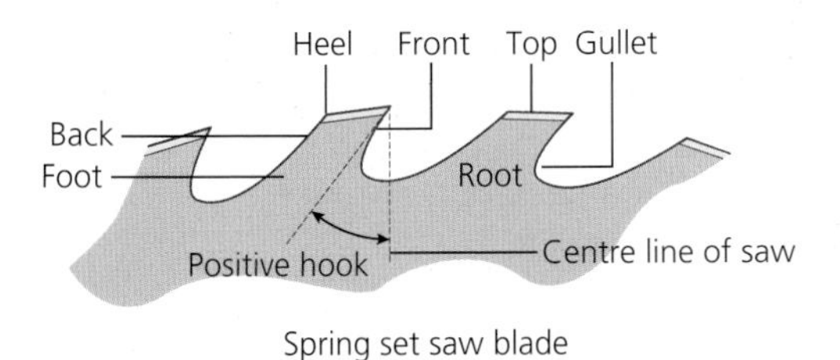

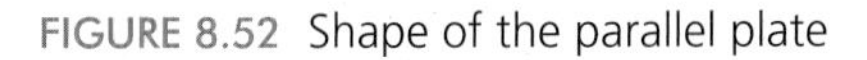

FIGURE 8.52 Shape of the parallel plate

FIGURE 8.53 TCT blade

Spring set blades are rarely used in industry nowadays because they quickly lose their sharp edge, especially when abrasive hardwoods and sheet materials are being cut. Tungsten carbide tipped (TCT) saw blades have now replaced spring set blades, and are commonly used in the construction industry. Tungsten is heat fused to the ends of the teeth on a saw blade because it is extremely hard; it also maintains its cutting edge and set for long periods of time.

SAW TEETH

The shape of the teeth is determined by the material on which they are going to be used.

RIP SAW BLADES

- If the 'hook angle' of the teeth is steep, it will have a deep gullet to collect the waste and be very effective in cutting through softwood. Although the edge on the teeth is extremely sharp, it is very weak and could break if used to cut hardwood.
- The 'hook angle' should be reduced to 20° for multi-purpose cutting of hardwood, softwood and panel materials. The lower hook angle means that the speed at which the material is fed into the saw must be reduced (feed speed).
- Hook angles of 10° should be used to cut panelled materials, including wood veneered boards. This type of circular saw will commonly have a minimum of 60 teeth on a 350 mm diameter blade.

PERIPHERAL SPEEDS

The 'peripheral speed', or 'rim speed', is the speed at which the circumference of a saw blade runs. If the peripheral speed of a circular saw blade is high, it will result in a poor-quality finish and compromise the health and safety of the machine operator. If the speed is too slow, it will make the process of cutting the material difficult and will normally require increased or excessive force to push the timber or sheet material through the saw, 50 metres per second is considered to be the most effective. The peripheral speed for a saw can be established by the calculation below. (*Remember* – the circumference of a circle is equal to 3.142 times the diameter.)

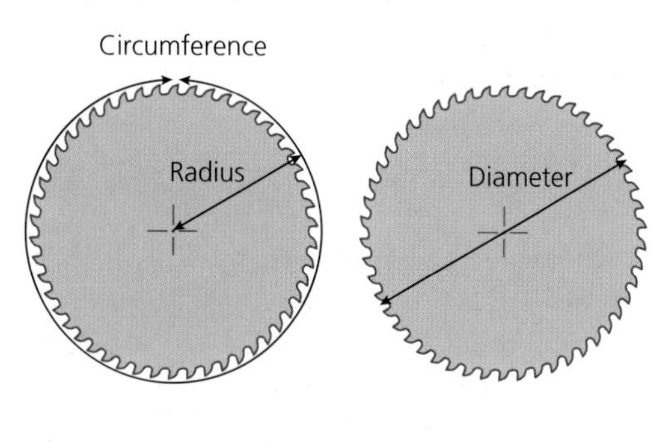

FIGURE 8.54

Example calculation

Circular saw circumference × spindle speed

= peripheral speed in metres per minute ÷ 60

= peripheral speed recorded in metres/second

If you are in any doubt as to the spindle speed and the suitable peripheral speeds for a circular saw, these can be found in the manufacturer's handbook.

ACTIVITIES

Activity 32 – Changing saw blades

Read through the following questions and answer them as fully as you can to help you develop your underpinning knowledge of this subject area.

1. How often should circular saws be serviced?
2. Who should carry out servicing on a circular saw?
3. What item needs to be removed from the bed on a circular saw to gain access to the riving knife and blade?
4. What is the minimum length of a push stick?
5. How can you identify faults on a circular saw blade?

While preparing to cut sections of solid timber on a circular saw, you must analyse the shape of each piece and look for timber defects, these may include dead knots, pockets of sap and splits. It is highly unlikely that longer lengths of natural timber will be perfectly 'true' (straight) or without an element of 'cupping', especially if the material is sawn. Consideration for the shape of the timber before cutting will save timber, avoid mistakes and protect the operator from the results of the timber 'kicking back' and 'snatching'. Snatching is the result of the timber being unsupported as it is fed into the saw. For example, the timber being used may have a bow; if this is fed into the saw with the bow side up (the 'crown'), it will be unsupported across a large area of the timber. As the teeth on the saw blade begin to cut into the timber, this will force the timber downwards onto the saw bed; this is known as 'snatching'.

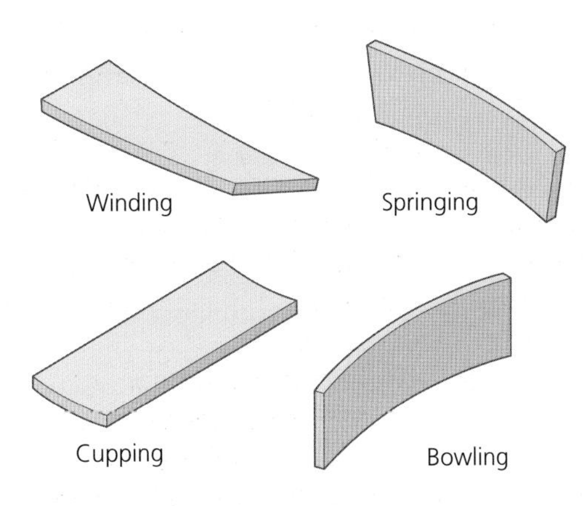

FIGURE 8.55 Defected timber

FREQUENTLY ASKED QUESTIONS

What is 'case hardening'?

Case hardening is a timber defect that occurs during the drying out of the timber after it has been converted. It is difficult to identify the defect until it has been recut after seasoning; it normally results in the saw cut (kerf) closing as it passes through the saw blade, and jamming on the blade. This is due to the timber being dried out too quickly and leaving a high level of moisture in the centre of the timber.

JIGS

Circular rip saws can be used for a number of operations besides ripping and cross-cutting. They are extremely versatile woodworking machines and, when they are used in conjunction with 'jigs', can accurately produce repetition work, with consistent results. Care should be taken when producing jigs to avoid using materials that could damage the saw blade if contact is made during its operation. Made-made sheet materials such as plywood and medium-density fibreboard (MDF) are usually used to manufacture jigs because they are both durable and stable.

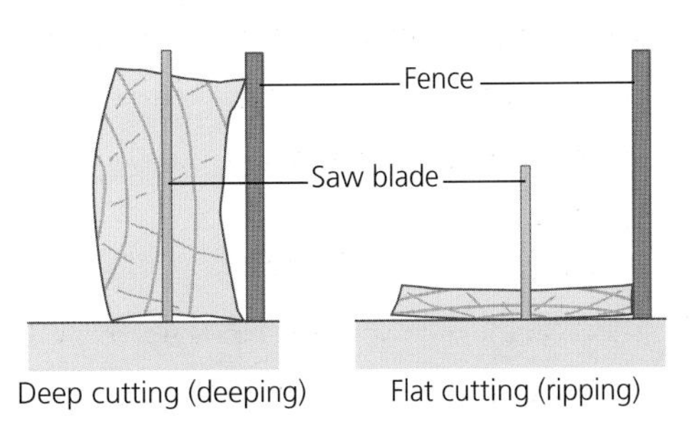

FIGURE 8.56

RECAP

SAFE WORKING PRACTICE ON CIRCULAR SAWS

- *Always* adjust guards to the lowest possible height.
- *Always* use the safety devices provided (e.g. push sticks).
- *Always* use dust extraction.
- *Always* report faults or defects found.
- *Always* isolate the power from the machine before setting up or adjusting.
- *Never* use machinery under the influence of alcohol or drugs.
- *Never* use a machine with loose clothing, jewellery or long hair.
- *Never* adjust, alter or walk away from the machine while the saw blade is still in motion.
- *Never* use a machine without the correct training, information and authorisation.
- *Never* use woodworking machinery without the correct personal protective equipment (PPE).
- *Never* use a machine without being authorised to do so.
- *Never* distract or talk to an operative (skilled worker) while they are operating machinery.
- *Never* use the machine in an unsafe condition.

ACTIVITIES

Activity 33 – Cutting timber and sheet materials

Read through the following questions and answer them as fully as you can to help you develop your underpinning knowledge of this subject area.

1. What is a 'jig' used for?
2. Explain the meaning of 'case hardening'.
3. Sketch the following timber conversions: 'flatting' and 'deeping'.

MULTIPLE-CHOICE QUESTIONS

1 When a circular saw is switched off, within how many seconds must the blade stop?
   a 5
   b 10
   c 15
   d 20

2 The guarding of circular saws is a requirement of which one of the following?
   a WaHR
   b PPER
   c PUWER
   d RIDDOR

3 A push stick should be used for the last
   a 200 mm of the cut
   b 300 mm of the cut
   c 400 mm of the cut
   d 500 mm of the cut

4 At table-top level the distance between the riving knife and the saw blade should not exceed:
   a 4 mm
   b 8 mm
   c 12 mm
   d 16 mm

5 A machine with a broken guard should be
   a sold
   b used with care
   c labelled unsafe for use
   d used if no other machine is available

6 Which **one** of the following hook types is required for a circular saw blade designed to rip timber?
   a Bill
   b Barbed
   c Positive
   d Negative

7 A second small-diameter circular saw, used on the underside of a cut to prevent breakout, is called a

a kerf
b crown
c raking
d scoring

8 The aid required to cut wedges safely and to a consistent shape is called a

a jig
b box
c saddle
d harness

9 Before changing a saw blade, the machine must be

a cleaned
b wedged
c isolated
d serviced

10 What percentage should the riving knife be thicker than a circular saw?

a 10 per cent
b 20 per cent
c 30 per cent
d 40 per cent

INDEX

ANSWERS TO MULTIPLE-CHOICE QUSTIONS

Chapter 1
Safe Working Practices

Q	a	b	c	d
1			/	
2		/		
3				/
4		/		
5				/
6		/		
7			/	
8			/	
9		/		
10			/	

Chapter 2
Information Quantities and Communication with Others 2

Q	a	b	c	d
1	/			
2				/
3		/		
4	/			
5	/			
6				/
7		/		
8			/	
9	/			
10		/		

Chapter 3
Building Methods and Contruction Technology 2

Q	a	b	c	d
1		/		
2		/		
3		/		
4	/			
5			/	
6				/
7			/	
8				/
9			/	
10				/

Chapter 4
First Fixing

Q	a	b	c	d
1	/			
2	/			
3			/	
4		/		
5		/		
6			/	
7			/	
8		/		
9			/	
10				/

Chapter 5
Second fixing

Q	a	b	c	d
1			/	
2		/		
3			/	
4			/	
5				/
6	/			
7				/
8			/	
9				/
10			/	

Chapter 6
Erect Structure Carcassing

Q	a	b	c	d
1		/		
2		/		
3			/	
4		/		
5	/			
6				/
7				/
8			/	
9				/
10			/	

Chapter 7
Maintenance

Q	a	b	c	d
1				/
2				/
3		/		
4	/			
5			/	
6	/			
7				/
8			/	
9		/		
10	/			

Chapter 8
Circular saws

Q	a	b	c	d
1		/		
2			/	
3		/		
4		/		
5			/	
6			/	
7				/
8	/			
9			/	
10	/